APPLIED CHEMISTRY

APPLIED CHEMISTRY

K. Bagavathi Sundari

Reader, Department of Chemistry
Guru Nanak College
Chennai

MJP PUBLISHERS

Chennai 600 005

Cataloguing-in-Publication Data

Bagavathi Sundari, K (1952 –   ).
  Applied Chemistry / K.
Bagavathi Sundari. – Chennai : MJP
Publishers, 2006.
    xxiv, 510p. ; 21 cm.
    Includes references and Index.
    ISBN 81-8094-025-X (pbk.)
    1. Chemistry, Applied II. Title.
    660 dc22   BAG   MJP 022

ISBN 81-8094-025-X
Copyright © Publishers, 2006
All rights reserved
Printed and bound in India

MJP PUBLISHERS
A unit of Tamilnadu Book House
47, Nallathambi Street
Triplicane, Chennai 600 005

Publisher : J.C. Pillai
Managing Editor : C. Sajeesh Kumar
Project Coordinator : P. Parvath Radha

Edited and Typeset at [logo] Editorial Services, Chennai, eserve@rediffmail.com
Cover : R. Shankari   CIP : Prof K. Hariharan

To

My beloved Father
for his selfless sacrifices and guidance

PREFACE

This book on Applied Chemistry is an exhaustive source of information on a variety of areas recommended under application-oriented chemistry. The fundamental concepts and applications of Pharmaceutical, Biological, Leather, Dairy, Polymer and Soil Chemistry have been discussed in detail.

The topics on biological chemistry and pharmaceutical chemistry are made cogent by presenting the concepts with adequate illustrations. This would provide a better platform for students lacking exposure in biological sciences. To instigate interest and understanding, *activities for classroom discussions* have been included to widen the horizons of learning the concepts from differing points of view.

The topics that fall under Industrial Chemistry include leather chemistry, dairy chemistry, and polymer chemistry. The knowledge of these is the fulcrum behind the growth of industries in any developing nation. The chapter on leather chemistry focuses on definitions of certain terminology used in the industry and detailed discussions on leather processing. The chapter on dairy chemistry deals amply with the science associated with milk and other dairy products. Polymer chemistry occupies fairly a larger volume of text in view of its indispensable and imperative role in life. In this chapter, apart from dealing with rudiment levels of information, more focus is given to conductivity polymers, polymers in optical lenses light-emitting polymers, which are gaining importance both in research and industries. A chapter on biodegradable polymers is to hint the students about emerging trends in polymers.

The chapters on soil chemistry reveal that soil is an embodiment of innumerable concepts of science. It is approached systematically by presenting its definition, systematic classification, and chemical and physical properties. The emphasis on colloidal properties of soil would help the students to understand the soil science in depth and to exploit it judiciously.

This book would make the learning process complete by facilitating students to grasp the actual application of science towards the betterment of man's life. It is hoped that this book receives the attention and enthusiasm of the academic community.

K. Bagavathi Sundari

ACKNOWLEDGEMENT

My acknowledgements are due to my colleagues from the departments of Biological Sciences for their invaluable suggestions on the chapters dealing with biological chemistry and soil chemistry. I thank all my colleagues of the Department of Chemistry, Guru Nanak College, for their constant support and encouragement without which this contribution could not have been possible. For any effort to bear fruit, the home front plays a major role. It is not an exaggeration to acknowledge the role of my daughter who played a pivotal role in all my endeavors. The idea of writing this book was seeded by my son and the support rendered by my husband is inestimable. My earnest thanks are due to MJP Publishers who have been extremely helpful and supportive throughout this task.

K. Bagavathi Sundari

CONTENTS

PART I   PHARMACEUTICAL CHEMISTRY

PART II   BIOLOGICAL CHEMISTRY

PART III   INDUSTRIAL CHEMISTRY

PART IV   AGRICULTURAL CHEMISTRY

PART I

PHARMACEUTICAL CHEMISTRY

INTRODUCTION

Medicinal or pharmaceutical chemistry is a scientific discipline at the intersection of chemistry and pharmacy involving designing and developing pharmaceutical drugs. Medicinal chemistry also focuses on identification, synthesis and development of new chemical entities suitable for therapeutic use. It also includes the study of existing drugs, their biological properties, and their quantitative structure–activity relationship (QSAR).

Pharmaceutical chemistry focuses on the development, production and delivery of drugs used to prevent, cure and relieve symptoms of diseases. Pharmaceutical chemists may synthesize new drugs, or modify older drugs so that the drugs have improved therapeutic value, less toxicity and improved stability. The need to demonstrate safety, bioavailability and effectiveness of all new drugs, as mandated by the FDA, places unique quality control requirements on all aspects of drug manufacturing and distribution process. To meet these requirements, pharmaceutical chemists also develop improved analytical techniques for monitoring the levels of drugs in the body and to ascertain the safety and potency of the drugs.

Terminology in Drug Chemistry

Some of the important terminology that are used in the chemistry of drugs are:

Medicinal chemistry   This field applies the principles of chemistry and biology in the creation of newer therapeutic agents. The basic understanding in medicinal chemistry lies in the awareness of relationships between the chemistry of a particular compound/compounds and their interaction with the body which is known as structure–activity relationship (SAR) and the mechanism by which the compound influences the biological system is known as mode of action. Medicinal chemistry aims at improving the therapeutic effect of drugs by minimizing their undesirable side effects.

Pharmacodynamics   It is the response of living organisms to chemical stimuli in the absence of disease. This is a quantitative study of biological and therapeutic effects of drugs. Such studies also elucidate the mechanism of action of drug and may bring a correlation between the action and chemical structure of drugs. Thus, knowledge of pharmacodynamics is essential for pharmacotherapy.

Pharmacophore   The physiological activity of drugs is due to the presence of certain functional groups or structural units. These are called as pharmacophores.

Pharmacodynamic agents   These are the drugs that stimulate or depress various functions of the body so as to provide relief from the symptoms of discomfort. However, they are not specific remedies for any particular disease, since they have no action on infective organisms, e.g. sedatives, analgesics, anaesthetics, etc.

The substances that take part in cellular metabolic reactions are called as metabolites. The chemicals that block the activities of metabolites are called antimetabolites. They act either by preventing the metabolite from combining with its specific enzyme, probably due to its close structural similarities or the antimetabolite may itself combine directly with the enzyme to form a compound which is metabolically inactive. Thus, these chemicals are referred to as structural antimetabolites.

Bacteria A group of unicellular microorganisms possessing rigid, complex, proteinous cell wall and capable of living freely, saprophytically or as parasites. Bacteria are of two types namely gram-positive and gram-negative. Those bacteria that retain the stain when stained with agents such as crystal violet in combination with iodine are known as gram-positive bacteria. And those bacteria that do not retain the stain are known as gram-negative bacteria.

Virus They are ultramicroscopic infectious agents. They replicate only within cells of living hosts; many are pathogenic causing diseases such as chickenpox, measles, mumps, rubella, pertussis and hepatitis. These obligate intracellular parasites consist of DNA or RNA enclosed in a protein coat. They range in diameter from 20 to 300 nm (1 nanometre is $1/1000000000$ ($1/10^9$) of a metre). Some animal viruses are also surrounded by a membrane. Inside the infected cell, the virus uses the synthetic capability of the host to produce progeny virus. Viruses are not affected by antibiotics, the drugs that are used to kill bacteria.

Pharmacy Pharmacy is the science of identification, selection, preservation, compounding and dispensing of medicinal substances.

Pharmacokinetics This describes the study of absorption, distribution, metabolism and excretion of drugs and their relationship to pharmacological response.

Therapeutics It is a branch of medicine dealing with the cure or relief of diseases with the help of drugs.

Pharmacopoeia This is an official code, containing a selected list of established drugs and medicinal preparations with the description of their physical properties and tests for their purity and potency. It defines the standards which any preparation must meet and their average doses for any individual. Few well-known pharmacopoeia are:

- Indian Pharmacopoeia (IP)
- British Pharmacopoeia (BP)
- United States Pharmacopoeia (USP)
- European Pharmacopoeia (EP)

Vaccine   Vaccine (named after "vaccinia", the infectious agent of cowpox, which when administered, provides protection against smallpox) is a biological preparation that is used to enhance the response of the immune system by producing specific antibodies or altered cells. Vaccines help in preparing the immune system to defend the body against a specific pathogen, usually a bacterium, a virus or a toxin, thus protecting against subsequent infections. It may also help the immune system to fight against degenerative (cancer) cells. Vaccines are administered through needle injections, orally or by aerosol. Depending on the infectious agent, the vaccine that is prepared can be a weakened bacterium or virus.

Assay   Estimation of the potency of the pure active principle in unit quantity of a medicinal preparation is called as assay.

Potency   The measure of the biological activity of a drug is its potency.

Parenteral route   This is the method of administering a drug through different routes other than the alimentary tract, e.g. inhalation, injection, etc.

Enteral route   This is the method of administering a drug orally.

MEDICINAL FLORA
OF INDIA

OVERVIEW

The medicinal flora of India is *comparable* to the
whole world's flora. It is very rich and highly
diverse. From ancient time, the Indian medicinal
flora and the traditional knowledge about these
plants have attracted the researchers, herbalists
and scientists around the world. In ancient Indian
history, it is mentioned that many such researchers
have visited India in search of this valuable
knowledge. When did ancient man start using the
medicinal plants and what was the name of the
first plant? These questions are still unanswered.
In addition, there is no scientific document
available regarding this. In India, Rigveda is the
most authentic document in this regard. In
Rigveda, the "Soma" is described as the first
medicinal plant used by ancient man. The
scientific name of Soma is not yet decided. In
Indian medicine, generally medicines of plant
origin are preferred than medicines of animal
origin. One possible reason for this may be the
presence of natural flora in abundance.

In Ayurveda, it is clearly mentioned that
any patient can be cured with the help of herbs

present in the environment. It is also mentioned that the "herb talks or expresses". According to ancient literature, by observing a particular herb minutely, one can understand its utility for different human ailments. From their deep studies, ancient Indian herbalists have found that the shape and size of different parts of herbs resembling different human organs, are useful in treatment of the diseases related to that particular organ. For example, Karela or bitter gourd (*Mimordica charantia*) fruits look like the pancreas of human body (diabetes is a result of disturbed activities of pancreas). In Ayurveda, it is mentioned that Karela is the best remedy for diabetes mellitus. Today the whole world is recognizing the medicinal properties of Karela fruits. Similarly, the seed of Akhrot (Walnut) resembles the structure of human brain. Ancient Indian herbalists have already mentioned the use of walnut to increase the activities of human brain. There are thousands of such examples mentioned in ancient literature. Another example is that of Aak, *Calotropis gigantea*, a valuable medicinal plant which has a reputed position in almost all branches of medicine. Like all herbs, the plant of Aak also "talks and expresses" its utility for human beings in Ayurveda. The structure of Aak flower resembles that of a patient, bowed down and whose spinal cord is stiff. This figure resembles the patient suffering from rheumatism sitting in front of a herbalist and expressing his/her problems. Aak is one of the best remedies for this disease.

According to the World Health Organization (WHO), more than one billion people rely on herbal medicines to some extent. The WHO has listed 21,000 plants which have reported medicinal uses around the world. India has a rich medicinal flora of about 2500 species. Of these, at least 150 species are used commercially on a fairly large scale. These traditional Indian healers have been appreciated globally. In India, these traditional healers are losing their popularity due to the "quick relief" Western systems of medicine like allopathy. Ultimately, the common Indian is shifting away from traditional systems.

SOME "INDIAN HEALERS" AND THEIR SIGNIFICANCE

Man is unaware about many "healers" that he is encountering in his day-to-day activities. Their significance and utility are not appreciated. Mentioned below are some of the prominent medicinal plants that express a wide range of significant utilities.

Neem

India has an impressive list of medicinal plants, almost all of them native to the soil. Topping the list is the neem plant (margosa). All parts of this ubiquitous tree are bitter and are used in medicine. A decoction of neem leaves helps cure fever, particularly malarial fever, liver problems such as hepatitis, boils, and all kinds of skin diseases. An extract of neem is a powerful insecticide, poisonous to insects and parasites. The medicinal use of neem can be traced back to the ancient Harappan civilization, 4500 years ago. From that time, neem was used widely to treat a wide range of ailments. As of now the plant plays an essential role in the Indian system of natural healing, Ayurveda. The neem tree is sacred in India because of its life-protecting properties and is also called as "The village pharmacy" in India. Neem is one of the most ancient and widely used herbs in the world. In fact, herbalists in ancient India have documented the healing qualities of this remarkable tree long before Western civilizations discovered the analgesic qualities of the willow tree from which aspirin is derived.

The neem is a powerful antibacterial and antifungal herb that is extremely bitter, with powerful detoxifying chlorophyll. It acts as an invaluable skin and blood cleanser. It is very effective for normalizing gut bacteria. It should be used up to 2 weeks for intestinal bugs and up to 2 months for inflamed skin. Neem products are helpful for conditions such as shingles, thrush, gum disease, scabies, muscle pain and against parasites such as head lice, etc. Common man finds the use of young twigs as toothbrushes that prevents pyorrhoea. The neem oil obtained from the seeds finds wide applications such as in healing wounds and in rheumatic pain. Neem is used as an antiseptic for treating various infections. The leaves are used in viral infections such as smallpox and measles, and as antineoplastics. Flowers and tender leaves are used as antihelminthics. Neem soap made from the neem oil helps to calm and soothe the irritated skin when used regularly.

Holy Basil

It is an aromatic herb also known as Tulsi in India, which is invaluable in traditional Ayurvedic medicine, and is considered sacred by the Hindus. It has long been used for its health-promoting properties. Traditionally it was used as an adaptogenic tonic in ayurvedic medicine. Holy basil extracts support the immune system and is used as an antioxidant and anti-inflammatory substance. It has been known to be useful as a diaphoretic, expectorant and mild sedative. There are three Holy Basil plant varieties—Rama Tulsi (*Ocimum canum*), Krishna Tulsi (*Ocimum sanctum*) and Vana Tulsi (*Ocimum gratissimum*).

Tulsi leaves contain a bright yellow volatile oil, which is useful against insects and bacteria. The principal constituents of this oil are eugenol, eugenol methyl ether and carvacrol. The oil is reported to possess antibacterial properties and it acts as an insecticide. It inhibits the *in vitro* growth of *Mycobacterium tuberculosis* and *Micrococcus pyogenes* var. *aureus*. It has marked insecticidal activity against mosquitoes. The juice of leaves and/or a concoction called jushanda, a kind of tea, is used to treat common cold, fever, bronchitis, cough, digestive complaints, etc. When applied topically, it helps in eradicating ringworms and other skin diseases. Tulsi oil is also used as eardrops in case of pain. The seeds are used in curing urinary problems and when mixed with ginger juice, black pepper and honey, it cures catarrh. It is a tonic for the heart and has been found effective in the first stages of many cancers. Tulsi purifies the air and is an insect repellent. No wonder the word "tulsi" itself means "matchless". Aphrodisiac virtue has been attributed to it, where powdered tulsi root with clarified butter (ghee) is prescribed. In addition to *Ocimum sanctum proper*, there are other species of this plant that go by the same name, viz. Tulsi. These plants are *Ocimum canum* (Rama Tulsi or Kali Tulsi), *Ocimum basilicum* or bobai tulsi, *Ocimum kilmand*, *O. scharicum* or camphor tulsi, etc. The medicinal effect of all these varieties is nearly similar, if not the same. The Indian scientists are actively involved at the threshold of finalizing their discovery of a reliable medicine against cancer out of Tulsi plant.

Mango

Mango is used as food in all stages of its development. Green or unripe mango contains a large portion of starch, which gradually changes into glucose, sucrose and maltose as the fruit begins to ripen. The starch disappears completely when the fruit is fully ripe. Green mango is a rich source of pectin, which gradually diminishes after the formation of the stone (kernel of mango seed). Unripe mango is sour in taste because of the presence of oxalic, citric, malic and succinic acids. The nutritional value of mango is summarized in Table 2.1.

Table 2.1   Nutritional value of mango

Component	Nutritional food value (%)*
Moisture	81.0
Protein	0.6
Fat	0.4
Minerals	0.4
Fibre	0.7
Carbohydrates	16.9
Total	100

Component minerals and vitamins	Nutritional food value (mg)*
Calcium	14
Phosphorus	16
Iron	1.3
Minerals	0.4
Vitamins C and B complex	16

Calorific value = 74

*All the above values are for 100 g of edible portion.

The raw mango is a valuable source of vitamin C. It contains more vitamin C than half-ripe or fully ripe mangoes. It is also a good source of vitamin B_1 and B_2 and contains sufficient quantity of niacin. These vitamins differ in their concentration in various varieties during the stages of maturity and environmental conditions. The ripe fruit is wholesome and nourishing. The chief food ingredient in mango is sugar. The acids contained in the fruit are tartaric and malic acid, besides a trace of citric acid. These acids are utilized by the body and they help to maintain the alkali reserve of the body.

Natural benefits and curative properties   Mango is well known for its medicinal properties both in unripe and ripe states. The unripe fruit is acidic, astringent and antiscorbutic. The skin of the unripe fruit is astringent and stimulant tonic. The bark is also astringent and has a marked action on mucous membranes. The ripe mango is antiscorbutic, diuretic, laxative, invigorating, fattening and astringent. It tones up the heart muscle, improves complexion and stimulates appetite. It increases the seven body nutrients, called "dhatus" in Ayurveda. These seven nutrients are food, juice, blood, flesh, fat, bone marrow and semen. The fruit is beneficial in curing liver disorders, gaining of weight and other physical disturbances.

Mangoes are found to be significant in curing various disorders. Some of them are summarized below.

Heat stroke   Unripe mangoes protect humans from the adverse effects of hot and scorching winds. The drink prepared from unripe mango by cooking it in hot ashes and mixing it with sugar and water is an effective remedy for heat exhaustion and heat stroke. Eating raw mango with salt quenches thirst and prevents the excessive loss of sodium chloride and iron during summer due to excessive sweating.

Gastrointestinal disorders   Unripe green mangoes are beneficial in the treatment of gastrointestinal disorders. Eating one or two small tender mangoes in which the seed is not fully formed with salt and honey is found to be a very effective medicine for summer diarrhoea, dysentery, piles, morning sickness, chronic dyspepsia, indigestion and constipation.

Bilious disorders Unripe mangoes are an excellent remedy for bilious disorders. The acids contained in the fruit increase the secretion of bile and act as intestinal antiseptics. Therefore, eating green mango daily with honey and pepper cures biliousness, food putrefaction, i.e., the process in which proteins are decomposed by bacteria, urticaria and jaundice. It tones up the liver and keeps it healthy.

Blood disorders Unripe mango, due to its high vitamin C content, is valuable in treating blood disorders. It increases the elasticity of the blood vessels and helps in the formation of new blood cells. It aids the absorption of food-iron and prevents bleeding tendencies. It increases body resistance against tuberculosis, anaemia, cholera and dysentery.

Scurvy The amchur, a popular diet in Indian houses, consists of green mangoes skinned, stoned, cut into pieces and dried in the sun. Due to its citric acid content 15 g of amchur is believed to be equivalent to 30 g of good lime. It is valuable in the treatment of scurvy.

Eye disorders Ripe mangoes are highly beneficial in the treatment of night blindness caused by vitamin A deficiency. It is very common among children who are malnourished due to poverty. Liberal use of mangoes during the season will be very effective in such conditions. Eating mangoes liberally also prevents development of refractive errors, dryness of the eyes, softening of the cornea, itching and burning in the eyes.

Gain of weight The mango–milk cure is an ideal treatment for gain of weight. For this mode of treatment, ripe and sweet mangoes are selected. They are taken thrice a day—morning, noon and evening. The mangoes should be taken first followed by milk. The mango is rich in sugar but deficient in protein. On the other hand, milk is rich in protein but deficient in sugar. The deficiency of the one is compensated by the other. Mango, thus combines very well with milk and exclusive mango–milk diet taken for at least one month, leads to improvement in health, vigour and gain in weight. The quantity of milk and of the mangoes to be consumed in this mode of treatment should be carefully regulated according to the condition of the patient.

Diabetes   The tender leaves of the mango tree are useful in the treatment of diabetes. An infusion is prepared from fresh leaves by soaking them overnight in water and squeezing them well in water before filtering it in the morning. This infused water is taken every morning to control early diabetes. As an alternative to infusion, leaves can be dried in the shade, powdered and preserved. Half a teaspoonful of this powder is taken twice a day, in the morning and evening.

Diarrhoea   The mango seeds are valuable in treating diarrhoea. The seeds are collected during the mango season, dried in the shade, powdered, and stored for use as medicine. It is given in doses of about 1½–2 g with or without honey. Juice of fresh flowers when taken with one or two teaspoonsful of curd is also helpful in treating diarrhoea.

Female disorders   Mango seeds are considered useful in certain gynaecological disorders such as menorrhoea and leucorrhoea.

Throat disorders   The mango bark is very efficacious in the treatment of diphtheria and other throat diseases. Its fluid is locally applied and used as a gargle. The gargle is prepared by mixing 10 ml of the fluid extract with 125 ml of water.

Scorpion bites   The juice which oozes out at the time of plucking the fruit from the tree gives immediate relief to pain when applied to the area of scorpion bite or the sting of a bee. The juice can be collected and kept in a bottle.

Infections and mango   Bacteria invade the body tissues by crossing the epithelial barrier. Any damage or poor formation of epithelium, the tissue that covers the external surface of the body may lead to this invasion and cause infection. Liberal use of mangoes during the season contributes towards formation of healthy epithelium, thereby preventing frequent attacks of common infections such as cold, rhinitis and sinusitis. This is attributed to high concentration of vitamin A in mangoes.

Precautions to be taken care   Unripe mangoes should not be eaten in excess. Their excessive intake may cause throat irritation, indigestion, dysentery and abdominal colic. One should therefore, not consume more than one or two green mangoes daily. Water should not be drunk immediately after eating the green mango because it coagulates the sap and makes it more irritant. Sap or milky juice, which comes out on breaking the stalk of the green mango, is irritant and astringent. Eating green mangoes without draining the sap may cause mouth, throat, and gastrointestinal irritations. The sap should therefore, be fully squeezed out or the skin should be peeled before using raw mango. Excessive use of mangoes produces ailments like constipation, eye infections, blood impurities and seasonal fever. Children who consume the fruit in excess during the season generally suffer from skin disease.

Adathoda vasica

Malabar nut commonly known as "vasaka" is indigenous to India, where it is found in sub-Himalayan regions. It has traditionally been used for the treatment of various diseases and disorders, particularly for the respiratory tract ailments. The major phytoconstituents are alkaloids. The leaf extract has been used for treatment of bronchitis and asthma for many centuries. It relieves cough and breathlessness. It is also prescribed commonly in Ayurveda for bleeding due to idiopathic thrombocytopenic purpura, local bleeding due to peptic ulcer, piles, menorrhagia, etc.

Large doses of fresh juice of leaves have been used in tuberculosis. Its local use gives relief from pyorrhoea and bleeding gums. The leaves, roots and flowers are also used as blood purifiers, and in antispasmodic dysuria.

Amla (Emblica officinalis)

Amla has been hailed as a nugget of vitamin C in heat-stable form. One amla fruit is said to contain more vitamin C than a dozen oranges. It is used for treating respiratory complaints and for rejuvenation of both body and hair. According to Charaka, a physician of yore and father of Ayurveda, regular intake of amla or amla-based preparations is a sure method of stalling the ageing process.

Shoe flower (*Hibiscus rosasinensis*)

The shoe flower is commonly called as "semparuthi" and "jab pushp" in Tamil and Hindi respectively. This shrub has great cooling properties. Crushed leaves of hibiscus, applied to the scalp in summer, prevent dandruff and provide lustre to the hair. Dried and powdered with henna leaves and made into a paste, it soothes rashes, particularly eczema. The paste applied before a regular shampoo makes the hair soft and silky.

Turmeric

The dried and powdered rhizome of the turmeric plant is a powerful antiseptic for treating external wounds as well as intestinal infections. A level teaspoon of turmeric in a glass of hot milk, taken at bedtime, is said to stave off an attack of flu. Laced with honey, this combination even helps in a case of trauma. Flowers are used as emollient, refrigerant and aphrodisiac. The decoction of flowers is effective in treating bronchial catarrh.

Kizhanelli (*Phyllanthus amarus*)

An ayurvedic herb, *Phyllanthus* has been used in connection with secondary hepatitis and other diseases in ayurvedic medicine for over 2000 years. It has a large number of traditional uses such as employing the whole plant for jaundice, gonorrhoea, frequent menstruation and diabetes. It is topically used as a poultice for skin ulcers, sores, swelling and itching. The young shoots of the plant are administered in the form of an infusion for the treatment of chronic dysentery. *Phyllanthus* species are also found in other countries, including China, Philippines, Cuba, Nigeria and Guam. In one study, administering this herb for 30 days appeared to eliminate the hepatitis B virus in 22 of 37 cases. However, other studies have failed to confirm a beneficial effect of *Phyllanthus amarus* against hepatitis B. *Phyllanthus* has been studied primarily in carriers of the hepatitis B virus, as opposed to those with chronic active hepatitis.

Thoothuvalai

Thoothuvalai is another valuable Indian plant used in daily life for its medicinal value. The entire plant has therapeutic value. It is used in throat infections and in congestive conditions to extricate phlegm. It is also used in the treatment of snake bite by administering the powder of the dried leaves as a 'snuff.

3

DRUGS AND DISEASES

INTRODUCTION

The word disease can be interpreted in many ways. A few such interpretations are given below:

- Literally, "disease" means "lack of ease", that is, a pathological condition that presents as a group of symptoms peculiar to it and which establishes the condition as an abnormal entity different from other normal states or pathological body states with discontinuous effect.

- An impairment of health or a condition of abnormal functioning.

- Any abnormal condition of the body or mind that causes discomfort, dysfunction or distress to the person affected or those in contact with the person. Sometimes, the term is used broadly to include injuries, disabilities, syndromes, symptoms, deviant behaviour and atypical variations of structure and function, while in other contexts these may be considered in distinguishable categories.

- An abnormal condition of body structure and function, usually indicated by symptoms.

HOW ARE DISEASES TRANSMITTED?

Diseases are caused by germs. Germs multiply rapidly in warm and moist places. Before looking at ways to prevent the spread of diseases, it is necessary to know their different modes of transmission.

Droplet Contact

Also known as the respiratory route, it is the typical mode of transmission of many infectious agents. If an infected person coughs or sneezes the microorganisms suspended in warm, moist droplets, may enter the body of another person in the same environment through the nose, mouth, or eye surfaces. Diseases that are commonly spread by coughing or sneezing include.

- Bacterial meningitis
- Chickenpox
- Common cold
- Influenza
- Mumps
- Strep throat
- Tuberculosis
- Whooping cough

Airborne diseases spread, when droplets of pathogens are expelled into the air due to coughing, sneezing or talking. Airborne diseases of concern to emergency responders include:

- Smallpox
- Anthrax
- Meningitis
- Chickenpox
- Tuberculosis (TB)
- Influenza

Many of these diseases require prolonged exposure for infection to occur, posing only minimal threat to emergency responders. However, there are preventive measures, such as wearing masks or maximizing ventilation, that help reduce these risks.

Faecal–Oral Transmission

Direct contact is rare in this route, for humans at least. More common are the indirect routes; transmission of gastrointestinal viruses, bacteria or other parasites occur through faeces. Foodstuff or water become contaminated and the people who eat and drink them become infected. This is the typical mode of transmission for the infectious agents of:

- Cholera
- Diarrhoea
- Hepatitis A
- Polio
- Rotavirus

Oral Transmission

Diseases that are transmitted primarily by oral means may be caught through direct oral contact, or by indirect contact such as by sharing a drinking glass.

- Cytomegalovirus infections
- Herpes simplex virus (especially HSV-I)
- Mononucleosis

Transmission by Direct Contact

Diseases that can be transmitted by direct contact are called contagious (contagious is not the same as infectious; although all contagious diseases are infectious, not all infectious diseases are contagious). These diseases can also be transmitted by sharing a towel or items of clothing in close contact with the body such as socks. For this reason, contagious diseases often break out in schools, where towels are shared and personal items of clothing accidentally get swapped in the changing rooms.

Some diseases that are transmissible by direct contact include:

- Athlete's foot
- Impetigo and ringworm
- Syphilis
- Warts

Vertical Transmission

This is from mother to child, often *in utero,* as a result of the exchange of bodily fluids (mostly blood) during childbirth or through breast milk.

Some diseases that can be transmitted in this way include:

- AIDS
- Hepatitis B
- Syphilis

Transmission by Indirect Contact

Indirect contact vectors or fomites also allow transmission of disease agents. The term vector is used in a broad sense to signify anything that allows the transport and/or transmission of pathogen. However, according to a strict, ecological definition, **vector-borne transmission** occurs when a living creature, because of its ecological relationship to others, acquires a pathogen from one living host and transmits it to another. Thus, vector-borne transmission is a form of indirect horizontal transmission in which a biological intermediary, often an arthropod, carries a disease agent between animals.

Vectors may be either biological or mechanical. A biological vector is a vector that supports replication of the pathogen. The disease agent and the biological vector are considered to have a long-standing ecological relationship. Biological vectors are usually persistently infected with the disease agent and may even be a required part of that organism's life cycle. A mechanical vector, on the other hand, is a vector that carriers the pathogen but the pathogen is not altered while on the vector. Infection in mechanical vectors tends to be short-lived and a mechanical vector is considered little more than a flying fomite.

Iatrogenic Transmission

This is the transmission due to medical procedures, such as injection or transplantation of infected material.

Some diseases that can be transmitted iatrogenically include:

- Creutzfeld–Jakob disease by injection of contaminated human growth hormone.
- MRSA infections is often acquired as a result of a stay in hospital.

Waterborne Diseases

Waterborne diseases are spread by contamination of drinking water systems. Contamination of water systems may occur due to flood waters, water run-off from landfills, septic fields, and sewer pipes. Symptoms of waterborne diseases depend upon the infecting agent; symptoms may include diarrhoea, abdominal cramps, nausea, vomiting and low-grade fever.

A few waterborne diseases are listed below:

- Shigellosis
- Cholera
- Salmonellosis/Enteric fever
- Dengue
- Malaria
- Campylobacteriosis
- Hepatitis A
- Giardiasis

Some bacterial, viral or parasitic infections are contagious even before symptoms appear. Therefore, it is important to take the necessary steps to prevent the spread of these communicable diseases. Precautions necessary to prevent the spread of germs are the same for all diseases and should be followed regularly and consistently.

COMMON TYPES OF COMMUNICABLE DISEASES

A communicable disease is any bacterial, viral or parasitic infection in the body that can be spread from one individual to another. This varies from common cold and flu to more uncommon diseases like meningitis. Some of the common diseases that occur in young children include:

- Cold
- Chickenpox
- Diarrhoea
- Ringworm/head lice
- Impetigo

Infectious diseases of a more serious nature include:

- Meningitis
- Hepatitis
- HIV/AIDS

The transmission of cold virus from one person to another does not carry with it the life-threatening implications of a disease like hepatitis or AIDS. On the other hand, some common diseases can be life-threatening if they are not treated appropriately. For example, bacteria and parasites that cause gastrointestinal illnesses, with symptoms of vomiting and diarrhoea, can be quite serious in young children. Diarrhoeal diseases can even be fatal if a child becomes severely dehydrated.

Some of the common human diseases are discussed in detail in the following sections.

CHOLERA

Cholera also called Asiatic cholera is a waterborne disease caused by the bacterium *Vibrio cholerae* (Figure 3.1), which are typically ingested by consuming contaminated water or food. Raw or undercooked seafood such as shellfish may be the source of infection in areas where cholera is prevalent. A series of six pandemics of cholera, originating in the Bengal basin, ravaged the world in the 19th and early 20th centuries killing thousands of people. Cholera, like plague, became a disease of fear in this part of world. Cholera is endemic in India and South-East Asia. There are references to deaths due to dehydrating diarrhoea dating back to Hippocrates and Sanskrit writings. It was first described in a scientific manner by the Portuguese physician Garcia de Orta in the 16th century. John Snow, a London physician, proved the mode of transmission of cholera by water in 1849. In 1883, Robert Koch successfully isolated the cholera *Vibrio* from the intestinal discharges of cholera patients and proved conclusively that is was the agent of the disease.

Transmission

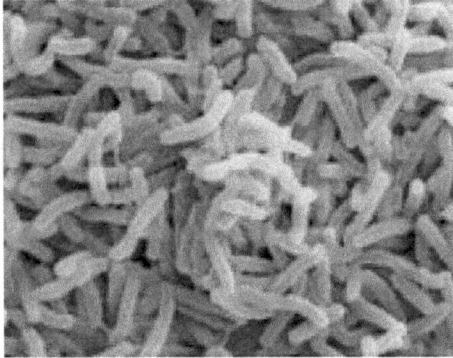

Figure 3.1   *Vibrio cholerae*—the bacterium that causes cholera (SEM image)

Cholera a severe diarrhoeal disease, is characterized by sudden onset of effortless vomiting and profuse watery diarrhoea. The bacterium *V. cholerae* is pathogenic only for humans.

Vibrio cholerae are

- Comma-shaped
- Curved gram-negative rods
- Motile

V. cholerae produces cholera toxin, the model for enterotoxins, whose action on the mucosal epithelium is responsible for the characteristic diarrhoea of the disease. In its extreme manifestation, cholerae is one of the most rapidly fatal illnesses known. *V. cholerae* can resist brackish water and coastal water environment with high salinity.

Cholera is transmitted through ingestion of food contaminated with the cholera bacterium. The contamination usually occurs when untreated sewage is released into waterways, affecting the water supply, or when any food is washed in the water, or when shellfish live in the affected waterway. It is rarely spread directly from person to person. The resulting diarrhoea allows bacteria to spread to other people under poor sanitary conditions.

Symptoms

The symptoms include those of general gastrointestinal tract upset: profuse diarrhoea (symptoms are the enterotoxins that *V. cholerae* produces). The cholera toxin interacts with G proteins and cyclic AMP in the intestinal lining to open ion channels. As ions flow into the intestinal lumen, water flows out through osmosis. This loss of fluid leads to dehydration, anuria, acidosis and shock. The watery diarrhoea is speckled with flakes of mucus and epithelial cells ("rice-water stool") and contains enormous number of vibrios. The loss of potassium ions may result in cardiac complications and circulatory failure.

The common symptoms include:

- Sudden onset of watery diarrhoea, up to litre (quart) per hour
- Diarrhoea has a "rice water" appearance—the stool looks like water with flakes of rice in it
- Diarrhoea has a "fishy" odour
- Dehydration can occur rapidly
- Rapid pulse (heart rate)
- Dry skin

- Dry mucous membranes or dry mouth
- Excessive thirst
- Glassy eyes or sunken eyes
- Lack of tears
- Lethargy
- Unusual sleepiness or tiredness
- Low urine output
- Sunken "soft spots" (fontanelles) in infants
- Abdominal cramps
- Nausea
- Vomiting

However the symptoms can vary from mild to severe and person to person.

Diagnosis

- Gram stain and stool culture show cholera
- Blood culture may show cholera

Treatment

- Cholera can be treated by adequate administration of oral rehydration salts (ORS) to replace the lost fluids and salts.
- Severely dehydrated persons should be given ORS intravenously. The World Health Organization (WHO) has developed an oral rehydration solution that is cheaper and easier to use than the typical intravenous fluid. This solution of sugar and electrolytes is now being used internationally.
- Use antibiotics like tetracycline; for tetracycline-resistant strains, cotrimoxazole, erythromycin, doxycyclin, chloramphenicol and furazolidone are administered to reduce the volume and duration of diarrhoea and the period of *Vibrio* excretion. However, tetracycline is usually not prescribed for children until after all the permanent teeth have come in, because it can permanently discolour teeth that are still forming.

Prevention

Simple sanitation is usually sufficient to stop an epidemic. There are several points along the transmission path at which the spread may be halted.

- Proper disposal and treatment of waste produced by cholera victims.
- Sewage: Treatment of general sewage before it enters the waterways.
- Sources: Warnings about cholera contamination posted around contaminated water sources.
- Sterilization: Boiling, filtering, and chlorination of water before use.
- Filtration and boiling is the most effective means of halting transmission. Cloth filters, though very basic, have greatly reduced the occurrence of cholera when used in poor villages in Bangladesh that rely on untreated surface water. In general, education and sanitation are the limiting factors in prevention of cholera epidemics.
- Sanitary disposal of faeces.
- Handwashing after defaecation using soap.
- Consuming uncontaminated food especially seafood and vegetables.
- Sanitary preparation of food.
- Maintaining cleanliness.

Routine treatment of community with antibiotics, or "mass chemoprohylaxis", has no effect in controlling the spread of cholera. Limited stocks of two oral cholera vaccines that provide high-level protection for several months against cholera caused by cholera O1 have recently become available in few countries.

MALARIA

Approximately 300 million people worldwide are affected by malaria and between one and 1.5 million people die from it every year. Previously extremely widespread, malaria is now mainly confined to Africa, Asia and Latin America. Malaria, the potentially fatal disease in developing countries, affects people of any age travelling to these countries. The problems of controlling malaria in these countries are aggravated by inadequate health structures and poor socioeconomic conditions. The situation has become even more complex over the last few years with the increase in resistance to the drugs normally used to combat the parasite that causes the disease.

Malaria is caused by protozoan parasites of the genus *Plasmodium*. Four species of *Plasmodium* can produce the disease in its various forms:

- *Plasmodium falciparum*
- *Plasmodium vivax*
- *Plasmodium ovale*
- *Plasmodium malariae*

P. falciparum is the most widespread and dangerous of the four. When left untreated it can lead to fatal cerebral malaria. Malarial parasites are transmitted from one person to another by the female anopheline mosquito. The males do not transmit the disease as they feed only on plant juices. There are about 380 species of anopheline mosquitoes, but only 60 or so are able to transmit the parasite. Like all other mosquitoes, the anophelines breed in water, each species having its preferred breeding grounds, feeding patterns and resting place. Their sensitivity to insecticides is also highly variable. Plasmodium grows in the gut of the mosquito and is passed on to the saliva of the insect, which is then transmitted to the victim. The parasites are then carried by blood to the victim's liver where they invade the cells and multiply. After 9–16 days they return to the blood and penetrate the red blood cells, where they multiply again, progressively breaking down the red blood cells.

The transmission of malaria can be outlined as follows:

1. The cycle begins when a mosquito bites an infected person. The parasite enters the feeding mosquito through the infected person's blood and develops into the next stage.

2. When the infected mosquito bites again, the parasites are injected into the person being bitten, spreading the infection.

3. Once inside a human, the parasites multiply in the liver. Within an average of 2 to 4 weeks, they mature and are released into the blood.

4. The parasites enter red blood cells (RBCs) and multiply again. Eventually, the RBCs rupture, releasing new parasites to enter new RBCs, repeating the cycle. This phase of the disease occurs in cycles of approximately 48 hours.

This induces bouts of fever and anaemia in the infected individual. In cerebral malaria, the infected red cells obstruct the blood vessels in the brain. Other vital organs can also be damaged often leading to the death of the patient.

The life cycle (Figure 3.2) of the malarial parasite comprises of

- The exogenous sexual phase (sporogony) in female anopheline mosquitoes.
- An endogenous asexual phase (schizogony) in human host.
- When a female anopheline mosquito bites a human host harbouring the malarial parasite, it ingests malarial parasites along with the blood meal.
- The asexual forms are digested, while the mature forms called gametocytes undergo further development.
- In the male gametocytes, the nucleus divides into 4–8 nuclei, each forming thread-like structures called microgametes.
- Female gametocytes undergo maturation process and form macrogametes.
- In the stomach of the mosquito, microgamete fuses with the macrogamete (fertilization) resulting in a product called a zygote.
- Within 18–24 hours, the zygote develops into a long mobile worm-like form called the ookinete $(18-24\,\mu)$.

- The ookinete passes between epithelial cells to the outer surface of the stomach of the mosquito and rounds up into a small sphere called an oocyst.

- The oocyst increases in size and the nucleus divides repeatedly to form (40–80 μ) sporoblast.

- The divided nuclei of sporoblast form elongated sporozoites (10–15 μ) and are released into the body cavity.

- These sporozoites migrate to the salivary gland of the mosquito and are now ready for transmission to human host.

- When the mosquito feeds on blood, sporozoites are released into the bloodstream of the human host.

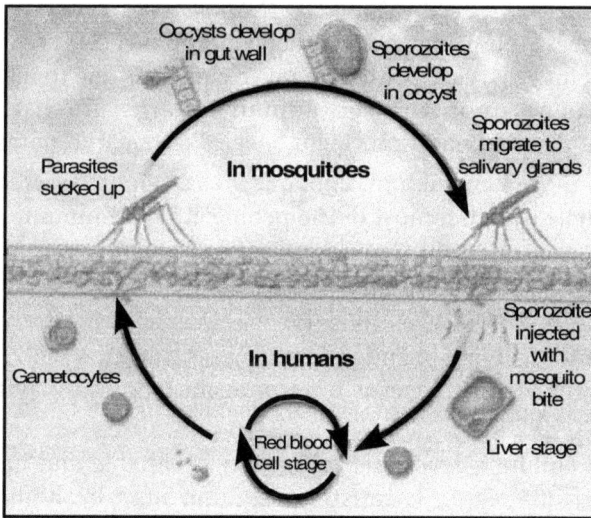

Figure 3.2   The malaria cycle (*Plasmodium* life cycle)

Ruptured blood cells release free parasites (gametocytes) into the host's bloodstream. The human host shows the classic malarial symptoms at this stage. The gametocytes are sucked up by a feeding mosquito and the cycle begins again.

Symptoms

The symptoms of malaria may be mild at first and are similar to the flu, making them difficult to diagnose. A person with malaria may experience the following symptoms.

- The classical malarial fever has three stages.

  Cold stage (Shivering) lasting for 2 hours,
  Host stage lasting for 3–4 hours,
  Sweating lasting for 2–4 hours.

- These symptoms occur at regular intervals, i.e., they show periodicity. Periodicity becomes apparent only after the first few days of illness and depending upon the species it could be tertian (every 2nd day for *vivax*, *falciparum* and *ovale*) or quartan (every 4th day for *malariae*)

- Malaria especially caused by *P. falciparum* is a very variable disease mimicking many other conditions such as typhoid, meningitis, gastroenteritis, etc. Symptoms due to malaria depend upon the age, immune status, intensity of transmission and prevalent species of malarial parasite.

- In those living in non-endemic areas, symptoms are very vague because, most of the people are non-immune. The disease often presents like flu. Presence of rigours and a rapid increase in temperature should alert the physician to the possibility of malaria.

- In those living in endemic areas, the attacks are modified by immunity. Fever is often intermittent and may have periodicity.

- In children, fever due to malaria is variable and has no periodicity. It is mostly irregular, may be high and continuous or low-grade.

- Other manifestations in children are pallor, nausea, vomiting, lethargy restlessness, headache, diarrhoea, muscle aches, abdominal pain and cough.

- Severe forms of malaria are seen with *P. falciparum* species. Some of the manifestations are severe vomiting and diarrhoea, cerebral malaria, algid malaria, hepato-renal syndrome, black water fever and malaria shock lung.

Malarial symptoms usually appear within 7 to 21 days of the mosquito bite. Pregnant women are also highly susceptible since the natural defense mechanisms are reduced during pregnancy.

Malaria caused by the *P. falciparum* parasite is the only type known to be life-threatening. *P. falciparum* malaria can lead to:

- Liver failure
- Kidney failure
- Fluid in the lungs
- Convulsions
- Coma

Diagnosis

Malaria is diagnosed by staining blood smears with Giemsa and examining through microscope using 100× oil immersion objective.

- Microscopy is sensitive, can detect densities as low as 5–10 parasites/ml blood.
- When parasites are found, their species and circulating stage can be found.

Treatment

The WHO through malaria eradication movement has eradicated malaria from the developed parts of the world. This was done by improving the living conditions, use of antimalarial drugs and use of pesticides for destroying the breeding of mosquitoes. However, the developing countries are yet to record their achievements. The treatment for malaria depends upon the geographic area to which the victim belongs. Different areas of the world have malarial types that are resistant to certain medications particularly to chloroquine.

Some of the antimalarial drugs used are

1. Quinine sulphate
2. Mefloquine
3. Chloroquine phosphate

4.   Pamaquine plasmoquin

5.   Mepacrine

6.   Proguanil

Prevention

Transmission of malaria can be barred by

- Preventing proliferation of mosquitoes; water should not be allowed to stagnate since stagnant waters are breeding grounds for mosquitoes.

- Preventing mosquito bites by using mosquito repellants, bed nets, and clothing that covers most of the body.

- Chemoprophylaxis (preventive medications).

LYMPHATIC FILARIASIS (ELEPHANTIASIS)

Lymphatic filariasis, also known as elephantiasis, is best known from dramatic photos of people with grossly enlarged or swollen arms and legs. The disease is caused by parasitic worms, including *Wuchereria bancrofti*, *Brugia malayi* and *B. timori*, all transmitted by mosquitoes. Lymphatic filariasis currently affects 120 million people worldwide, and 40 million of these people have serious disease. When an infected female mosquito bites a person, it may inject the worm larvae, called microfilariae, into the blood. The microfilariae reproduce and spread throughout the bloodstream, where they can live for many years.

Often disease symptoms do not appear until years after infection. As the parasites accumulate in the blood vessels, they can restrict circulation and cause fluid accumulation in surrounding tissues. The most common, visible signs of infection are excessively enlarged arms, legs, genitalia and breasts. Medicines to treat lymphatic filariasis are most effective when used soon after infection, but they do have some toxic side effects. In addition, the disease is difficult to detect at an early stage. Therefore, improved treatments and laboratory tests are needed. A vaccine is not yet available.

Filariasis is an infection with any of the several round, thread-like parasitic worms. The most common type of filariasis is infection with a parasitic worm that lives in the human lymph system. This is called lymphatic filariasis. At least 120 million people in 73 countries worldwide are estimated to be infected with filariasis parasites. The most widespread is *Wuchereria bancrofti*, which affects about 100 million people in Africa, India, South-East Asia, the Pacific islands, South America and the Caribbean. The *Brugia malayi* and *Brugia timori* parasites affect about 12 million people in South-East Asia. Lymphatic filariasis is a disease of the tropics.

Filariasis is spread from infected persons to uninfected persons by mosquitoes. Adult worms live in an infected person's lymph vessels. The females release large numbers of very small worm larvae, which circulate in an infected person's bloodstream. When the person is bitten by a mosquito, the larvae are introduced. These develop into adult worms and can spread to other people via mosquito bites. After a bite, the larvae pass through the skin, travel to the lymph vessels, and develop into adults, which live for about 7 years. Then the cycle begins again. Most of the signs and symptoms of filariasis are caused because of the adult worms living in the lymphatic system. Tissue damage caused by the worms restricts the normal flow of lymph fluid. This results in swelling, scarring, and inflammations. The legs and groin are most often affected. Filariasis larvae can sometimes be detected in blood. Lymphatic filariasis is rarely fatal, but it can cause recurring infections, fevers, severe inflammation of the lymph system, and a lung condition called tropical pulmonary eosinophilia (TPE). In about 5% of infected persons, a condition called elephantiasis causes the legs to become grossly swollen. This can lead to severe disfigurement, decreased mobility, and long-term disability. Testicular hydrocele is the disfiguring enlargement of the scrotum.

Treatment

1. Administrating medicine to kill circulating larvae and adult worms.
2. Soap and water and skin care to prevent secondary infections.

3.   Elevation, exercises, and, in some cases, pressure bandages to reduce swelling.

There is no vaccine for filariasis. Prevention centres on mass treatment with anti-filariasis drugs to prevent ingestion of larvae by mosquitoes, public health action to control mosquitoes and individual action to avoid mosquito bites.

To avoid being bitten by mosquitoes:

- If possible, staying inside between dusk and dark. This is when mosquitoes are most active in their search for food.
- When outside, wearing long pants and long-sleeved shirts.
- Spraying exposed skin with an mosquito repellent.

JAUNDICE

Jaundice is not a disease but rather a sign that manifests itself in many different diseases. Jaundice comes from the French word *jaune*, which means yellow. Jaundice is the yellow colouration of the skin, the mucous membranes or the eyes (sclerae, the whites of the eyes) caused by high levels of the chemical, bilirubin, in blood. The colour of the skin and sclerae vary depending on the level of bilirubin. When the bilirubin level is slightly elevated, they are yellowish. When the bilirubin level is high, they tend to be brown.

What Causes Jaundice?

Bilirubin comes from red blood cells. When red blood cells get old, they are destroyed. Haemoglobin, the iron-containing chemical in red blood cells that carries oxygen, is released from the destroyed red blood cells after the iron it contains is removed. The chemical that remains in the blood after the iron is removed becomes bilirubin. One of the functions of liver is to produce and secrete bile into the intestine to help digest dietary fat. Another function is to remove toxic chemicals or waste products from the blood. Since bilirubin is a waste product, the liver removes bilirubin from the blood. After the bilirubin has entered the liver cells, the cells conjugate (attaching to other chemicals, primarily glucuronic acid) to the bilirubin, and then secrete the bilirubin–glucuronic acid complex into bile. The complex that is secreted in bile is called conjugated bilirubin. The conjugated bilirubin is eliminated in the faeces. (Bilirubin is what gives faeces its brown colour.) Another condition called Gilbert's syndrome is a benign, hereditary condition in which mild jaundice develops. It is caused by low levels of some bilirubin-processing enzymes in the liver. This condition, once recognized, requires no further treatment or evaluation. There are other more rare hereditary conditions caused by elevated bilirubin levels.

Physiological jaundice is the name for normal jaundice commonly seen in babies. Pathological jaundice is the name given when jaundice presents a health risk, either because of its degree or its cause. Pathological jaundice can occur in children or adults.

It arises due to many reasons, including blood incompatibilities, blood diseases, genetic syndromes, hepatitis, cirrhosis, bile duct blockage, other liver diseases, infections or medications. The term also applies to physiological jaundice exaggerated by dehydration, prematurity, difficult delivery or other reasons.

Jaundice occurs when there is:

- An excess of bilirubin production. For example, patients with haemolytic anaemia have an abnormally rapid rate of destruction of red blood cells that releases large amounts of bilirubin into the blood.
- A defect in the liver that prevents bilirubin from being removed from the blood, and converted to bilirubin–glucuronic acid complex (conjugated) or secreted in bile.
- Blockage of the bile ducts that decreases the flow of bile and bilirubin from the liver into the intestine. For example, the bile ducts can be blocked by cancers, gallstones or inflammation of the bile ducts.

The decreased conjugation, secretion or flow of bile that results in jaundice is referred to as cholestasis. However, cholestasis does not always result in jaundice.

The possible reasons for jaundice include heavy use of alcohol (alcoholic liver disease), and use of illegal, injectable drugs (viral hepatitis). Recent investigations with administering modern drugs suggests drug-induced jaundice. Episodes of abdominal pain associated with jaundice suggests blockage of the bile ducts usually by gallstones.

The physical examination of the abdomen of a jaundiced patient showing masses (tumours) in the abdomen suggests cancer infiltrating the liver (metastatic cancer) as the cause of the jaundice. An enlarged, firm liver suggests cirrhosis. A rock-hard, nodular liver suggests cancer within the liver.

Measurement of bilirubin can be helpful in determining the causes of jaundice. Markedly higher concentration of unconjugated bilirubin relative to that of conjugated bilirubin in the blood suggests haemolysis (destruction of red blood cells).

Marked elevations in liver tests (aspartate aminotransferase (AST) and alanine aminotransferase (ALT) suggest inflammation of the liver (such as viral hepatitis). Elevations in other liver tests, e.g. alkaline phosphatase, suggest diseases or obstruction of the bile ducts. Ultrasonography, and magnetic resonance imaging (MRI) are some of the diagnostic measures used. Liver biopsy is particularly good for diagnosing inflammation of the liver and bile ducts, cirrhosis, cancer, and fatty liver.

Treatment

With the exception of the treatments for special cases of jaundice, the treatment of jaundice usually requires a diagnosis of the specific cause of jaundice and treatment directed at the specific cause, e.g. removal of a gallstone blocking the bile duct. Patients suffering from jaundice are advised rest, intake of plenty of oral fluids such as sugar cane juice and other fruit juices, and fat-free diet.

ANAEMIA

In a healthy person, the number of red blood cells and the percentage of haemoglobin depend on a delicate balance between fresh formation and destruction. This balance may be upset in disease or other conditions thus resulting in deficiency in the oxygen-carrying capacity of blood, which is termed as anaemia. The anaemic deficiencies are mainly due to the inadequate concentration of iron, vitamin B_{12} or folic acid.

Iron

In the human body, approximately 3 g of iron is present. About 1 g of it is present in haemoglobin, and 1 g is stored in liver, spleen and bone marrow. Nearly 50 mg of iron is released in the body daily by the destruction of red cells but most of it is reused. Iron is effective only in the treatment of anaemias caused by iron deficiency. The approximate daily requirement of iron is 8–18 mg for children, 15–20 mg for women and 10–15 mg for men. In a carrying mother, the foetus accumulates 200–400 mg of iron. Iron is lost during childbirth and lactation. During the menstrual cycle, women regularly lose nearly 25 mg of iron. Therefore, it is necessary to supplement the loss of iron through either dietary factors and/or medicinal preparations.

Some of the iron drugs are:

- Dried $FeSO_4$ 200 mg tablets
- $FeSO_4$ syrup containing 60 mg/5 ml
- $FeSO_4$ paediatric drops with 125 mg/ml
- Ferrous gluconate as 300 mg tablets and syrup of 36 mg/5 ml
- Ferrous fumarate as 200 mg tablets. This is resistant to oxidation
- Various other preparations like iron citrate, ferrous chelates are available. However, they offer no remarkable advantage over the other preparations. Actually, the iron absorption is much less with iron citrate, tartrate, pyrophosphate, etc.

(In all the preparations ferrous sulphate is mixed with reducing reagents like glucose or ascorbic acid to prevent its oxidation.)

The side effects noticed with ferrous sulphate preparations are gastric disturbances, diarrhoea and constipation. If a patient is given iron preparations for a longer duration they may suffer from haemochromatosis, a condition in which the skin becomes brown coloured due to the deposition of increased amount of iron in the tissue. Among children large doses of iron may produce metal poisoning. Such conditions are treated if deferoxamine is given.

Vitamin B$_{12}$

The deficiency of vitamin B$_{12}$ or cyanocobalamine, a cobalt complex, causes pernicious anaemia (deadly anaemia). In the early days, pernicious anaemia was considered as incurable. When liver extracts are administered, an improvement in the condition was observed. Later on, vitamin B$_{12}$ was isolated from liver extracts and fermentation liquors of *Streptomyces griseus*. Vitamin B$_{12}$ is also present in cow's milk. This vitamin is not present in higher plants but can be synthesized by certain microorganisms.

Folic Acid

Folic acid is otherwise called pteroyl-L-glutamic acid. It is a water-soluble B vitamin and is made up of pteridine, PABA and glutamic acid moieties. Other names for folic acid are folacin, vitamin B$_c$, vitamin B$_9$ and *Lactobacillus casei* factor. It is present in leaves, foliages, spinach, mushrooms, liver, yeast, bone marrow, soyabean, etc. Folic acid deficiency causes anaemia and leucopenia, a condition where a decrease in the number of white blood cells occurs.

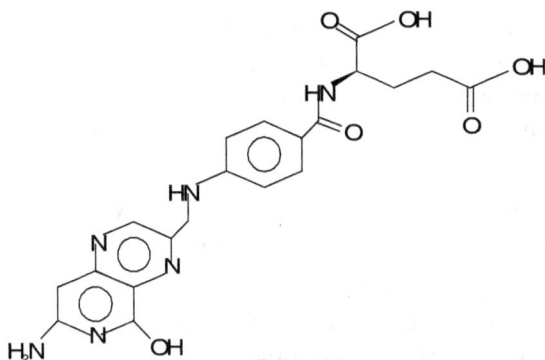

Folic acid

Folinic acid is the natural factor in the digest of liver, and not folic acid. It is an effective antanaemic agent for human beings, a growth factor for chicks. It is prepared from folic acid on formylation, reduction reactions, etc. Folic acid is formed from folinic acid during the isolation procedure.

4

DRUG METABOLISM AND ACTION

The word "Drug" is derived from the French word *drogue* which means dry herb. In a general way, drugs may be defined as substances which are made available by chemical synthesis or from natural resources, and when administered, can cause a reaction, which can help to prevent, relieve, cure or recognize an illness or disorder in human beings or animals.

REQUIREMENTS OF AN IDEAL DRUG

The basic requirements for any drug to be "ideal" are:

1. When administered to the ailing individual or host, its action should be localized at the site where it is desired to act. In actual practice, there is no drug, which can function so. It normally distributes itself to many tissues of the host.

2. It should act on systems with safety and efficiency.

3. It should have minimum side effects.

4. It must not harm the host tissue or the physiological processes.

5. It should not have any toxic effects.

6. The cells should not acquire tolerance or resistance to the drug when administered repeatedly.

Very few drugs satisfy all the above conditions. Thus, the search for ideal drugs continues. Important sources of drugs include synthetic, plant-derived, animal-derived, minerals, etc. The beginning of therapy by chemicals has lost in antiquity, because it preceded recorded medical history. Early success in the quest for chemicals effective against diseases was predominantly found for infective diseases rather than the illness accompanying the ageing process. The earliest references about medicinal preparations in writing came from India (Rig Veda) and from China in 2500–3000 BC. It was the Greek physician Hippocrates in 450 BC, who laid the foundation for modern medicine. It is only in the 18th century that Paul Ehrlich did outstanding works in medicinal chemistry and is called as the Father of Chemotherapy.

An antimicrobial drug is a chemical substance that destroys disease-causing microorganisms with minimal damage to host tissues. Chemotherapeutic agents include chemicals that combat disease in the body. Paul Ehrlich developed the concept of chemotherapy to treat microbial diseases. He predicted the development of chemotherapeutic agents that would kill pathogens without harming the host. The modern era of antimicrobial chemotherapy began in 1929 with Fleming's discovery of the powerful bactericidal substance, penicillin. In the early 1940s, spurred partially by the need for antibacterial agents in the World War, penicillin was isolated, purified and injected into experimental animals, where it was found to not only cure infections but also to possess incredibly low toxicity for the animals. This fact ushered into being the age of antibiotic chemotherapy and an intense search for similar antimicrobial agents of low toxicity to animals that might prove useful in the treatment of infectious disease. The rapid isolation of streptomycin, chloramphenicol and tetracycline soon followed, and by the 1950s, these and several other antibiotics were in clinical usage. Domagk's discovery in 1935 of synthetic chemicals (sulphonamides) with broad antimicrobial activity could also be viewed as an important episode in medicinal chemistry.

The most important property of an antimicrobial agent, from the point of view of the host, is its selective toxicity, i.e., the agent acts in some way that inhibits or kills bacterial pathogens but has little or no toxic effect on the host. This implies that the biochemical processes in the bacteria are in some way different from those in the host cells, and that the advantage of this difference can be utilized in chemotherapy.

In its non-oncological use, the term may also refer to antibiotics (antibacterial chemotherapy). In that sense, the first modern chemotherapeutic agent was Paul Ehrlich's arsphenamine, an arsenic compound discovered in 1909 and used to treat syphilis. This was later followed by sulphonamides discovered by Domagk and penicillin-G discovered by Alexander Fleming.

Other uses of cytostatic chemotherapy agents are the treatment of autoimmune diseases such as multiple sclerosis and rheumatoid arthritis, the treatment of some chronic viral infections such as hepatitis, and the suppression of transplant rejections.

Most microbiologists distinguish two groups of chemotherapeutics used in the treatment of infectious disease: antibiotics, which are invariably natural substances produced by certain groups of microorganisms, and sulphadrugs, which are chemically synthesized. A hybrid substance is a semi-synthetic antibiotic, wherein a molecular version produced by the microbe is subsequently modified by the chemist to achieve desired properties.

The majority of chemotherapeutic drugs can be divided into: alkylating agents, antimetabolites, anthracyclines, plant alkaloids, topoisomerase inhibitors and antitumour agents. All these drugs affect cell division or DNA synthesis.

In addition, some drugs may be used which modulate tumour cell behaviour without directly attacking those cells. Hormone treatments fall into this category of adjuvant therapies.

DRUG METABOLISM

To be useful, a drug must not only enter the body reliably and reach the site of action, but it must also be eliminated in a reasonable time. Drugs which are relatively lipid-insoluble and ionized, are excreted unchanged by the kidney. However, the volatile anaesthetics are highly lipid-soluble and are eliminated due to their volatile nature. In general, the highly lipid-soluble, unionized drugs are reabsorbed from the glomerular filtrate by diffusion across the renal tubule epithelium. By this, the drugs remain in the body for a longer time unless altered suitably. To be eliminated, these drugs must be converted into lipid insoluble and ionized metabolites, and this is the most important function of drug metabolism.

> Brodie suggested that there are drug-metabolizing enzymes present in the body during the course of evolution and these enzymes are non-specific. He also demonstrated a situation by estimating the rate of disappearance of drugs from the body if there were no drug-metabolizing enzymes. He assumed that if a less lipid-soluble drug is evenly distributed throughout the body, it requires more than 5 h to reduce the drug to half its the concentration and for a lipid-soluble drug it requires about 30 days. The use of such substances in industry or in agriculture that are little or never metabolized little or not would pose an environmental threat.

The main site of drug metabolism is the liver, although other tissues may also participate. Therefore, drugs have greater or less than expected effects in liver diseases. Metabolism of drugs involves two main chemical changes:

- Phase I—Non-synthetic reactions such as oxidation, reduction and hydrolysis.
- Phase II—Synthetic reactions such as conjugation with glucuronic acid, acetic acid, glycine, etc.

Some drugs pass through phase I first and then phase II, while others pass only through any one of the two phases.

All transformations of metabolism make the drug metabolic products more polar than the parent drug. This is of significance in the clearance of drugs and other metabolic products chiefly by kidney and biliary excretion. When a drug is absorbed from the intestine, all of it must pass through liver. A drug that is readily metabolized by the liver is required relatively in high doses to get adequate concentration in the systemic blood. It is for this reason that certain drugs are injected in fairly smaller doses than the oral doses. Drugs or substances having high lipid/water partition coefficient will have very low renal clearance. This leads to prolonged stay of drug in the body. If such drugs are metabolized producing polar metabolic products possessing low lipid/water partition coefficient they are cleared readily. For instance, a drug containing a methyl group in it, on metabolism involving oxidation reaction will be converted into a more polar metabolite, carboxylic acid. Thus, a single metabolic alteration could reduce the biological half-life of the drug from many hours to few minutes. This is also achieved by conjugating a relatively non-polar drug with certain anions such as sulphate ion. However, the decreased lipid solubility does not necessarily lead to increased water solubility. When the antibacterial sulphonamides are metabolized to more polar, less lipid-soluble acetyl derivatives, some of them are less water-soluble than the parent drug.

For example, sulphathiazole on metabolism is transformed to acetylsulphathiazole, which is less water-soluble than the parent compound.

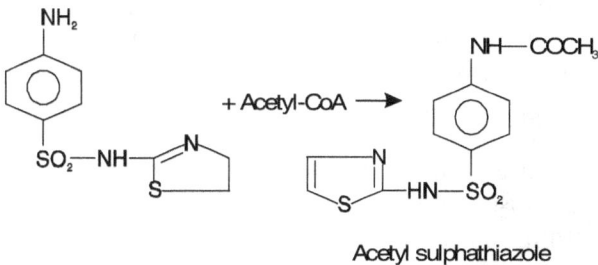

Acetyl sulphathiazole

This reduced water solubility of acetylsulphathiazole leads to serious toxicity such as precipitation in renal tubules. It is also known that certain polar drugs are metabolized to less polar products. An example is the deacetylation of acetanilide to aniline. Although the metabolic transformations tend to reduce the drug action, it is inappropriate to view the drug metabolism as "detoxification." In many instances, metabolites have greater biological activity than the drug itself. In certain cases, the drugs themselves are inert but the metabolites are responsible for the pharmacological action. There are drugs whose metabolites are wholly or in part responsible for the toxic side effects.

CYTOCHROMES IN DRUG METABOLISM

Cytochromes are conjugate proteins containing haeme as the prosthetic group and associated with electron transport and redox processes. They participate in the phase of biochemical respiration called oxidative phosphorylation. The terminal electron transport chain of oxidative respiration involves at least five different cytochromes. It has been customary to assign the cytochromes to the groups a, b, and c according to the nature and mode of binding of the haeme prosthetic group; group d was introduced later by the IUB. The various cytochrome subclasses are organized by the type of haeme and wavelength range of their reduced alpha-absorption bands. Cytochromes act as carriers of hydride ions (sometimes are considered to be the equivalent of electron pairs) in the series of complex enzymes known as the electron transport chain. The hydride ions or their equivalents travel along the electron transport chain. Each cytochrome is in turn reduced by accepting a hydride ion or pair of electrons and then oxidized by donating the hydride ion or pair of electrons to the next acceptor in the chain; in the process, the iron atom in the cytochrome haeme shuttles between the ferrous and ferric states. Formally, this redox change involves a single-electron, reversible equilibrium between the Fe(II) and Fe(III) states of the central iron atom. Cytochromes are thus capable of performing oxidation and reduction. Because the cytochromes (as well as other complexes) are held within membranes in an organized manner, the redox reactions are carried out in the proper sequence for maximum efficiency. The haeme group is a highly conjugated ring system (its electrons are very mobile) surrounding an iron ion, which readily interconverts between the Fe^{2+} (reduced) and Fe^{3+} (oxidized) states.

Cytochrome P450 was discovered in 1955 by Axelrod and Brodie *et al.* who identified an enzyme system in the endoplasmic reticulum of the liver which was able to oxidize xenobiotic compounds. A CO binding pigment in liver microsomes which had an absorption maximum at 450 nm (Figure 4.1) was demonstrated to be a haemoprotein and was named cytochrome P450. In addition, although found mainly in the liver, P450s have been identified in many other organs. P450s are responsible for the metabolism of numerous endogenous compounds as well as an enormous range of xenobiotic compounds including many toxins and carcinogens as well as drugs.

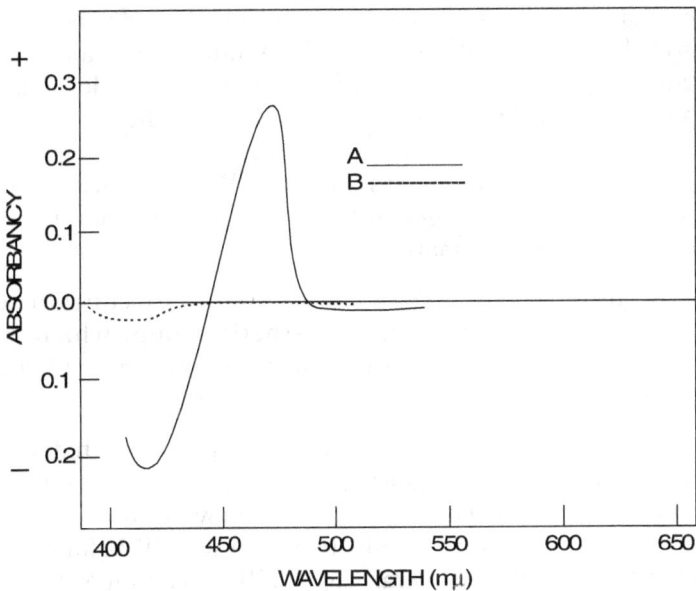

Figure 4.1   UV absorption spectrum of P450

The cytochrome P450 (CYP) haemoprotein superfamily is the major group of enzymes that participates in oxidative biotransformation in most organs. CYPs are located in the endoplasmic reticulum (microsomal fraction) or mitochondrial fraction of mammalian tissues. Cytochrome P450(CYP) enzymes catalyse the oxidative conversion of drugs and other lipophilic compounds to hydrophilic metabolites. Thus, CYPs play a dominant role in the elimination of drugs from the body.

Inhibitory interactions occur when drugs compete for oxidation by specific CYPs whereas certain drugs increase the capacity for oxidative biotransformation by inducing the synthesis of new CYPs. Most drugs and other lipophilic chemicals undergo oxidative biotransformation to form hydroxylated metabolites that are either eliminated or metabolized further by conjugation with a range of endogenous moieties, such as glucuronic acid or sulphate. Thus, metabolic oxidation has a critical role in the termination of drug action and influences the residence time of toxic chemicals in the body.

Drug-metabolizing CYPs are expressed at high level in hepatic microsomal fractions, with lower levels in lung, kidney and the gut. Other CYPs, that have specialized physiological roles, may be inactive in drug biotransformation. During the 1970s, it became apparent that mammalian liver contained a number of different CYPs. Typically, one or more CYPs were purified to homogeneity from detergent-solubilized microsomes from differently pretreated animals.

As of now all CYPs have been found to contain a ferroprotoporphyrin IX haeme prosthetic group, which is involved in the activation of molecular oxygen, and an apoprotein portion.

A systematic nomenclature based on amino acid relatedness was developed in 1987. Regular updates and refinements have been made since then. This nomenclature may be summarized as follows. The particular family to which a CYP belongs is indicated by an arabic numeral, e.g. CYP1. A capital letter is then used to indicate CYPs that belong to subfamilies, e.g. CYP1A or CYP1B. Another arabic numeral then signifies the individual CYP gene product, e.g. CYP1A1, CYP1A2 or CYP1B1. CYPs within the same family are at least 40% similar while those in subfamilies are at least 55% identical at the amino acid level. The nomenclature was developed to indicate the evolutionary relationships between CYPs; this relationship has been extended recently. There are 43 CYP families in the animal kingdom but some families tend to segregate into clusters termed CYP clans.

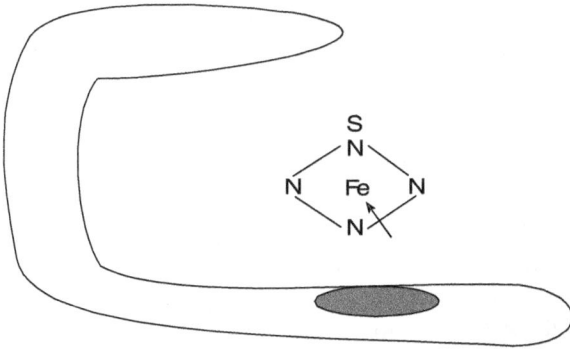

Figure 4.2   Schematic diagram of the CYP holoenzyme

Figure 4.2 shows how the apoprotein is bonded to the haeme prosthetic group via a cysteinyl residue. The arrow indicates the sixth axial position of the CYP haeme iron at which molecular oxygen is bound and activated for substrate oxidation; the substrate-binding site is the hatched region shown opposite the haeme moiety.

In the native microsomal membrane, the haeme iron of CYP is in the ferric state and is able to bind drugs or other xenobiotics to generate a binary enzyme–substrate complex. This complex undergoes reduction by an electron from NADPH, which is transferred to CYP via the flavoprotein NADPH-CYP-reductase (Figure 4.4). Molecular oxygen is subsequently coordinated at the sixth ligand position of the reduced (ferrous) CYP opposite the cysteine thiolate ligand at the fifth axial position of the haeme group. This oxyform of CYP is reduced by a second electron from NADPH or, in some cases, from NADH delivered via NADH-cytochrome b_5 reductase. Heterolytic cleavage of the dioxygen bond releases the distal atom of oxygen as a water molecule, while the proximal atom remains coordinated to the CYP haeme and constitutes the reactive oxidant. Although the nature of the activated oxyform remains uncertain, current evidence favours an oxyperferryl species. Oxygen activation by CYPs may be circumvented in the presence of appropriate exogenous oxygen sources, including peroxides, iodosylbenzene and periodates. Two-electron oxo-transfer from such species to CYP generates the pro-oxidant form of the cytochrome and supports substrate oxidation directly. This pathway has been termed the "peroxide shunt."

In an *in vitro* system containing liver enzymes, various drugs can be oxidized if the components NADPH, O_2 and Mg^{2+} are available. The requirement of NADPH-reduced pyridine nucleotide and molecular oxygen evinces the involvement of the enzyme monoxygenase. One molecule of oxygen is consumed in the reaction for the oxidation of each molecule of substrate/drug. One atom of oxygen is introduced into the substrate, while the other is reduced to form a molecule of water. The overall reaction is summarized in the equations given.

A-P450 cytochrome is in the oxidized form.

AH_2 is the reduced form of P450.

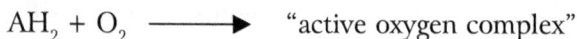

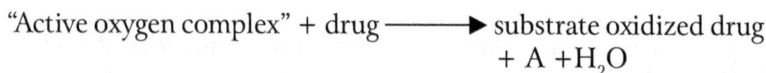

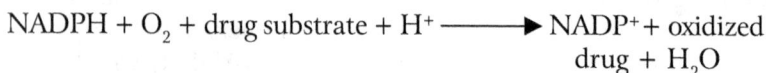

$$NADPH + A + H^+ \longrightarrow AH_2 + NADP^+$$

$$AH_2 + O_2 \longrightarrow \text{"active oxygen complex"}$$

$$\text{"Active oxygen complex"} + drug \longrightarrow \text{substrate oxidized drug} + A + H_2O$$

$$NADPH + O_2 + drug\ substrate + H^+ \longrightarrow NADP^+ + \text{oxidized drug} + H_2O$$

The principal physiological role of the P450 superfamily of enzymes is that of a monoxygenase.

The catalytic reaction can be summarized as follows.

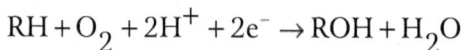

$$RH + O_2 + 2H^+ + 2e^- \rightarrow ROH + H_2O$$

where RH can be one of a large number of possible substrates.

A simplified depiction of the proposed "activated oxygen" cytochrome P450 substrate complex is shown in Figure 4.3. The figure shows the simplified apoprotein portion the haeme (protoporphyrin IX) portion of cytochrome P450 and the close proximity of the substrate RH undergoing oxidation.

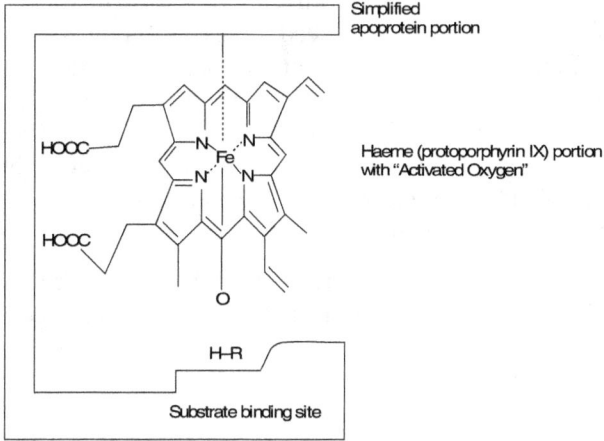

Figrue 4.3   Activated oxygen–cytochrome P450 substrate complex

THE P450 CATALYTIC CYCLE

The catalytic cycle of cytochrome P450 is shown in Figure 4.4. The intermediate states enclosed in square brackets, [], are only hypothetical.

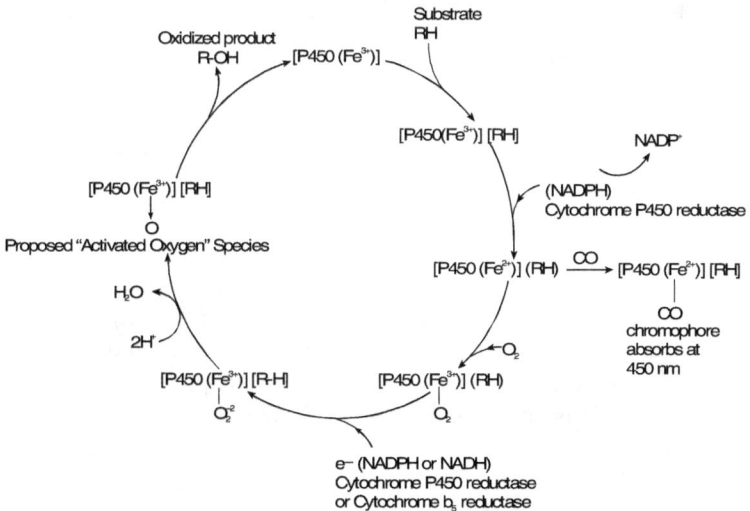

Figure 4.4   The CYP electron transfer cycle

Substrate (RH) binds to oxidized CYP to form the binary complex Fe(III) RH which is reduced to Fe (II) RH by an electron from NADPH. Molecular oxygen enters the cycle and a second electron is delivered to the complex. After loss of water molecules, the putative activated CYP supports metabolite formation and regenerates ferric CYP. The peroxide shunt, in which the activated oxyform of CYP is generated without the requirement for electron transfer and oxygen activation, is indicated.

The proposed catalytic reaction cycle involving cytochrome P450 in the oxidation of xenobiotics may be summarized as in Figure 4.5.

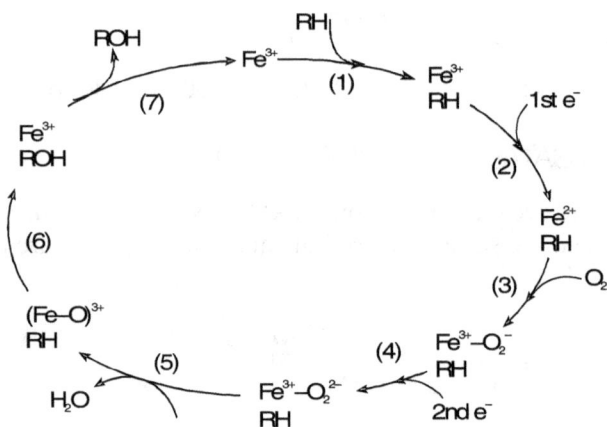

Figure 4.5   Different stages involved in the cytochrome P450 catalytic cycle

Stage 1: Substrate Binding

The binding of a substrate to a P450 causes a lowering of the redox potential by approximately 100 mV, which makes the transfer of an electron favourable from its redox partner, NADH or NADPH. This is accompanied by a change in the spin state of the haeme iron at the active site. It has also been suggested that the binding of the substrate brings about a conformational change in the enzyme which triggers an interaction with the redox component.

Stage 2:   First Reduction

The next stage in the cycle is the reduction of the Fe^{3+} ion by an electron transferred from NADPH via an electron transfer chain.

Stage 3:   Oxygen Binding

An O_2 molecule binds rapidly to the Fe^{2+} ion forming Fe^{2+}—O_2. There is evidence to suggest that this complex then undergoes a slow conversion to a more stable complex Fe^{3+}—O_2^- .

Stage 4:   Second Reduction

A second reduction is required by the stoichiometry of the reaction. This has been determined to be the rate-determining step of the reaction. A comparison between the bond energies of O_2, O_2^- and O_2^{2-} suggest that the Fe^{3+} complex is the most favourable starting point for the next stage of the reaction to occur. However, evidence from resonance Raman spectroscopy indicates the presence of a superoxide (O_2^-) complex.

Stage 5: O_2 Cleavage

The O_2 reacts with two protons from the surrounding solvent, breaking the O–O bond, forming water and leaving an $[Fe—O]^{3+}$ complex.

Stage 6:   Product Formation

The Fe-ligated O atom is transferred to the substrate forming a hydroxylated form of the substrate.

Stage 7:   Product Release

The product is released from the active site of the enzyme which returns to its initial state. The structures of the transitional states following processes (4) and (5) have never been directly observed and are hypotheses based on analogy with other haemoproteins.

- A polar drug on metabolism may yield a less polar product. This can be illustrated by the deacetylation of acetanilide to aniline.

- The reduction of chloral hydrate, a hypnotic drug to another pharmacologically active trichloroethanol involves participation of NADH.

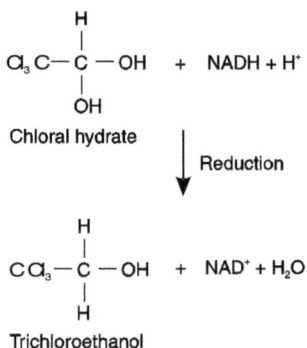

$$Cl_3C-\underset{\underset{OH}{|}}{\overset{\overset{H}{|}}{C}}-OH \quad + \quad NADH + H^+$$

Chloral hydrate

$\downarrow$ Reduction

$$CCl_3-\underset{\underset{H}{|}}{\overset{\overset{H}{|}}{C}}-OH \quad + \quad NAD^+ + H_2O$$

Trichloroethanol

- A pharmacologically inactive drug on metabolism is transformed into an active substance. This includes conversion of an inactive drug talampicillin into active metabolite, ampicillin, benorylate into paracetamol and aspirin.

Activity   The student may attempt to draw chemical reactions for the above metabolic transformations as an exercise.

METABOLIC TRANSFORMATION OF HALOTHANE

The metabolic transformation of halothane, one of the volatile anaesthetic agents is as shown in Figure 4.6. Fe^{3+} and Fe^{2+}, the oxidized and reduced haeme of cytochrome P450; FPT, NADPH-cytochrome P450 reductase; cyt.b_5, cytochrome b_5; FPD, and NADH-cytochrome b_5 reductase are the different species involved.

(a) Halothane   2-Chloro-1,1-   2-Chloro-1,1-
                       trifluoroethane   difluoroethylene

(b)

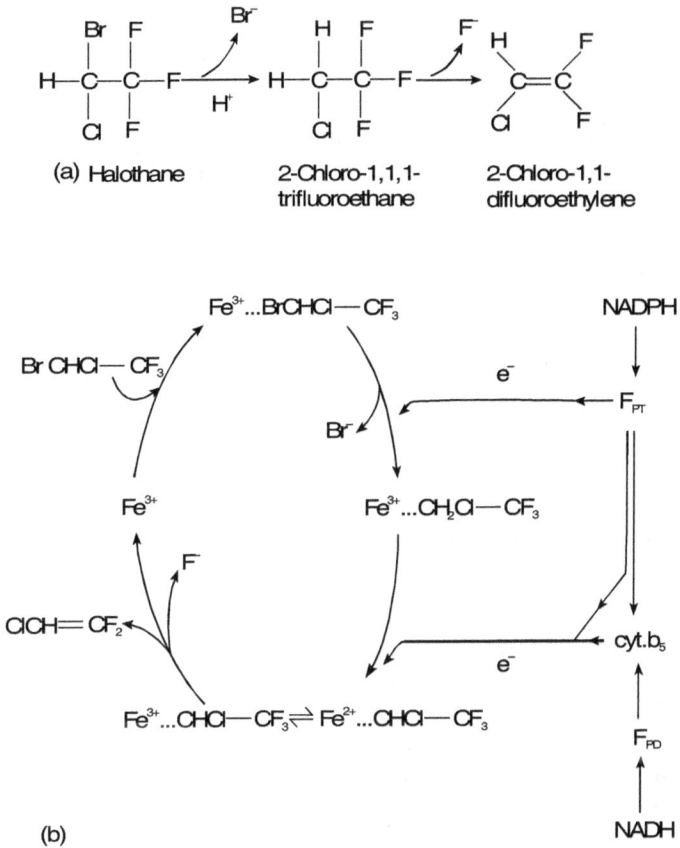

Figure 4.6   Metabolic transformation of halothane. (a) Halothane metabolism forming 2-chloro-1,1,1-trifluoroethane and 2-chloro-1,1-difluoroethylene. (b) Proposed reaction scheme of cytochrome P450 in the dehalogenation of halothane.

PHASE I—NON-SYNTHETIC REACTIONS

Non-synthetic reactions include oxidation, reduction and hydrolysis.

Oxidation Reactions

Typical examples of the oxidative transformations catalysed by the liver microsomal enzyme system include:

Aliphatic side chain oxidation   The principal metabolites of pentobarbitol are alcoholic derivatives, formed by the side chain oxidation.

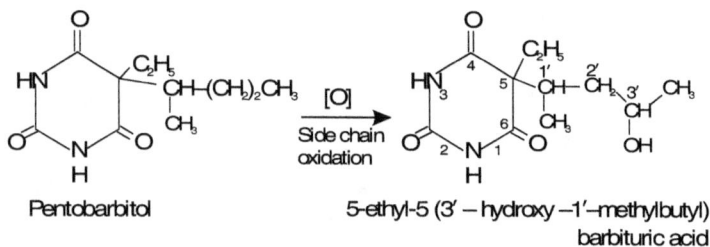

Pentobarbitol

5-ethyl-5 (3' – hydroxy –1'–methylbutyl) barbituric acid

Hexobarbitol undergoes oxidation of cyclohexenyl ring attached to C5 position to a keto derivative.

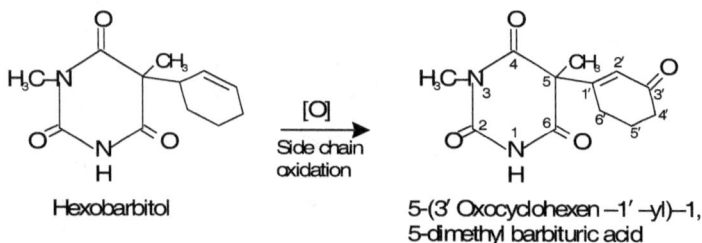

Hexobarbitol

5-(3' Oxocyclohexen –1' –yl)–1, 5-dimethyl barbituric acid

Aromatic hydroxylation   The conversion of acetanilide to *p*-hydroxyacetanilide.

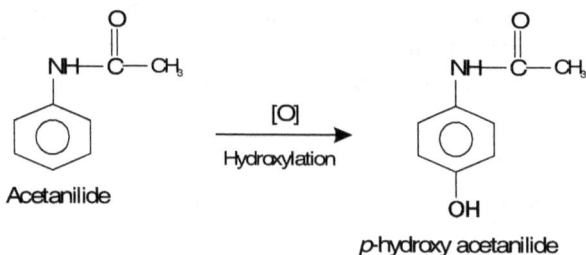

Acetanilide

p-hydroxy acetanilide

N-Dealkylation   N-methyl, ethyl and other alkyl groups are oxidatively converted to aldehydes.

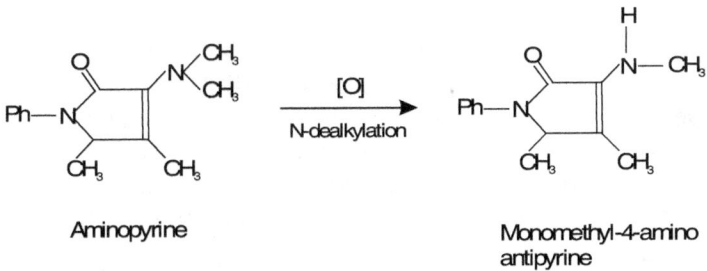

Aminopyrine   Monomethyl-4-amino antipyrine

O-Dealkylation   Aromatic ethers are cleaved.

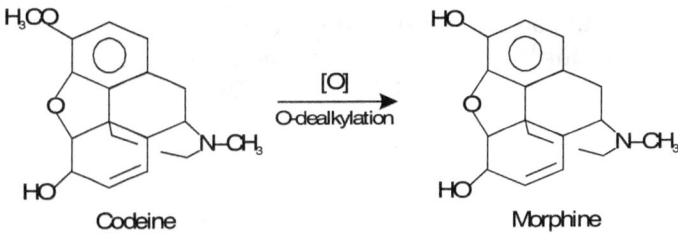

Codeine   Morphine

Sulphoxide formation   4-phenylthioethyl analogue of phenylbutazone (antirheumatic, analgesic, antipyretic), was known to be metabolized into a sulphoxide–sulphinpyrazone which possesses potent uricosuric action.

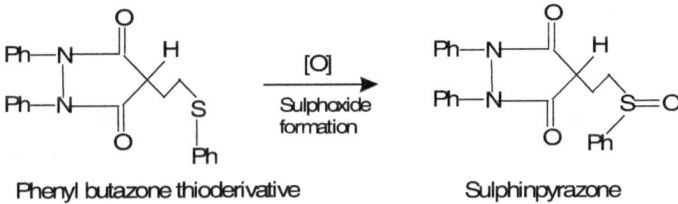

Phenyl butazone thioderivative   Sulphinpyrazone

N-oxidation and N-hydroxylation   This is illustrated by 2-acetylaminofluorene which when administered into the system, produces cancer and the N-hydroxy metabolite was found in the urine indicating that this oxidation product is the actual carcinogen.

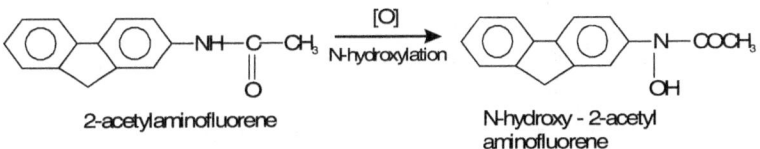

2-acetylaminofluorene   N-hydroxy - 2-acetyl aminofluorene

Several naturally occurring amines and a number of drugs containing amino groups are oxidatively deaminated mainly by two enzyme systems monoamine oxidase (MAO) and diamine oxidase (DAO). The reaction products are initially aldehydes, which on further oxidation produce carboxylic acids. Substrates such as tyramine, dopamine, epinephrine, norepinephrine, phenylethylamine, etc. undergo metabolism by this reaction.

Dehalogenation Certain halogenated insecticides, anaesthetics and other halogenated compounds can undergo dehalogenation in the living system. The reaction includes displacement of hydrogen halide by a hydroxyl group. The metabolism of DDT in DDT-resistant houseflies illustrates the dehydrohalogenation to the non-toxic derivative DDE.

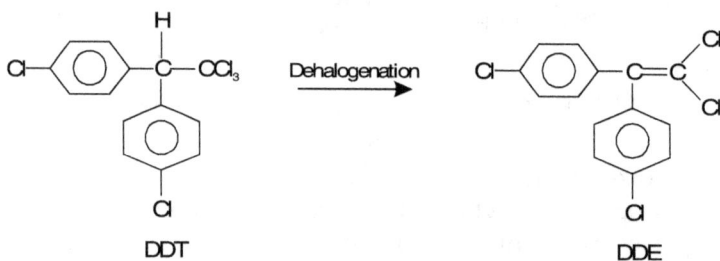

DDT   DDE

Hydrolysis

Drug metabolism by hydrolysis is restricted to esters and amides. The numerous hydrolytic enzymes esterases and amidases found in the blood plasma and other tissues including liver bring about this reaction. The hydrolysis of local anaesthetic procaine by liver and plasma choline esterase is as follows:

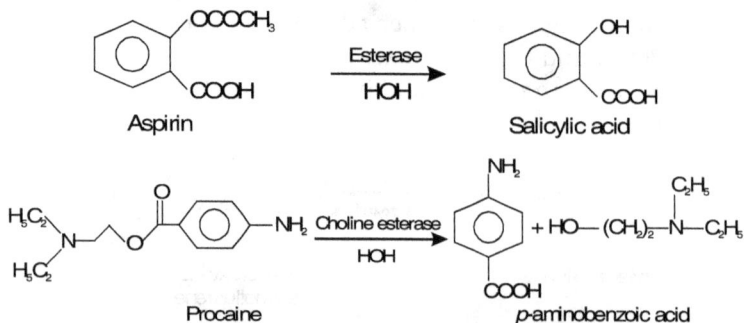

Aspirin   Salicylic acid

Procaine   *p*-aminobenzoic acid

Reduction Reactions

Reduction reactions are illustrated by

Reduction of azo and nitro groups The era of antibacterial chemotherapy began with the introduction of prontosil, an azo dye, used in the treatment of streptococcal and pneumococcal infections. Later on it was established that metabolite sulphanilamide was the active drug and not prontosil.

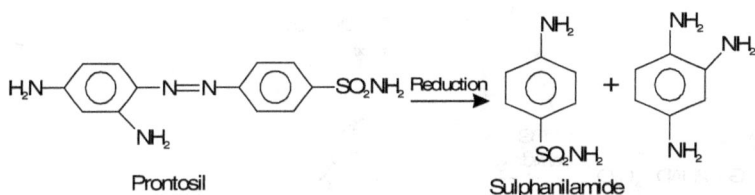

Prontosil   Sulphanilamide

The reduction reaction of a drug can be illustrated by choosing the antibacterial agent chloramphenicol, which is transformed into an amino compound by nitro reductase enzymes.

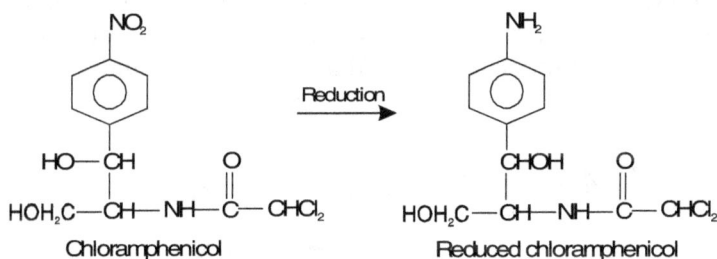

Chloramphenicol   Reduced chloramphenicol

PHASE II—SYNTHETIC REACTIONS

Synthetic reactions include conjugation with glucuronic acid, acetic acid, glycine, etc.

Conjugation Reactions

There are several small molecules present in the body capable of reacting with drugs or drug metabolites.

Glucuronic acid (formed by the oxidation of $-CH_2OH$ group in glucose moiety) combines with groups such as phenols, aromatic amines, alcohols and carboxylic acids to produce glucuronides.

Synthesis of uridine diphosphate-glucuronic acid (UDPGA)

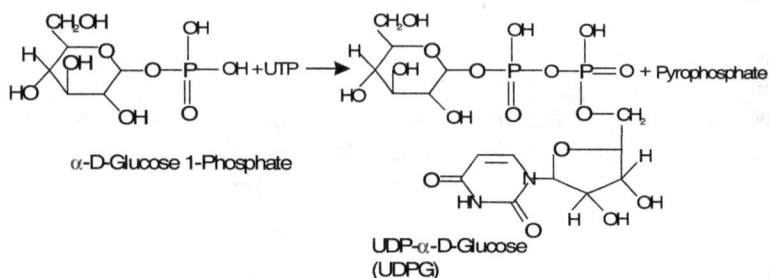

α-D-Glucose 1-Phosphate

UDP-α-D-Glucose
(UDPG)

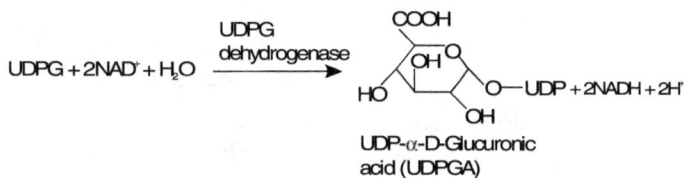

$$UDPG + 2NAD^+ + H_2O \xrightarrow{\text{UDPG dehydrogenase}}$$

UDP-α-D-Glucuronic
acid (UDPGA)

UDPG + 2NAD⁺ + H₂O → + 2NADH + 2H⁺

UDPGA serves as a good donor of glucuronic acid to various acceptors containing phenolic hydroxyl, aliphatic alcohols to form hemiacetal glucuronide often called as "ether glucuronides'.

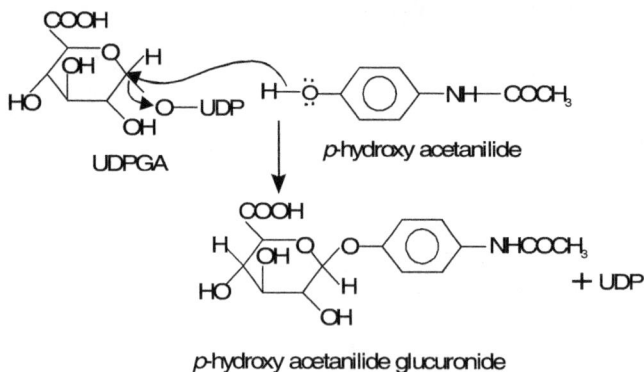

UDPGA

p-hydroxy acetanilide

p-hydroxy acetanilide glucuronide

+ UDP

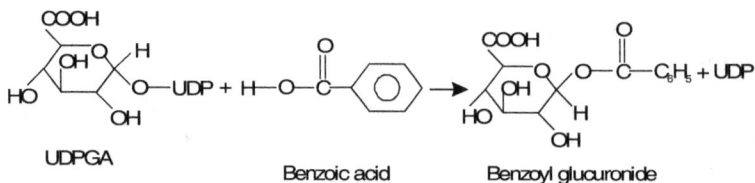

UDPGA   Benzoic acid   Benzoyl glucuronide

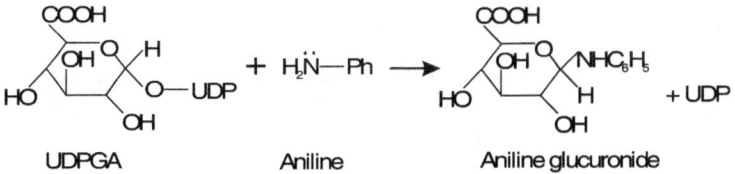

UDPGA   Aniline   Aniline glucuronide

In all the above reactions, there is a nucleophilic attack of the electron-rich oxygen atom, nitrogen, etc. of the substrate on the C1 atom of glucuronic acid of UDPGA. This results in Walden inversion on glucuronic acid part of the glucuronide.

Similarly, addition of ribose and phosphates to heterocyclic bases like purine and pyrimidine analogues yield nucleosides and nucleotides.

Acylation Reactions

This reaction occurs commonly at the amino groups. It is effected chiefly by the enzyme acyl transferase. It is worth mentioning about coenzyme A (CoA), a coenzyme of the acetylation reaction, discovered during the investigation of acetylation of sulphanilamides. The carboxylic acids in their active form react with the CoA through its sulphydryl group to form the acyl-CoA derivative. The latter then transfers the acyl group to the suitable substrates such as amines.

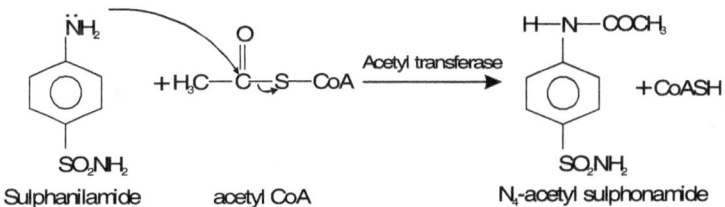

Sulphanilamide   acetyl CoA N_4-acetyl sulphonamide

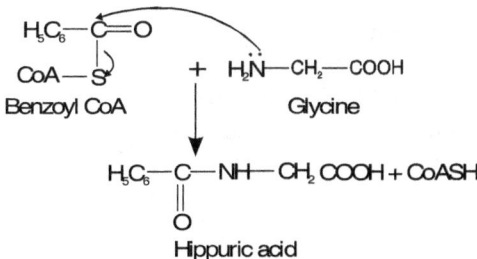

Benzoyl CoA   Glycine

Hippuric acid

Alkylation Reactions of N- and O-

Generally, it is the methylation reaction that occurs about these centres. The methyl donor is S-adenosyl methionine. A number of methyl transferase enzymes bring about these transformations, e.g. catechol *o*-methyl transferase (COMT). It can catalyse transfer of methyl group to a phenolic —OH group of nephrine, norepinephrine, etc. Methylation occurs in meta position and is dependent on the presence of magnesium ions.

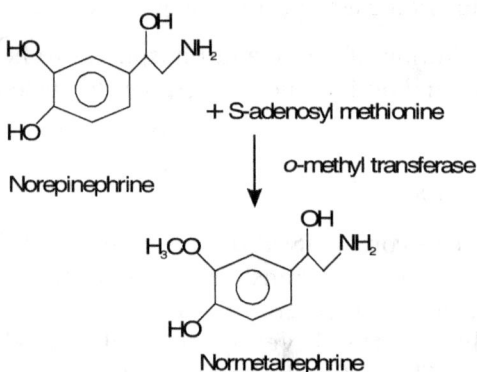

Norepinephrine

+ S-adenosyl methionine

o-methyl transferase

Normetanephrine

N-methylation reactions of a number of amines are also known. These enzymes are more specific. Enzyme phenyl ethanolamine N-methyltransferase (PNMT), methylates phenyl ethanolamines, e.g. norepinephrine is converted into epinephrine.

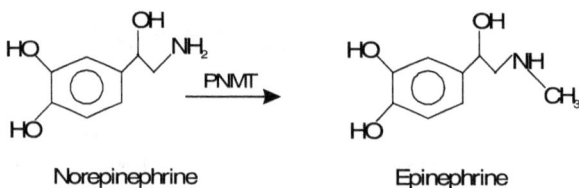

Norepinephrine   Epinephrine

PNMT

One of the mechanisms for solubilizing natural and synthetic chemicals in the body and particularly those that are electrophilic, or those that can be metabolized to an electrophilic form, involves reaction with nucleophilic amino acids. Conjugation with amino acids involves the classical activation of carboxyl groups to the thioester followed by amide formation with the amino group of an amino acid.

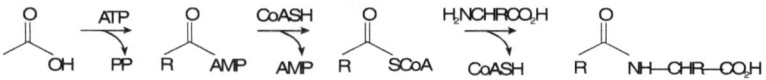

Glutathione is the tripeptide Gly-Cys-Glu with a central cysteine group:

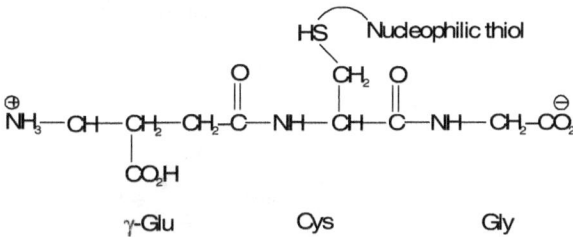

Through its nucleophilic thiol group it is used metabolically as a trap for various electrophilic substrates. It reacts by addition to conjugated alkenes, $ArSN_2$ substitution of aromatic halogen groups as well as ring opening of epoxides. The schematic structure shown below picturizes how naphthyl group in a drug is metabolized to 1-naphthyl mercapturic acid. After metabolic oxidation, the critical step in this sequence is the reaction of the glutathione with the 1, 2-epoxide of naphthyl group.

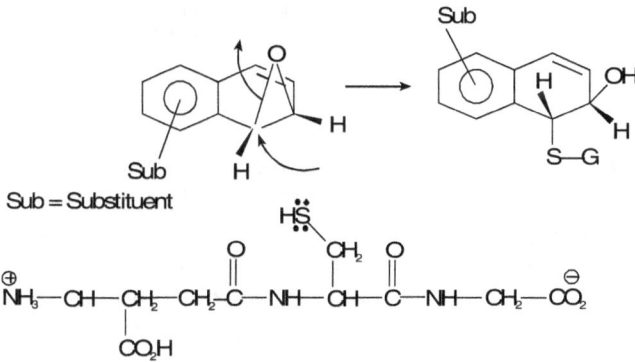

Sub = Substituent

METABOLISM OF COMMON DRUGS

The phases I and II in the metabolism of common drugs like paracetamol and prontosil are discussed below.

Prontosil

The activity of prontosil depends upon metabolism for its activation. Sulphanilamide is released on bioreduction. This active constituent is then rendered inactive by conjugation.

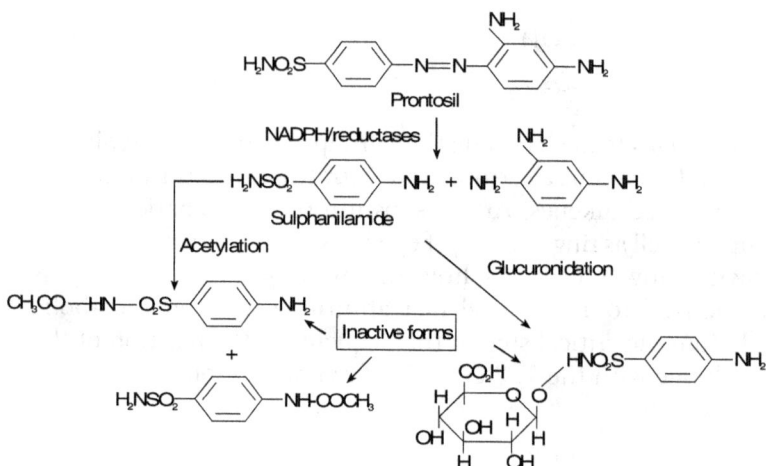

Paracetamol

Paracetamol is administered in its active form. Metabolism leads to its inactivation. It is conjugated and detoxified directly by the sulphonic acid (path I) and the glucuronide (path II). However, a small proportion is activated through the mixed function oxidases, (path III) to a harmful N-hydroxyl compound which, as an electrophilic iminoquinone has the potential to attack nucleic acids or proteins. However, glutathione reacts with the iminoquinone structure safely removing it through solubilization and excretion.

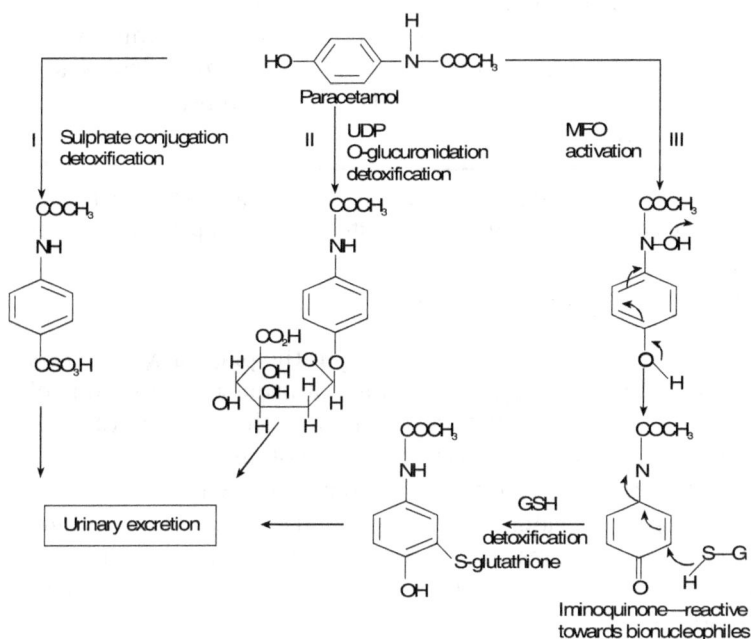

Iminoquinone—reactive towards bionucleophiles

ROLE OF KIDNEY IN DRUG EXCRETION

Metabolism, storage and excretion are the three mechanisms by which the drugs are ultimately removed from their sites of action. Excretion at the kidneys, biliary system, and sometimes by lungs account for most of the elimination. Renal excretion is the most important of all. In renal excretion, the following factors are of importance:

- Extent of plasma protein binding of drug
- Glomerular filtration rate
- Amount of back diffusion from filtrate influenced by the urine pH
- Active renal tubular secretion and reabsorption

About 130 ml of plasma water is filtered every minute by the glomerular membranes amounting to 190 L per day. Of this, only about 1.5 L are excreted as urine and the remaining are reabsorbed.

The glomerular capillaries contain large pores that are readily visible in electron micrographs. The filtration constants, when experimentally determined, show that the glomerules are permeable to solutes and only the free drug and not the plasma-bound protein is filtered.

A simple quantitative method known as "clearance" is used in the determination of excretion of certain volume of plasma water per minute.

Clearance = UV_u/tc ml min^{-1}

If c is the concentration of a drug in the plasma water, U its concentration in urine, V_u the volume of urine and t the time of urine collection in minutes and UV_u/t is the amount of drug excreted per minute. Then UV_u/tc is the volume of plasma water in which this amount of drug is present. In other words, such a volume of plasma water is actually cleared of its drug per minute. The term UV_u/tc represents the clearance of drug in ml per minute. It is also stated to be UV/P and UV is defined as the urinary excretion per minute.

Carbohydrate insulin is not bound to plasma proteins and is filtered at the glomeruli. Its clearance is taken as a measure of the glomerular filtration rate, which is about 130 ml/min. The clearance of glucose under normal conditions is zero, because it is completely reabsorbed in the tubules.

5

SULPHA DRUGS AND ANTIBIOTICS

INTRODUCTION

A drug is defined as a substance that affects the function of living cells, and is used in medicine to diagnose, cure and prevent the occurrence of diseases and disorders, and prolong the life of patients with incurable conditions.

Drugs can be classified in many ways: by the way they are dispensed over the counter or prescription; by the substance from which they are derived—plant, mineral or animal; by the form they take—capsule, liquid or gas; and by the way they are administrated—by mouth, injection, inhalation or direct application to the skin (absorption). Drugs are also classified by their names.

Another way to categorize drugs is by the way they act against diseases or disorders: chemotherapeutic drugs attack specific organisms that cause a disease without harming the host, while pharmacodynamic drugs alter the function of bodily systems by stimulating or depressing normal cell activity in a given system. The most common way to categorize a drug is by its effect on a particular area of the body or a particular condition.

Anti-infective drugs are classified as antibacterials, antivirals or antifungals depending on the type of microorganism they combat. Anti-infective drugs interfere selectively with the functioning of a microorganism while leaving the host unharmed.

Antibacterial drugs—sulpha drugs, penicillins, cephalosporins, and many others—either kill bacteria directly or prevent them from multiplying so that the body's immune system can destroy invading bacteria. Antibacterial drugs act by interfering with some specific characteristics of bacteria. Antibiotics often cure an infection completely. However, bacteria can spontaneously mutate, producing strains that are resistant to existing antibiotics.

ANTIBACTERIALS AND SULPHONAMIDES

These are synthetic chemotherapeutic agents, containing sulphonamide group (SO_2NH_2) in their structure. These were the first effective chemotherapeutic agents to be widely used for the cure of bacterial infections in human beings. These are less effective against certain gram-positive and gram-negative cocci, bacilli, and protozoa. They are available at moderate cost, safe, administered orally, and their mechanism of action has given tremendous impetus to investigate other antimetabolites of therapeutic interest. At present, sulphonamides have been largely replaced by antibiotics, but they are still in use either alone or in combination with other antibiotics. While preparing some synthetic dyes, the preparation of sulphanilides was achieved during the year 1908. For nearly twenty-five years, nothing much was known about this compound. It was Paul Ehrlich's idea on dyestuff and their staining properties to the bacterial cell that stimulated Domagk in 1930 for testing certain dyes for their effect on germs in mice. His daughter who was suffering from infection in her finger, which could not be cured by even surgical methods was administered the oral dose of an azo dye called prontosil. To Domagk's surprise, she recovered miraculously. Later it was found that the dye was active against streptococci. This was prepared by diazotizing sulphanilamide followed by coupling with *m*-phenylenediamine.

In a shorter period, the drug became very popular and was used widely in the treatment of various bacterial infections. It was then known that the antibacterial activity was due to the degradation of prontosil into sulphanilamide *in vivo*. This led to the synthesis and testing of various sulphanilamides. This had been used in the treatment of pneumonia, meningitis, gonorrhoea, tonsillitis, sinusitis, etc. Sulpha compounds were proved to be effective and less toxic.

From the therapeutic point of view, the sulpha drugs are classified as:

- Compounds which are readily absorbed and excreted, for example, sulphadiazine, sulphamethazine, sulphamerazine, etc.

- Compounds which are readily absorbed, but excreted slowly, for example, sulphaphenazole, sulphadimethoxine, etc.

- Compounds which are poorly absorbed, for example, phthalyl sulphathiazole, sulphaguanidine, etc.

- Compounds with special indication, for example, sulphapyridine.

- Absolute compounds, for example, sulphanilamide and sulphathiazole.

According to the therapeutic utility they are further classified into:

I.   Sulphonamides for systemic infections

    i.   Short-acting sulphonamides

    ii. Long-acting sulphonamides

    iii. Intermediate acting sulphonamides

II. Sulphonamides for local gastrointestinal infections

SYNTHESIS AND APPLICATIONS OF FEW SULPHONAMIDES

Most sulphonamides are synthesized from *p-N*-acetyl benzene sulphonyl chloride as one of the substrates.

Sulphapyridine

It is prepared by condensing 2-aminopyridine with *p-N*-acetyl sulphonyl chloride in the presence of pyridine as a base followed by hydrolysis in an alkaline medium.

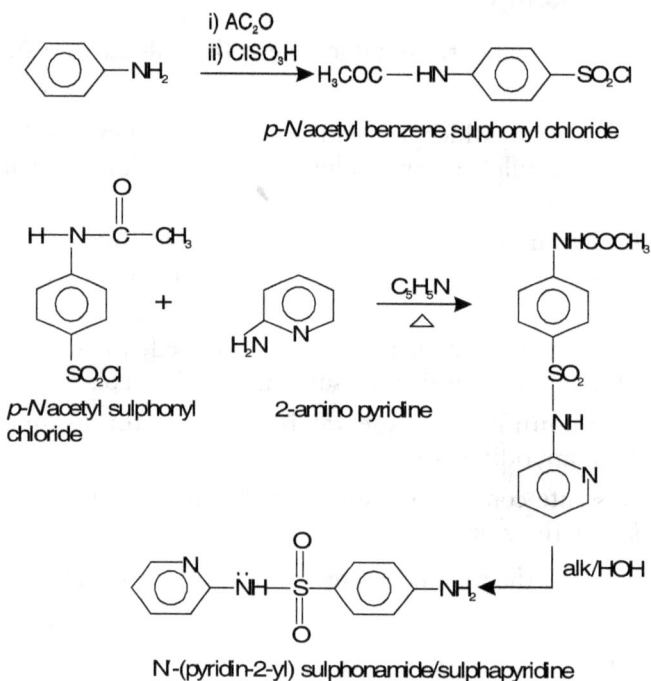

p-N acetyl benzene sulphonyl chloride

N-(pyridin-2-yl) sulphonamide/sulphapyridine

Sulphapyridine was the first drug to have an outstanding action on pneumonia. It is more potent than sulphonamide. Because of its high toxicity, it has been largely substituted by sulphadiazine and sulphamerazine. It causes nausea in many cases, and is readily acylated in the system. In a number of cases, renal damage has been reported due to the deposition of acetyl sulphapyridine crystals in the kidney.

Sulphadiazine

It is prepared by the condensation of 2-aminodiazine with *p-N*-acetyl sulphonylchloride and hydrolysing the product with an alkali.

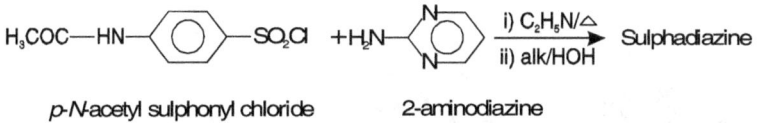

p-N-acetyl sulphonyl chloride   2-aminodiazine

It is used in the treatment of pneumonia, meningitis and influenza. *In vivo,* it is eight times as active as sulphonamide and exhibits fewer toxic reactions than the other members. It is absorbed readily and excreted slowly.

Phthalyl Sulphathiazole

It is prepared by heating sulphathiazole with phthalic anhydride.

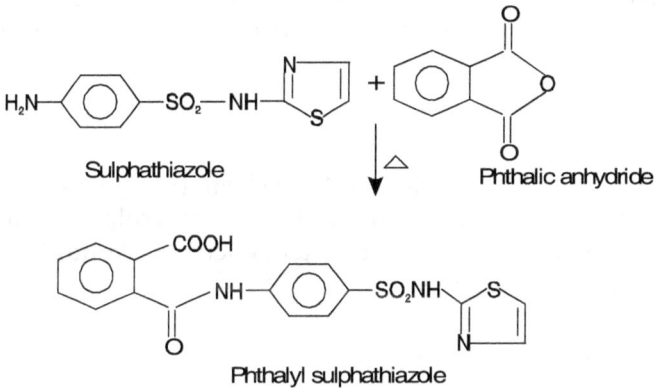

Sulphathiazole

Phthalic anhydride

Phthalyl sulphathiazole

It is used in intestinal infections such as bacillary dysentery and cholera. It is absorbed poorly in the intestinal tract and is more potent.

Sulphafurazole

It is also called as sulpha isoxazole.

It is prepared by condensing *p-N*-acetyl sulphonyl chloride with 3,4-dimethyl aminoisoxazole in the presence of pyridine followed by the alkaline hydrolysis.

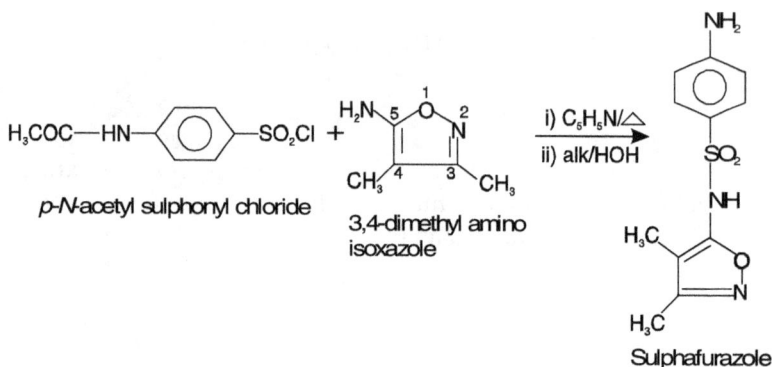

Sulphafurazole

It is used in the infections of sulphanilamide-sensitive bacteria and effective in the treatment of gram-negative urinary tract infections. Sometimes the acyl derivative of the drug is used which is given orally.

Prontosil

This is the first sulphonamide used as a chemotherapeutic agent. This breaks down in the physiological system into sulphonamide, which is responsible for its antibacterial activity. Prontosil S is more suitable than prontosil.

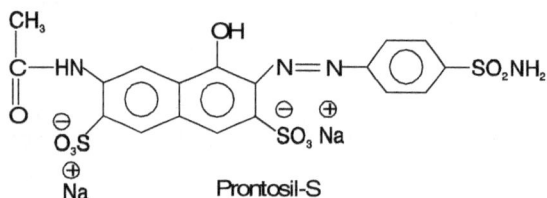

Prontosil-S

Prontosil is prepared by diazotising *p*-amino benzene sulphonamide into its diazonium salt and coupling it with *m*-phenylene diamine.

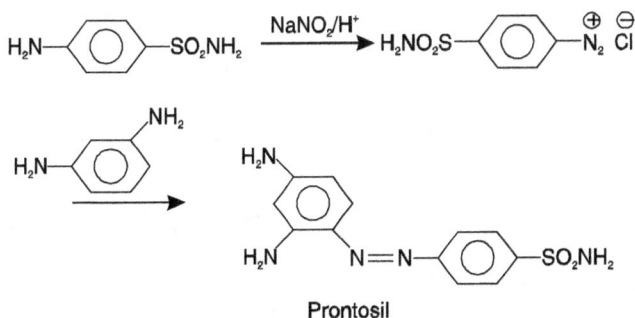

Prontosil

MECHANISM OF ACTION OF SULPHA DRUGS

It is well established that these drugs are not "-cidal" but "-static" in action. As early as 1940, it was found that the bacteriostatic property of sulphonamide was due to the structural resemblance of *p*-amino sulphonamide with PABA. The latter is important for the vital functions of the bacteria. PABA is necessary for the synthesis of folic acid and essential for the growth and multiplication of microorganisms. Folic acid is a group of substances. One of the important members of this group is pteroyl glutamic acid.

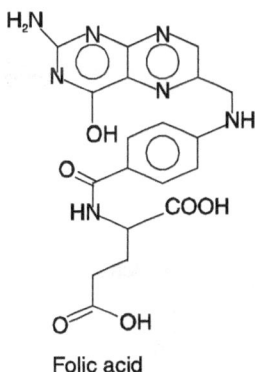

Folic acid

PABA

Sulphanilamide

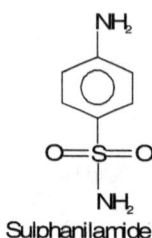

When sulpha drug is administered, it attaches itself to the enzyme which is responsible for the conversion of PABA into folic acid. By this, the synthesis of folic acid is blocked; consequently, the organisms grow weaker and lose their ability to multiply and finally get eliminated from the system. Thus, the sulphonamides are established to be acting as structural antimetabolites. They also interfere with other metabolic processes in addition to blocking the synthesis of folic acid. The antibacterial activity of sulphonamides is confined only to microorganisms which synthesize their own folic acid and is ineffective against those using folic acid from the environment. Also, sulphonamides are ineffective in the presence of pus and tissue broken-down products containing a large quantity of PABA. Similarly, these drugs are not effective when certain drugs administered are capable of producing PABA in the body on biotransformation, e.g. procaine antagonizes the action of sulphonamides.

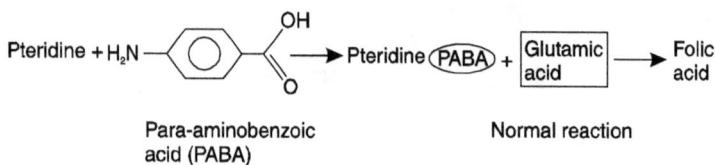

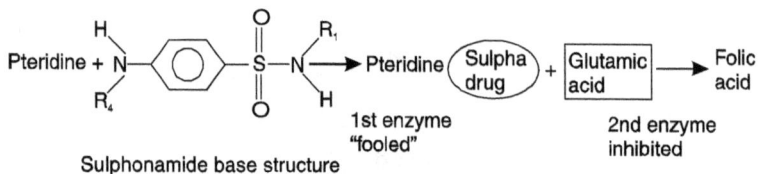

Para-aminobenzoic acid (PABA)

Normal reaction

Sulphonamide base structure

1st enzyme "fooled"

2nd enzyme inhibited

Sulpha drug as antimetabolite

In bacteria, normally the folic acid synthesis takes place in two steps. In the first step, the products containing pteridine and

PABA are produced. The next and final step is the reaction of PABA with glutamic acid to make folic acid, which is catalysed by the enzyme folic acid synthetase. If the sulpha drug is substituted for the PABA, then the final enzymatic reaction is inhibited and folic acid is not produced. Recent studies indicate that substituents on the N(I) nitrogen may play the role of competing for a site on the enzyme surface reserved for the glutamate residue in *p*-aminobenzoic acid glutamate in one of the following two ways:

1. Direct competition in the linking of PABA glutamate with the pteridine derivative.
2. Indirect interference with the coupling of glutamate to dihydropteroic acid.

ANTIBIOTICS

Antibiotics are the chemical substances produced by the microorganisms which in turn inhibit the growth of other microorganisms or destroy them. Because of their diversity in structure and functions, they are classified in many ways.

Antibiotics are classified as those derived from:

1. Amino acids for example, chloramphenicol derived from a single amino acid and penicillin from two amino acids.
2. Amino sugars amino glycosides, for example, streptomycin.
3. Acetates and propionates.
4. Miscellaneous.

Antibiotics are also classified based on their spectra of action as:

1. Effective on gram-positive bacteria only.
2. Effective on gram-negative bacteria only.
3. Effective on both gram-positive and gram-negative.
4. Effective on protozoans.
5. Effective on fungi.
6. Effective against malignancy, e.g. actinomycin, mithomycin, etc.

Streptomycin

This belongs to the class amino glycosides, and is a product of actinomycetes (soil bacteria) or is a semi-synthetic derivative of natural products. Compounds derived from *Streptomyces* are known as mycins. Streptomycin is isolated from the culture of *Streptomyces griseus*. This belongs to the group of antibiotics used against gram-negative infections. Its action is bacteriostatic in lower concentration and bacteriocidal in higher concentration. It interferes with the enzymatic action of the disease-causing microorganism.

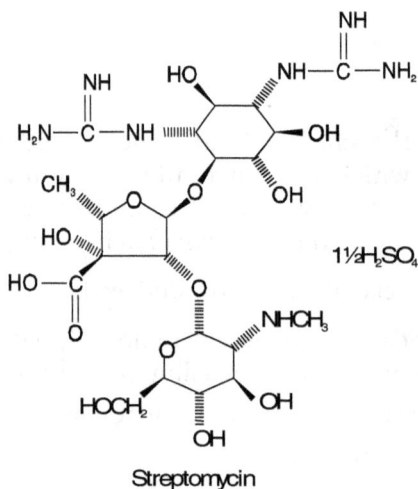

Streptomycin

It is used in the treatment of tuberculosis, meningitis, influenzae, plague and pneumonia.

The adverse effects of the usage of this drug include harmful effects on the 8th cranial nerve in the CNS, dizziness, perforation of ear drum, local irritation, rashes, etc.

Penicillin

Penicillin is the general name given to a class of compounds possessing molecular formula $C_9H_{11}O_4N_2SR$. There are six natural penicillins known so far. The most common is benzyl penicillin or penicillin-G. Penicillins belong to a class of antibiotics called lactam antibiotics. So far, only two types of β-lactam antibiotics have been found in nature—penicillins and cephalosporins. The fusion of lactam with the thiazolidine ring is called as the penam structure. The fused rings are not coplanar but folded along the joining axis.

Penicillin was initially derived from *Pencillium notatum.* Recently high yields of penicillin are obtained from the mould *P. chrysogenum*, the commercial source. A variety of semi-synthetic penicillins are produced by altering the composition of the side chain attached to 6-amino penicillanic acid (6APA).

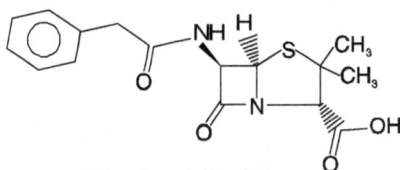

Benzyl penicillin-G

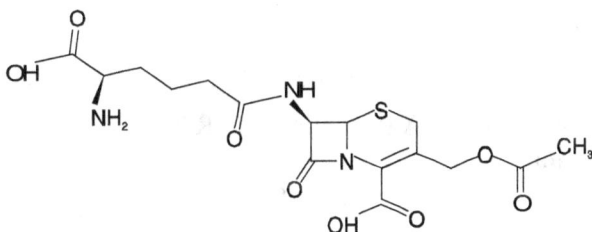

Cephalosporin C

The basic structure of penicillin is called as penam structure also known as β-APA (β-amino penicillanic acid) and attached to it are various side chains. Both the side chains and β-APA are essential for the antibacterial activity. The side chain is important in determining the stability against degradation by the gastric acid and enzyme penicillinase, which is produced by certain microorganisms. Penicillin is a bacteriocidal drug. It interferes with the synthesis of cell wall of gram-positive cocci. As the cell wall synthesis occurs during growth stage, the antibiotic is effective against the multiplication of microorganisms. Pencillins are effective in pneumococcal-infections such as meningitis, pneumonia, etc., and in streptococcal infections like scarlet fever, rheumatic fever, tetanus, diphtheria, gangrene, etc. Penicillin can be administered orally, parenterally or topically. In oral administration, higher concentration is required as it is inactivated by the lower pH of gastric juices and the presence of the enzyme penicillinase.

Local anaesthetic procaine is administered along with penicillin-G to prolong the action of the drug. In general, penicillins are safer drugs, but at times exhibit adverse effects. In certain cases, cardiovascular collapse is also noticed.

The important allergic reactions of penicillin are:

• Skin rashes

• Shock syndrome

• Renal disturbances

Natural penicillins   Natural penicillins are based on the original penicillin-G structure. Some of the natural penicillins are as follows:

1. Penicillin-I or -F (the side chain R is $—CH_2—CH=CH—C_2H_5$)

2. Penicillin-II or -G (the side chain R is $—CH_2—Ph$)

3. Penicillin-III or -X (the side chain R is $—CH_2—Ph—OH$ (p)-ampicillin)

4. Penicillin-IV or -K (the side chain is $—(CH_2)_6—CH_3$)

5. Dihydro-F-penicillin (the side chain is $—(CH_2)_4—CH_3$)

6. Penicillin-V (the side chain is $—CH_2—O—Ph$)

Disadvantages of natural penicillins   They are inactivated readily in the acid medium and duration of action is short. They have poor absorption in the intestine. Therefore, several synthetic penicillins have been prepared. All the synthetic analogues show the allergic reactions as that of natural ones.

Synthetic penicillins   Synthetic penicillins like α-amino benzyl penicillin has been found to be active against those microorganisms towards which benzylpenicillin is not active. 6-aminopenicillanic acid has been used as the starting material for the synthesis of various penicillins.

Semi-synthetic penicillins This includes oxacillin, cloxacillin, amoxycillin, etc. The semi-synthetic analogues are advantageous over the natural ones in being resistant to acid and the enzyme penicillinase.

The commercial preparations of penicillins contain one or more of penicillins in different proportions. To the culture medium various compounds containing benzyl group such as phenyl acetamide, and phenyl acetic acid are added to increase the total yield of penicillin and also the proportion of benzyl penicillin. Similarly, addition of compounds containing para hydroxy benzyl group to the medium increases the proportion of para hydroxy benzyl penicillin. When addition of compounds other than these are done a number of synthetic penicillins have been obtained.

Chloramphenicol/Chloromycetin

This is the first synthetic antibiotic. It is produced by *Streptomyces venezuelae*.

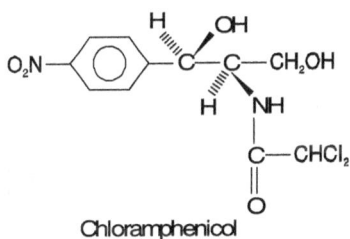

Chloramphenicol

The other names of chloramphenicol are

- D(–)-threo-2-dichloroacetamido-1-(4-nitrophenyl)-1, 3-propanediol

- D(–)-threo-2,2-dichloro-N-[β -hydroxy- α - (hydroxymethyl)- β -(4-nitrophenyl)ethyl] acetamide

- D(–)-threo-2,2-dichloro-N-[β -hydroxy- α - (hydroxymethyl)-4-nitrophenethyl] acetamide

Action of chloramphenicol D(–) threo isomer is active. It interferes with the protein synthesis of bacteria. They are used in the treatment of acute typhoid, and other infections caused by *Salmonella* bacteria, pneumonia, influenza, whooping cough, plague, etc. It is given orally. It is also used in the ophthalmic ointments and eye drops. The nitrobenzene moiety is responsible for the bone marrow toxicity.

All the structural features of chloramphenicol are essential for its pharmaceutical action.

1. A number of analogies of chloramphenicol have been prepared and found to be less effective to the parent compound in their action.

2. Replacing the phenyl ring for heterocyclic systems show less effect.

3. Replacing the para nitro group with other groups such as $-CN$, NH_2, amide group, etc. retained the biological activity.

4. The dichloro group is essential for its action. When it was replaced by bromine, it resulted in the reduction of the activity.

5. Its succinic ester has higher solubility and is used in the parenteral administration.

6. Its palmitate is tasteless and used in the pediatric preparations.

7. The stereochemistry is important. Out of the four stereoisomers, D(–) threo alone is physiologically active. The L-isomers are inactive. The D(+) isomer is toxic.

8. The presence of propanediol is essential for the activity.

9. Any change in the –OH group or the propane side chain decreases the activity.

10. Any other para-substituted derivatives are also as toxic as the parent compound.

Tetracyclines

Tetracyclines are broad spectrum antibiotics. They are derived from a species of *Streptomyces aureofaciens*. These compounds vow their name to their basic skeleton of four rings. All of them have a hydronaphthalene skeleton in their structure. This group includes tetracycline, chlorotetracycline, aureomycin, oxytetracycline (terramycin), doxycycline, etc.

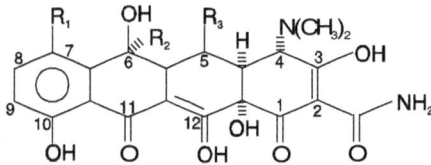

| Tetracyclines | R₁ | R₂ | R₃ |

Tetracyclines	R_1	R_2	R_3
1. Auromycin (7-chlorotetracycline)	Cl	CH₃	H
2. Terramycin (5-oxytetracycline)	H	CH₃	OH
3. 6-Demethyltetracycline	H	H	H
4. Parent tetracycline	H	CH₃	H

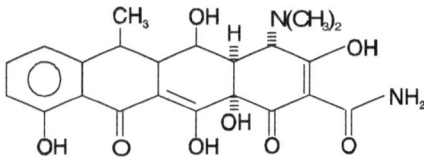

Doxycycline

Therapeutic uses   They are effective against both gram-positive and gram-negative microorganisms, acid-fast bacilli, etc. They are bacteriostatic in action. They are used in the treatment of pneumonia, urinary tract infections, plague, etc. The hydrochlorides of tetracyclines are administered orally as capsules. Terramycin is used in eye infections.

Macrolides

The macrolide antibiotics are derived from *Streptomyces* bacteria and their nomenclature is attributed to the fact that they all possess a macrocyclic lactone chemical structure. Macrolides are products of actinomycetes (soil bacteria) or semi-synthetic derivatives of them. They are weakly alkaline and high molecular weight compounds. The structural feature shows the presence of a large lactone ring, a few double bonds. There are few sugar residues; some of them are amino sugars. The classifications of macrolides are based on their nature and arrangement of sugar residues. The sugar residue is essential for the biological activity of the compounds. They are -"static" in lower concentration and -"cidal" in higher concentration. Some examples are erythromycin (*Streptomyces erythraeus*) and rifamycin (*Streptomyces mediterranis*), aminoactinomycin, etc.

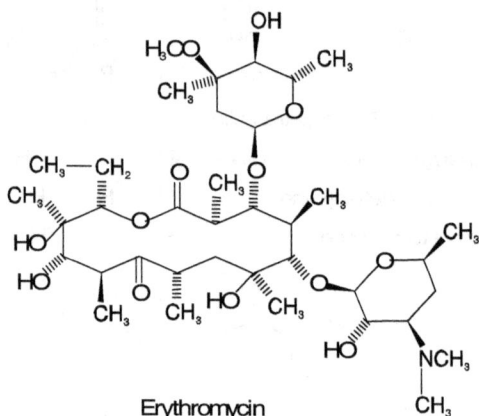

Erythromycin

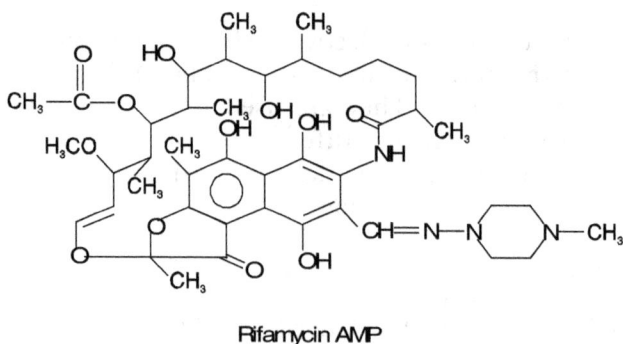

Rifamycin AMP

6

ANTISEPTICS AND DISINFECTANTS

INTRODUCTION

Antiseptics and disinfectants are generally nonselective, anti-infective agents that are applied topically. Their activity ranges from simply reducing the number of microorganisms to safe limits of public health interpretations (sanitization) to destroying all microbes (sterilization) on the applied surface. In general, antiseptics are applied on tissues to suppress or prevent microbial infection. Disinfectants are germicidal compounds usually applied to inanimate surfaces. Sometimes the same compound may act as an antiseptic or as a disinfectant, depending on the concentration, conditions of exposure, number of organisms, etc. To achieve maximum efficiency, it is essential to use the proper concentration of the anti-infective agent for the purpose intended.

Topical anti-infective agents are extensively used in surgery for antisepsis of the surgical area and surgeon's hands and to disinfect surgical instruments, apparel and the hospital premises. Other common uses are as disinfectants for home and farm premises, in water treatment, in public health sanitation, and as antiseptics in soaps, teat dips, dairy sanitizers, etc. Antiseptics have also been used for treating local infections. However, in most cases, systemic chemotherapeutic agents are preferred because they often penetrate better into the foci of infection and are less likely than the topical anti-infectives to lose their potency when in contact with body fluids and debris in the infected area.

Ideally, antiseptics and disinfectants should have a broad spectrum and potent germicidal activity, with rapid onset and long-lasting effect. They should withstand environmental factors (e.g. pH, temperature, humidity, etc.) and must retain activity even in the presence of pus, necrotic tissue, and other organic material. High lipid solubility and good dispersibility increase their effectiveness. Antiseptic preparations should not be toxic to the host tissues and should not impair healing. Disinfectants should be nondestructive to applied surfaces. Offensive odour, colour and staining properties should be absent or minimal. Most of these compounds exert their antibacterial activity by denaturation of intracellular proteins.

The term antiseptic includes the anti-infective agents which are applied to the living tissues. Hence, they are bacteriostatic and do not necessarily sterilize the surface under treatment. An antiseptic prevents the growth of microorganisms as long as it is in contact with the wound.

Antiseptics are milder and prolonged in action. An ideal antiseptic would destroy bacteria, spores, fungi and other infective agents without harming the host tissue. Most of them have limited spectrum of activity and their use is always local because of the toxicity they produce. This, therefore prevents their other routes of administration, which depend on absorption by the body.

The disinfectants are those applied to non-living surfaces or inanimate objects. The disinfectants are bacteriocidal in nature, kill all the microorganisms and produce irreversible lethal effects. The action of disinfectants is immediate and has a shorter duration.

STANDARDIZATION OF DISINFECTANTS

The method adopted to standardize the disinfectants is called as Rideal–Walker method. This method gives a measure of the biological activity of a chemical compound in relation to phenol. Phenol is taken as the standard because of its nonspecific nature of action. The germicidal activity of phenol is attributed to the denaturation of bacterial protein. In this method, the minimum concentration of phenol, which kills a 24-hour culture of β-typhus in a finite duration is first determined and then the concentration of unknown substance that produces the same effect under identical conditions is determined. This relationship is termed as phenol coefficient or Rideal–Walker coefficient.

$$\text{Rideal} - \text{Walker coefficient} = \frac{\text{Con. of test compound}}{\text{Con. of the phenol}}$$

This can be determined in the absence or presence of standard amount of organic matter.

SOME TYPICAL ANTISEPTICS

Some typical antiseptics are described below:

Formaldehyde

Either as a gas or in solution, formaldehyde is an excellent germicide and is as effective as phenol. Its volatility renders it more penetrating. Formaldehyde gas is employed in disinfecting rooms, excreta, instruments and clothing. 40% solution of formaldehyde in water or alcohol is applied as a hardener of skin to prevent excessive perspiration and to disinfect the hands or the site of surgery.

Nitrofurazones

Nitrofurazone possesses good bactericidal property. It is effective against a wide range of both gram-positive and gram-negative microorganisms. It is used in the infections of burns, ulcers, wounds and other skin diseases.

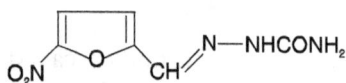

Nitrofurazone
semiarbazone

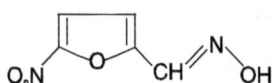

Nitrofuroxime

Phenol

Commonly called as carbolic acid, phenol is used as an antiseptic in surgery. It has no specific action and is called as protoplasmic poison, in injuring microorganisms and cells of body tissue. For this reason phenol is used as a standard against which the potency of other antiseptics are judged. Many alkylated and halogenated phenols are used as antiseptics. The antiseptic value of homologous phenols depends on the nature of the side chain. The value increases with the increasing chain length of up to five carbon atoms and thereafter it diminishes. Similarly, the introduction of chlorine atoms into the aromatic nucleus enhances the antiseptic value.

The alkylated homologous phenols used as antiseptics include the following.

Carvacrol   Carvacrol is a naturally occurring antiseptic. It is prepared from *p*-cymene by sulphonation followed by fusion with alkali. It has both antiseptic and antifungal activity.

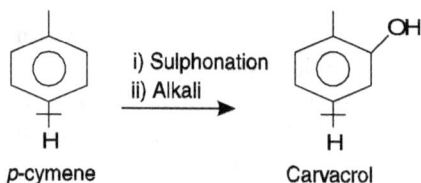

i) Sulphonation
ii) Alkali

p-cymene

Carvacrol

Thymol   Thymol is also a naturally occurring antiseptic. It is synthesized from *m*-cresol.

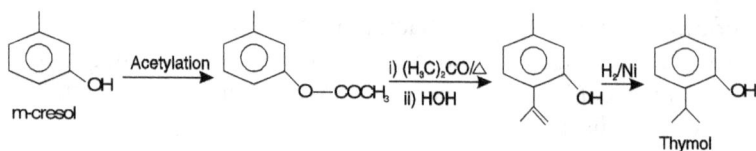

m-cresol

Acetylation

i) (H₃C)₂CO/Δ
ii) HOH

H₂/Ni

Thymol

It is used as a disinfectant and flavouring agent in most of the mouthwashes and gargles.

N-hexyl resorcinol   N-hexyl resorcinol is prepared by the condensation reaction of caproic acid chloride with resorcinol in the presence of anhydrous zinc chloride followed by Clemmenson's reduction. It is a general antiseptic and has irritant action on the skin. It is also used as an antihelminthic.

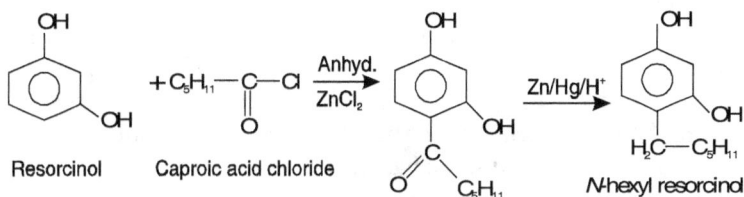

| Resorcinol | Caproic acid chloride | | N-hexyl resorcinol |

The chlorinated phenols include:

Chloro-m-cresol   Chloro-*m*-cresol is prepared by chlorinating *m*-cresol.

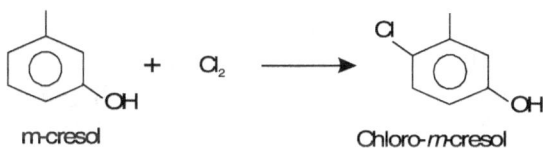

m-cresol   Chloro-*m*-cresol

Chloro-m-xylenol   Chloro-*m*-xylenol is an important antiseptic of wider application. The alcoholic solution of the compound is the major constituent of dettol.

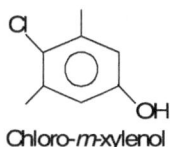

Chloro-*m*-xylenol

SURFACTANTS

These are surface-active agents and are used as antiseptics. They possess a hydrophobic core and hydrophilic ends. The hydrophobic core accumulated in the interface acts on the bacterial cell membrane that contains lipids and hence these agents behave as antimicrobials;

they alter the energy relationship at interfaces. Based on the charge on the amphiphile in the molecule, surfactants are classified as anionic or cationic.

Anionic Surfactants

Soaps are dipolar anionic detergents with the general formula $RCOO^-Na^+/K^+$, which dissociate in water into hydrophilic K^+ or Na^+ ions and the amphiphilic fatty acid ions. Most soap solutions are alkaline (pH 8–10) and may irritate sensitive skin and mucous membranes. Soaps emulsify lipoidal secretions of the skin and remove most of the dirt, desquamated epithelium, and bacteria, which are then rinsed away with the lather. The anionic surfactants are active against gram-positive bacteria. The antibacterial potency of soaps is often enhanced by inclusion of certain antiseptics like hexachlorophene, phenols, carbanilides, potassium iodide, etc. They are incompatible with cationic surfactants.

Cationic Surfactants

These are quaternary ammonium compounds (Figure 6.1), e.g. benzalkonium chloride, benzethonium chloride, cetylpyridinium chloride, triton, etc. They are active against both gram-positive and gram-negative bacteria. The major site of action of these compounds appears to be the cell membrane, where they become adsorbed and cause changes in permeability. Their activity is reduced by porous or fibrous materials (e.g. fabrics, cellulose sponges, etc.) that adsorb them. They are inactivated by anionic substances (e.g. soaps, proteins, fatty acids, phosphates). Therefore, they are of limited value in the presence of blood and tissue debris. They are effective against most bacteria, some fungi (including yeasts), and protozoa but not against viruses and spores. An aqueous solution of 1 : 1000–1 : 5000 has good antimicrobial activity, especially at slightly alkaline pH.

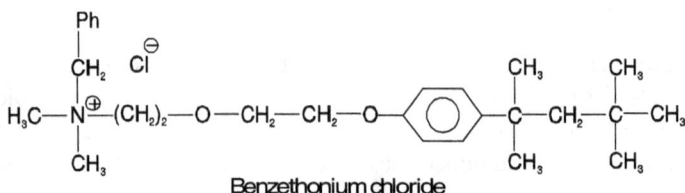

Benzethonium chloride

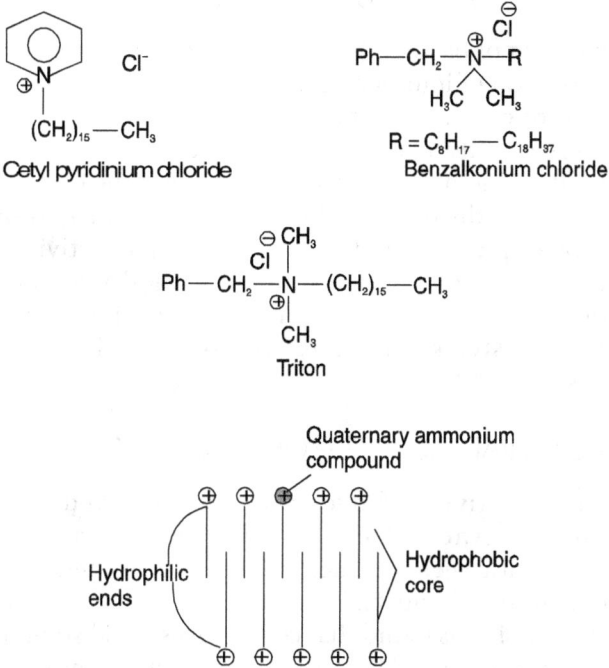

Cetyl pyridinium chloride

$R = C_8H_{17} - C_{18}H_{37}$
Benzalkonium chloride

Triton

Quaternary ammonium compound

Hydrophilic ends

Hydrophobic core

Figure 6.1   Cationic surfactants

Benzalkonium chloride solutions are rapidly acting anti-infective agents with a moderately long duration of action. They are active against bacteria and some viruses, fungi and protozoa. Bacterial spores are considered to be resistant. Solutions are bacteriostatic or bactericidal according to their concentration. Gram-positive bacteria are generally more susceptible than gram-negative. Activity is not greatly affected by pH, but increased substantially at higher temperatures and prolonged exposure times.

Antimicrobial activity of cationic surfactants Cationic surfactants are capable of replacing phospholipid molecules in the phospholipid bilayer in the cell wall, which disrupts the functionality of the cell wall, and ultimately, kills the organism.

Mechanism of Action of Surfactants

The mechanism of bacterial/microbicidal action is thought to be due to disruption of inter molecular interactions. This can cause dissociation of cellular membrane bilayers, which compromises cellular permeability controls and induces leakage of cellular contents. Other biomolecular complexes within the bacterial cell can also undergo dissociation. Enzymes, which finely control a plethora of respiratory and metabolic cellular activities, are particularly susceptible to deactivation. Critical intermolecular interactions and tertiary structures in such highly specific biochemical systems can be readily disrupted by cationic surfactants.

ANTIBACTERIAL ACTIVITY OF DYES

Antibacterial activity of dyes was first reported in 1913. Subsequently the discovery of sulphonamides as chemotherapeutic agents ensued from the antibacterial activity observed in the dye prontosil. Before the advent of sulphonamides and antibiotics, it was found that certain dyes could stain certain tissues and not others. These dyes also possess antiprotozoal properties, promote wound healing, are slow acting and have remarkable specificity. The cationic dyes are more active in the basic medium and show specificity to gram-positive bacteria. The anionic dyes are active in the acidic medium and show specificity to gram-negative bacteria. Some of these dyes possessing antiseptic property include:

1. Acridine dyes
2. Triphenyl methane dyes
3. Xanthine dyes
4. Azo dyes
5. Thiazine dyes

Acridine Dyes

Acridine dyes are also called as flavins because of their yellow colours. They are powerful antiseptics. The parent compound acridine has the following structure:

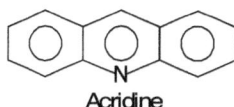

Acridine

The 3, 6-diamino derivative of acridine is of pharmaceutical interest as an antiseptic in ophthalmic surgery under the name proflavin.

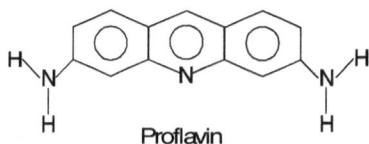

Proflavin

Acriflavin (3,6-diamino-10-methyl acridine chloride) came into prominence during the First World War as an antiseptic for wounds. It is used as an orange mixture of hydrochlorides of proflavin and acriflavin.

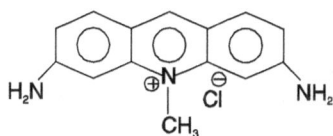

3,6-diamino-10-methyl
acridinium chloride/acriflavin

In recent years, 9-aminoacridine(aminacrine) has become an important antiseptic. It is advantageous over the other two above-mentioned dyes in not staining the fabrics or skin.

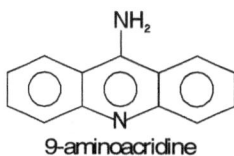

9-aminoacridine

Triphenyl Methane Dyes

The important members of this group include methyl violet, crystal violet, and malachite green.

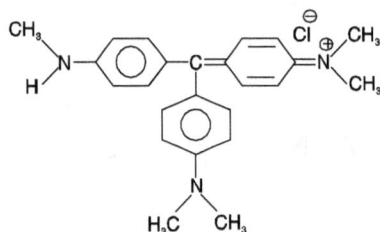

Methyl violet

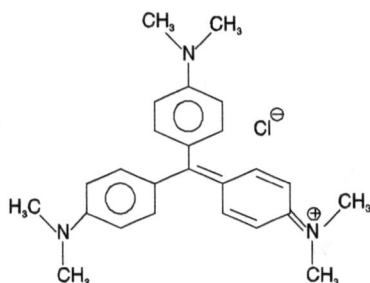

Crystal violet–gentian violet

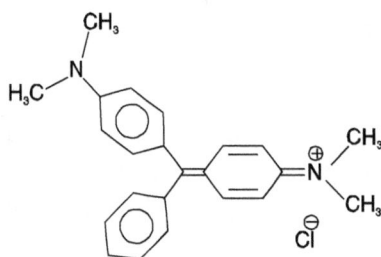

Malachite green

These members are effective against skin eczemas, ulcers, burns, etc. and are also used in washing the wounds.

Xanthine Dyes

An important member of this group is mercurochrome, which was the first organomercurial to be used. It is also known as merbromin. It is a mercurial fluorescein dye, non-irritant and a poor skin disinfectant when compared to mercurous chloride. The better mercurial antiseptic is merthiolate, which is used as a tincture solution.

Organomercuric compounds are the most important ones among the group of antiseptics. Mercuric bichloride is one of the early antiseptics which was later replaced by the less irritant and less toxic organic mercurials like merbromin, thiomersal, nitromersol, phenyl mercuric nitrate, mercurochrome, merthiolate, etc. At moderate concentrations, they are bacteriostatic and act by inhibiting the bacterial enzymes through their affinity for the sulphhydryl groups present in the enzymes. This effect can be reversed by sulphur-containing compounds like citrine and glutathione. Mercurials are not effective against spores. Due to potential persistence of mercury in the environment as a contaminant, the use of mercurial antiseptics or disinfectants has largely decreased.

Mercurochrome is used as a surgical solution on the wounds, on the skin and mucous surfaces. Thiomersal is used as both fungicide and bacteriocide. Mercurophen is used to sterilize the surgical instruments and applied on mucous membrane.

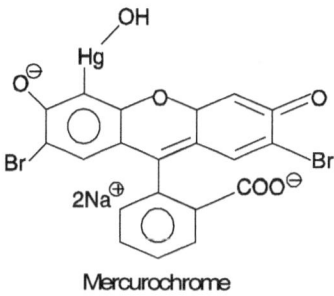

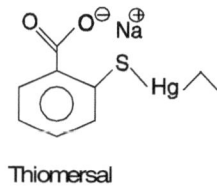

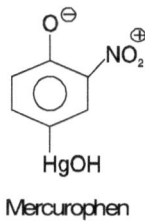

Mercurochrome

Thiomersal

Mercurophen

Azo Dyes

The important azo dyes like dimazon, scarlet red and phenazopyridine are most active in an acidic medium and effective against gram-negative bacteria. Scarlet red and dimazon are often used as a 5% ointment on sores, ulcers and wounds. They are of more use for their healing nature by stimulating tissue proliferation than for their antiseptic value.

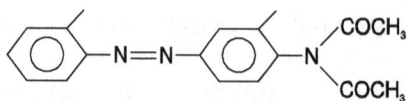

Dimazon

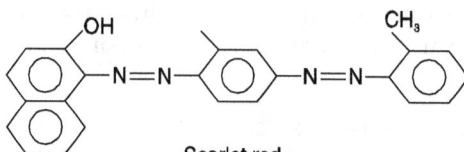

Scarlet red

Pyridium is a better azo antibacterial often incorporated as an analgesic with sulphonamides in the treatment of urinary tract infections and as a non-irritant wound disinfectant.

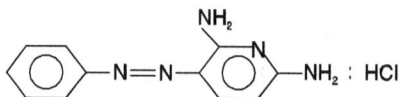

Pyridine hydrochloride

Thiazine Dyes

Methylene blue is a thiazine dye. It is a weak antiseptic, used as an intestinal and urinary antiseptic. It is also used to test the renal functions of kidneys by staining vital nerves and as an antidote of cyanide and nitrate poisoning.

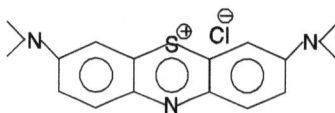

Methylene blue

SILVER COMPOUNDS

Silver compounds can have caustic, astringent and antibacterial effects. Silver ions combine with sulphhydryl, amino, phosphate and carboxyl groups, and thus precipitate proteins and also interfere with essential metabolic activities of microbial cells. A 0.1% aqueous solution is bactericidal and irritant, whereas 0.01% solution is bacteriostatic. A 0.5% solution is sometimes applied as a dressing on burns to reduce infection and induce rapid eschar formation. Colloidal silver compounds, which release silver ions slowly, are bacteriostatic and have a more sustained effect. They do not irritate the tissues and have little astringent or caustic effect. They are generally used as mild antiseptics and in ophthalmic preparations.

Biguanidines as Antiseptics

Chlorhexidine is the most popular antiseptic of this group.

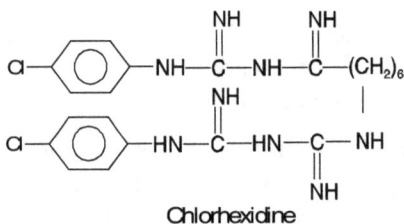

Chlorhexidine

It has potent antimicrobial activity against most gram-positive and some gram-negative bacteria but not against spores. A 0.1% aqueous solution is bactericidal against *Staphylococcus aureus*, *Escherichia coli*, and *Pseudomonas aeruginosa* in 15 sec. It is relatively ineffective against other gram-negative organisms, spores, fungi and most viruses. Nosocomial infections by *Pseudomonas* sp. have occurred from the use of contaminated chlorhexidine solutions in which the bacteria persisted. In susceptible organisms, it disrupts the cytoplasmic membrane. Its activity is enhanced by alcohols, quaternary ammonium compounds and alkaline pH, and is somewhat depressed by high concentrations of organic matter (pus, blood, etc.), hard water and contact with cork. It is incompatible with anionic compounds, including soap. It is one of the most commonly used surgical and dental antiseptics. A 4% emulsion of chlorhexidine gluconate is used as a skin cleanser, a 0.5% w/v in 70% isopropanol is used as a general antiseptic, and a 0.5% w/v in 70% isopropanol with emollients is used as a hand rinse. Chlorhexidine is incorporated in shampoos, ointments, skin and wound cleansers, teat dips, surgical scrubs, etc. for its antimicrobial properties. A 1% chlorhexidine acetate ointment is used as a topical antiseptic in treatment of external wounds in animals. Chlorhexidine has low potential for systemic or dermal toxicity.

HALOGENS AND HALOGEN-CONTAINING COMPOUNDS

Iodine and chlorine are used as topical antimicrobial agents. They owe their activity to their high affinity to protoplasm, where they are believed to oxidize proteins and interfere with vital metabolic reactions.

Elemental iodine is a potent germicide with a wide spectrum of activity and low toxicity to tissues. A solution containing 50-ppm iodine kills bacteria in 1 min and spores in 15 min. It is poorly soluble in water but readily dissolves in ethanol, which enhances its antibacterial activity.

Preparations   Iodine tincture contains 2% iodine with 2.4% potassium iodide (KI) dissolved in 50% ethanol; it is used as a skin disinfectant. Strong iodine tincture contains 7% iodine and 5% KI dissolved in 85% ethanol; it is more potent but also more irritating than tincture of iodine. Iodine solution contains 2% iodine and 2.4% KI dissolved in aqueous solution; it is used as a non-irritant antiseptic on wounds and abrasions.

Strong iodine solution contains 5% iodine and 10% KI in aqueous solution.

Iodophores, e.g. povidone-iodine are water-soluble combinations of iodine with detergents, wetting agents that are solubilizers, and other carriers. They slowly release iodine as an antimicrobial agent and are widely used as skin disinfectants, particularly before surgery.

They do not leave any malodour or stain. They are non-toxic to tissues but may be corrosive to metals. They are effective against bacteria, viruses, and fungi but less so against spores. Iodophore solutions retain good bacterial activity at pH <4, even in the presence of organic matter, and often change colour when the activity is lost. Phosphoric acid is often mixed with iodophores to maintain an acidic medium. They have been used in teat dips to control mastitis, as dairy sanitizers, and as a general antiseptic or disinfectant for various dermal and mucosal infections.

Chlorine exerts a potent germicidal effect against most bacteria, viruses, protozoa and fungi through formation of hypochlorous acid in water at acidic to neutral pH. It is effective against most organisms at a concentration of 0.1 ppm, but much higher concentrations are required in the presence of organic matter. Alkaline pH ionizes chlorine and decreases its activity by reducing its penetrability. Chlorine has a strong acid smell and is irritant to the skin and mucous membranes. It is widely used to disinfect water supplies and inanimate objects. The search for other chlorinated compounds provided a range of them possessing better therapeutic value, e.g. Calx chlorinata, Dakin's solution, chloramines, etc.

Including molecular chlorine all other chlorinated compounds act on the microorganisms for their ability to liberate hypochlorous acid in different concentration which in turn reacts with the proteinous part of the cell membrane of the microorganism. During this reaction, the nitrogen atom of the peptide is chlorinated leading to the denaturation of protein.

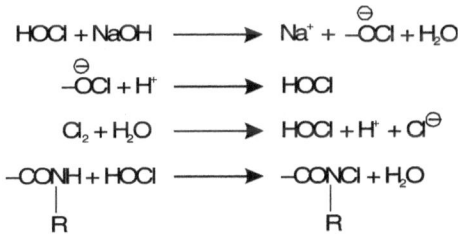

$$HOCl + NaOH \longrightarrow Na^+ + {}^-OCl + H_2O$$

$$^-OCl + H^+ \longrightarrow HOCl$$

$$Cl_2 + H_2O \longrightarrow HOCl + H^+ + Cl^-$$

$$\underset{R}{-CONH} + HOCl \longrightarrow \underset{R}{-CONCl} + H_2O$$

The organic chloramines received a special place as antiseptics for the following advantages:

- They are less irritant.
- They are stable in solid state and in solution liberate hypohalous acid of definite strength.
- The liberation of hypohalous acid is slow and therefore has better antiseptic value.
- They are effective over a longer period.

The organic chloramines include: chloramine-T, dichloramine-T, halazone and dequalinium chloride.

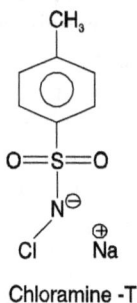

Chloramine -T

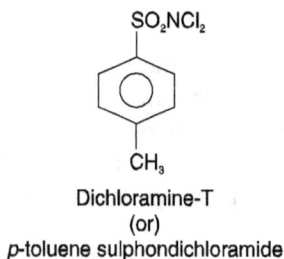

Dichloramine-T
(or)
p-toluene sulphondichloramide

Halazone
(or)
p-dichloro sulphanoyl
benzoic acid

Dequalinium chloride is a sparingly soluble solid in water. It is used as an antiseptic for mouth and throat infections in the form of lozenges and paint. This has a long chain of nonpolar aliphatic groups at one end and has a quaternary ammonium group at the other end, and thus behaves like a cationic surfactant. It has been established in general that quaternary ammonium salts have marked bactericidal activity if there is a long chain of 8–18 carbon atoms attached to the quaternary nitrogen.

Dequalinium chloride

Certain aliphatic chloramines also exhibit antiseptic property. This includes succinchlorimide, chloroazodin (azochloramide), etc.

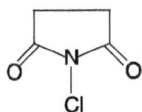

Succinchlorimide

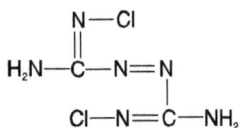

Chloroazodin

Succinchlorimide is used in sterilizing water. Chlorazodin called as modern antiseptic is non-toxic to tissues. Its action is relatively longer than chloramine-T. The advantage of using it is that the loss of activity is less in bloodstream. It is useful on wounds and as a packing for cavities.

7

ANALGESICS AND ANTIPYRETICS

ANALGESICS

Analgesics are the drugs which relieve pain without leading to the state of unconsciousness. Their role is to relieve or decrease the sensation of pain. Thus, it is a pharmacodynamic agent. The analgesics act by increasing the threshold of pain, which is defined as the lowest perceptible intensity of pain. The pain is induced by stimuli. The amount of stimulus is regarded as the measure of threshold of pain. If more stimuli are required, the threshold of pain is increased and vice-versa. When analgesics are used, the pain is not actually decreased but only the threshold of pain is increased. Thus the sensation of pain is not felt. From the nature of action, it can be said that the analgesics act on the CNS to decrease the sensation of pain.

Analgesics are of two types, namely narcotic analgesics and non-narcotic analgesics.

NON-NARCOTIC ANALGESICS

Non-steroidal anti-inflammatory drugs (NSAIDs) are drugs which act to relieve inflammation, but which are not structurally related to the corticosteroids. NSAIDs have four major activities.

1. Analgesia   The term refers to the relief of pain by a mechanism other than the reduction of inflammation (e.g. headache). These agents produce a mild degree of analgesia, which is much less than the analgesia produced by opioid analgesics such as morphine.

2. Antipyretic   These are drugs which lower elevated body temperature by their action on the hypothalamus. However, they will not reduce normal body temperature.

3. Anti-inflammatory   These drugs are used to treat rheumatoid disorders, and also in other inflammatory diseases and injuries. Their anti-inflammatory activity is due to their ability to inhibit the cyclooxygenase activity of prostaglandin synthase, an enzyme which mediates the production of prostaglandins from arachidonic acid. These drugs were developed as an alternative to the corticosteroids and their analogues, which have many side effects.

4. Uricosuric   Many of these agents cause the excretion of uric acid, and thus are useful in the treatment of gout.

Aspirin, naproxen (Alleve), and ibuprofen (Advil, Motrin) are examples of non-steroidal anti-inflammatory drugs (NSAIDs). These reduce inflammation caused by injury, arthritis or fever.

NARCOTIC PAIN RELIEVERS

The term narcotic is derived from the Greek word *narcotuc* meaning drowsiness. Narcotic pain relievers are also called opiate pain relievers, as their action is similar to that of opium. The most potent narcotic pain relievers are morphine, pethidine and methadone. However, due to their high addicting ability, these are reserved for acute emergencies and prescribed for short durations. Other opiates like codeine, dextropropoxyphene, pentazocine or ethoheptazine are less addictive and less potent. When used in combination with the non-narcotic pain relievers, they provide greater relief of pain than the individual drugs as the two act by different mechanisms.

(–)-Morphine   (+)-Morphine

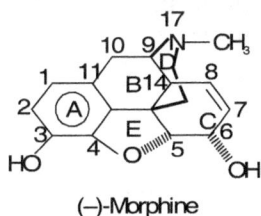

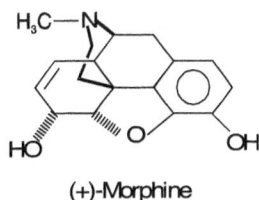

Examination of the morphine molecule reveals the following structural features important to its pharmacological profile:

1. A rigid pentacyclic structure consisting of a benzene ring (A), two cyclohexane rings (B and C), a piperidine ring (D) and a dihydrofuran ring (E). Rings A, B and C form the phenanthrene ring system providing very little conformational flexibility.

2. Two hydroxyl functional groups, a C3-phenolic OH and a C6-allylic OH.

3. An ether linkage between C4 and C5.

4. Unsaturation between C7 and C8.

5. A basic, $3°$-amine function at position 17.

6. There are five chiral centres (C5, C6, C9, C13 and C14) in morphine exhibiting a high degree of stereo selectivity of analgesic action. Only (–)-morphine is active.

Morphine contains an acidic phenolic group and a basic tertiary amine function. However, since the amine function is significantly more basic than the acidic phenol group, morphine as well as a majority of narcotic analgesics are functionally basic compounds both pharmaceutically (dosage forms) and physiologically. Hence, morphine exists as a cation at physiological pH, and readily forms salts with appropriate acids.

The opium class of drugs particularly morphine is the potent clinically useful analgesic. Nevertheless, they cause depression on the central nervous system. The other undesirable side effects of morphine as analgesics include: respiratory depression, nausea and addiction. Considerable efforts have been made to improve and maintain its analgesic properties while eliminating its undesirable side effects, which include:

1. Etherification of the phenolic group has been found to increase convulsant action. But this decreased the analgesic action by one-tenth.

2. Catalytic hydrogenation of the isolated double bond increases the analgesic property only slightly.

3. Modifying the alcoholic hydroxylic group increases the toxicity, convulsant action and analgesic properties.

4. Opening the ether bridge decreases the analgesic properties.

The attempt to synthesize morphine analogues possessing improved analgesic property yielded little or no improvement in side effects. The synthetic analogues include: diacetyl morphine commonly called as heroin is as potent as morphine but is more habit forming than morphine. Levorphanol tartrate is a potent narcotic analgesic having an action similar to that of morphine.

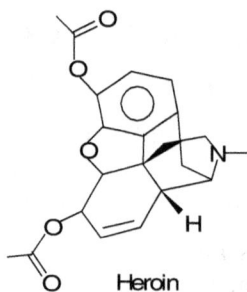
Heroin

Structure–Activity Relationships (SAR) of Morphine

Modifications at the 3- and 6-hydroxy groups

1. Inversion of configuration at C6 increases analgesia. This is the only chiral centre that when inverted, causes increase in analgesia.

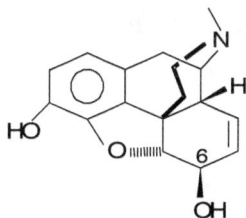
C6 isomer of morphine

2. Conversion of the 3-OH to a 3-OCH$_3$, gives codeine whose analgesic activity is reduced to 15% of morphine. Groups larger than a methoxy, when introduced, reduced the activity dramatically.

3. Conversion of the 6-OH to a 6-OCH$_3$ gave heterocodeine possessing six-fold increase in the activity.

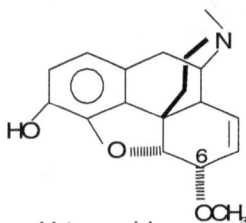
Heterocodeine

4. By the oxidation of the 6-OH to a ketone, the activity is reduced when the double bond is present (morphinone). Oxidation of the 6-OH to a ketone increases the activity when double bond is not present (dihydromorphinone).

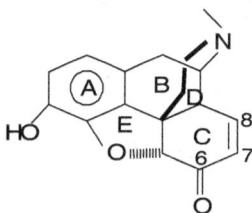
Dihydromorphinone

5. Acetylation of morphine gives 3,6-diacetylmorphine, heroin, which is 2–3 times more potent than morphine with increased lipid solubility and rapid CNS penetration.

6. If ether linkage is broken, with hydroxyl group on aromatic ring, activity is reduced to 90%.

Modifications at the 7,8-double bond

1. Reduction of the 7,8-double bond slightly increases the activity of dihydromorphine.

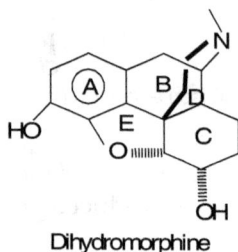

Dihydromorphine

2. The similar reduction slightly increases the activity in dihydrocodeine.

Modifications of the nitrogen substituent

1. Introduction of a methyl group maintains the optimal agonist activity of morphine.
2. Introduction of an N-phenylethyl group leads to 14-fold increase in activity over morphine. On quaternization of nitrogen, no analgesic activity was observed (N,N-dimethylmorphine).

Ring substitutions

1. In ring E (tetrahydrofuran ring), breaking up the ether linkage, reduces the catechol activity by 90%.

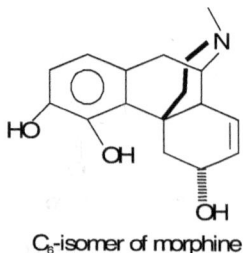

C_6-isomer of morphine

2. 14-β-OH in dihydromorphine series shows dramatic increase in activity of oxymorphone.

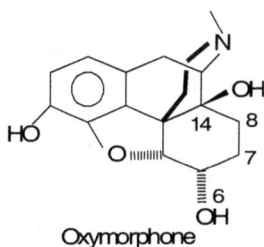

Oxymorphone

3. Oxycodone has an activity equal to morphine.

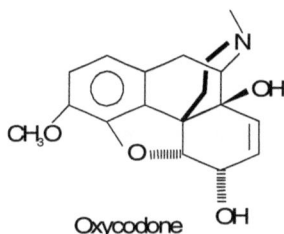

Oxycodone

Synthetic Narcotic Analgesics

More extensive structural modification of morphine has been carried out in an attempt to achieve more therapeutics—analgesics, antitussives—with minimal adverse effects—respiratory depression, sedation, and physical dependence reactions. Continued SAR studies of nuclear modified opiates suggested that a minimum "structural" pattern analgesiophore exists in the relatively complex morphine structure that is primarily responsible for narcotic analgesia produced by this agent. The analgesiophore consists of an aromatic ring linked to a quaternary carbon, which in turn, is connected via a two-carbon atom chain to a tertiary basic amine. These studies led to the design and development of narcotic analgesics of simpler structure, which is outlined below.

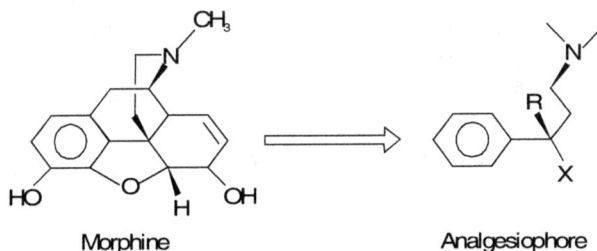

Morphine   Analgesiophore

Morphinans are structurally derived from morphine by removal of the ether oxygen (ring E) and therefore elimination of the E ring. The (–) morphinan analogue derived from such a modification is levorphanol (Levo-dromoran). It is approximately 5–8 times more potent than morphine as both an analgesic and respiratory depressant, but produces less nausea and vomiting. Levorphanol is more lipophilic than morphine and has a greater oral/parenteral potency ratio (2 : 1). It is available as a bitartrate salt, which is administered orally and by injection (IM and SC) and has duration of 3–6 hours.

Morphine   Remove E ring   Levorphanol

Dextromethorphan   The (+)-isomer of levorphanol with a 3-methylether group is dextromethorphan.

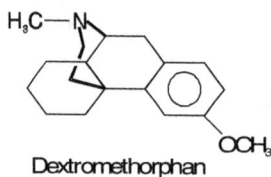

Dextromethorphan

Dextromethorphan is inactive as an analgesic, but has antitussive potency equal to that of codeine. It is about half to one-third as potent as codeine for the relief of pain. However, its action lasts for more than 24 h. It is available in the form of dextropropoxyphene hydrochloride salt of which a 65 mg dose is equal to 600 mg of aspirin or paracetamol for the relief of pain. Most of the drug combinations available contain 32.5–65 mg of the above-mentioned salt and 325–500 mg of paracetamol. It is also available with ibuprofen. For relief of pain, 2–3 tablets or capsules per day are required. The cost of these combinations is quite high.

The most common adverse effects are dizziness, drowsiness, nausea and vomiting. The less common adverse effects include constipation, abdominal pain and headache, although it is a pain reliever. A false sense of well-being (euphoria) or hopelessness (dysphoria) and some visual disturbances have been reported in rare cases. The addiction liability of propoxyphene is less than that of codeine. Replacement of the *N*-methyl moiety in the morphinan series with an *N*-allyl or *N*-cycloalkyl groups results in partial agonist activity.

Butorphanol tartrate (Stadol)

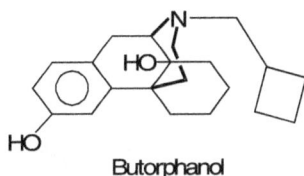

Butorphanol

As an analgesic (agonist) Butorphanol is five times more potent than morphine with a similar onset and duration of action. Benzomorphans such as metazocine are structurally derived from morphine by the cleavage of both rings C and E. Removal of these rings yields a benzomorphan ring system that has conformational flexibility unlike the morphines and morphinans, thus different ring conformational extremes may exist. Also, the presence of methyl

Remove C and E rings

Morphine   **Metazocine**

substituents at positions 5 and 9 in these compounds produces sites of asymmetry. Thus the benzomorphans can exist in different stereoisomeric and conformational forms. Different stereoisomeric forms of the benzomorphans differ in analgesic potency. In the case of the metazocine, the *trans*-5,9-dimethyl stereoisomers also termed as beta isomers are significantly more active as analgesics than the *cis*-alpha geometric isomers (Figure 7.1). Metazocine isomers have only about one-sixth the activity of morphine.

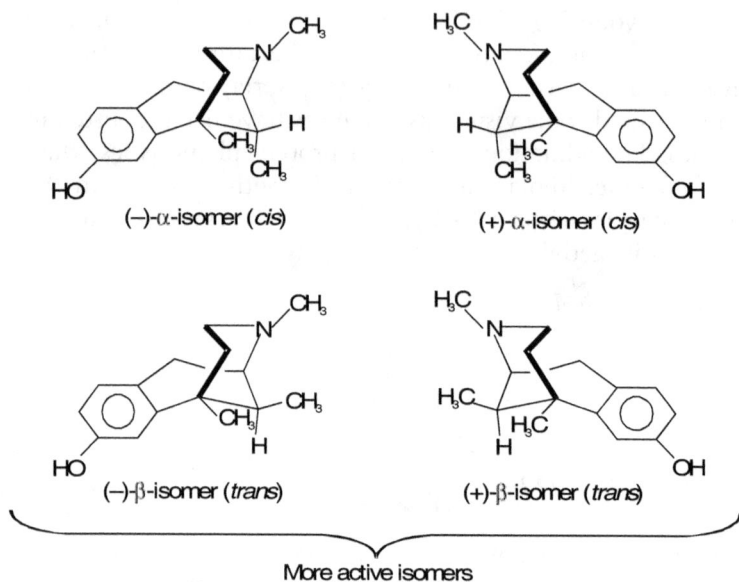

Figure 7.1   Stereoisomers of benzomorphan

Pentazocine   Replacement of the *N*-methyl of metazocine with an *N*-substituted allyl group yields pentazocine (Talwin).

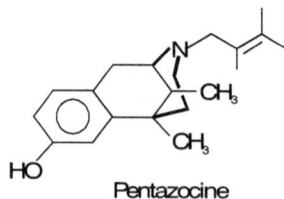

Pentazocine

It is more effective against moderate pain but tolerance and abuse potential do exist. The 4-phenylpiperidines can be regarded as analogues of morphine in which rings B, C and E have been removed as illustrated by the figure below with the example of meperidine.

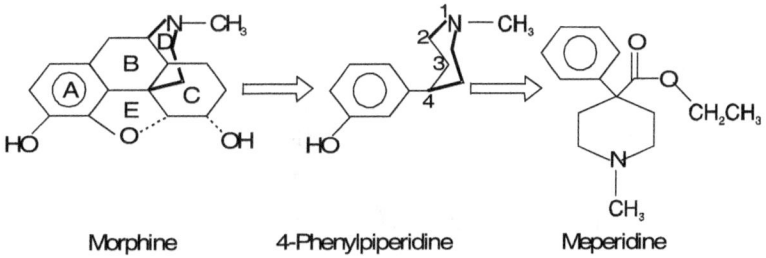

| Morphine | 4-Phenylpiperidine | Meperidine |

Meperidine is 5–10 times less potent than morphine. It has no antitussive activity and possesses the same degree of analgesia, sedation and respiratory depression as morphine and has a relatively short duration of action. Apomorphine produced by heating morphine with a strong acid has no morphine-like effects and it is used as an emetic agent.

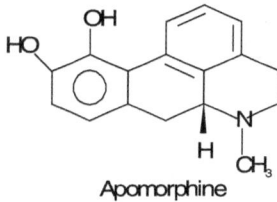

Apomorphine

Normorphine   Normorphine (7,8-didehydro-4,5-epoxymorphinan -3, 6-diol desmethylmorphine) is used as a narcotic analgesic.

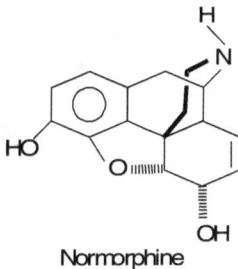

Normorphine

Oxymorphone   Oxymorphone [Numorphan] 14-hydroxy-7, 8-dihydromorphinone is 6–10 times more potent than morphine.

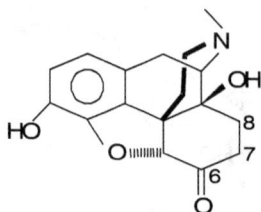

Oxymorphone

Buprenorphine   This is a derivative of thebaine, a semi-synthetic drug. As a partial agonist, it is 25–50 times more potent than morphine.

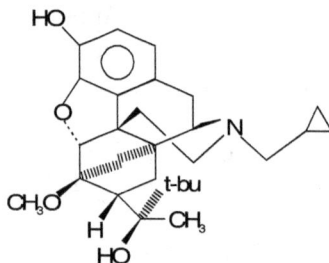

Buprenorphine

Nalorphine   It is also called as *N*-allylnormorphine and is an analgesic antagonist. It is a derivative of morphine used in the form of its hydrochloride as a respiratory stimulant. It acts to reverse the effects of morphine and other narcotics. It counteracts narcotic-induced nervous system and respiratory system depression but is not effective against depression induced by other sedatives such as barbiturates. Because nalorphine causes withdrawal symptoms in addicts, it is administered to apparent ex-addicts to determine if they have returned to drug use. Nalorphine is marketed under the trade name Nalline. Nalorphine is an unusual opiate. The low doses of nalorphine antagonize morphine analgesia but higher nalorphine doses are analgesic.

There are a number of totally synthetic compounds finding use as narcotic analgesics. This includes pethidine and methadone.

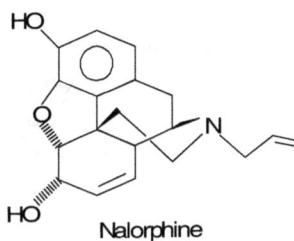
Nalorphine

Pethidine   It is a piperidine derivative having analgesic action as that of morphine, but with less habit-forming action.

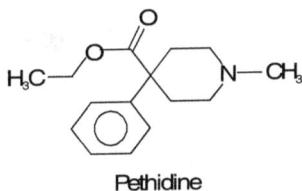
Pethidine

Pethidine is prepared by reacting benzyl cyanide with bis-2-chloro ethyl *N*-methyl amine in the presence of $NaNH_2$. This involves a double C-alkylation reaction. The product pethidine is obtained by hydrolysing the cyanide to carboxylic ester (Figure 7.2).

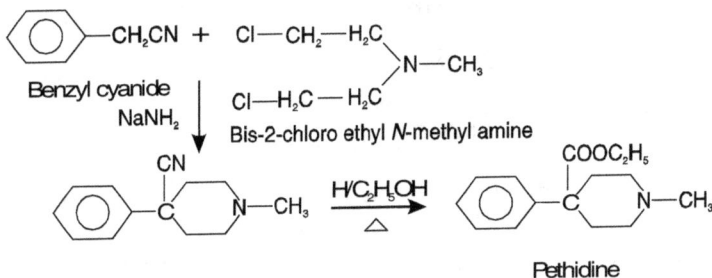

Figure 7.2   Synthesis of pethidine

Methadone   It is also known as amidone. It is obtained as a racemic mixture. Both the d and l enantiomers are active. The l-form is more active. Its analgesic activity is similar to that morphine. The synthesis of methadone is shown in Figure 7.3.

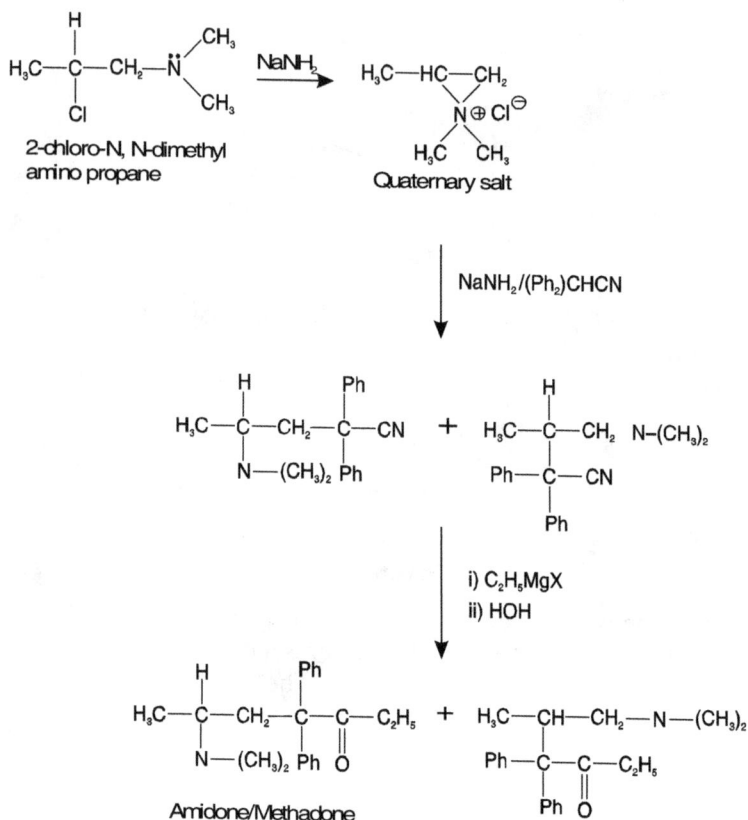

Figure 7.3   Synthesis of methadone

ANTIPYRETICS

The drugs responsible for lowering the temperature of a feverish organism to normal, without any effect on normal states are called as antipyretics. In the earlier days, the antipyretics studied and synthesized also possessed analgesic property. However, some compounds possess either one of the properties only.

MODE OF ACTION OF ANTIPYRETICS

To regulate the body temperature there should exist a balance between the heat produced and heat lost. Hypothalamus situated in the CNS plays an important role in maintaining this balance. For this reason, the hypothalamus is known as the thermostat of the body. During fever there exists a balance between the heat production and heat loss, but now the thermostat is set at a higher level. The antipyretics help to reset the thermostat to the normal level. During this, the heat production is not reduced but the heat loss is increased by increasing the peripheral blood circulation, which in turn increases the rate of perspiration. This results in loss of more heat and makes the body to return to the normal temperature.

COMMON ANTIPYRETIC ANALGESICS

Some of the most commonly used antipyretic analgesics are:

- Pyrazolone and pyrazolidones
- Aniline derivatives
- Derivatives of para amino phenols
- Quinoline derivatives
- Salicylic acid derivatives

A few of them are described here.

Pyrazolone and Pyrazolidones

Pyrrole on complete reduction yields pyrazolidine. Several derivatives of pyrazolidine are of pharmaceutical interest particularly 5-pyrazolone and 3,5-pyrazolidine dione are of paramount importance as analgesics, e.g. antipyrine, aminopyrine

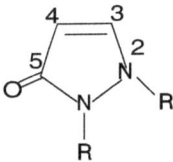

| General structure | Antipyrene (Felsol®) | Aminopyrene (Pyramidon®) |

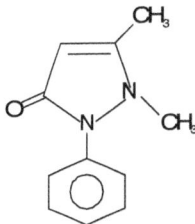

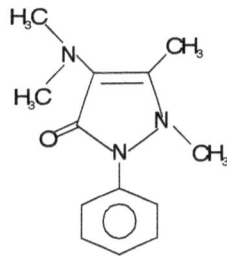

and analgine. Pyrazolinones which are or were clinically useful have the general structure shown in the figure and are numbered as indicated.

The pyrazolinone, antipyrene (Felsol®) as an analgesic is less effective compared to salicylates. It is a non-anti-inflammatory drug. Antipyrine is administered orally to reduce pain and fever in migraine, chronic rheumatism and other headaches.

The structurally related analogue aminopyrene (Pyramidon®) has similar activity to antipyrene, but is more potent. Aminoantipyrine and aminopyrine are the intermediates of analgin. Aminopyrine had also been used in the treatment of heat exposure and ache, but it is replaced by analgin because it reduces the count of leucocytes. They are more powerful than antipyrine. These drugs have been removed from the market due to adverse side effects and the availability of more efficacious agents.

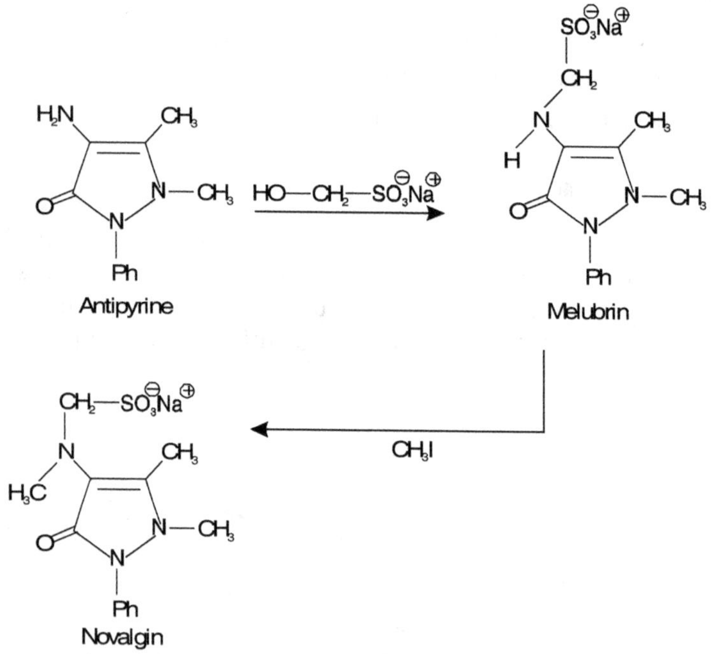

Figure 7.4   Preparation of novalgin

Novalgin is prepared from melubrin by methylation reaction (Figure 7.4). Both melubrin and novalgin have antipyretic action. They are used to reduce body pain and headache. Novalgin is twice as active as melubrin.

Derivatives of Pyrazolidine

Derivatives of this class are 3,5-pyrazolidinediones with the general structure as shown, and they are numbered like the pyrazolinones. These compounds are analgesic, antipyretic, anti-inflammatory (due to their weak acidity) and uricosuric. The acidity in these molecules is due to the presence of an enolizable hydrogen in the 4 position, and is pK_a-dependent. The most active analogues have a pK_a near 4.5. At lower pK_a values, uricosuric activity predominates. The most widely used analogues in this class are shown below.

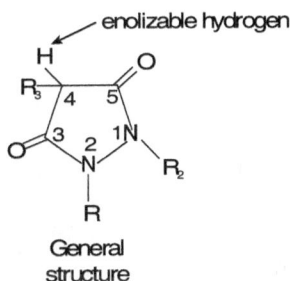

General
structure

Phenylbutazone (Azolid, Butazolidin)   This is an excellent anti-inflammatory drug, which has no inherent analgesic properties (except for analgesia due to the relief of inflammation). Phenylbutazone, sometimes known as benzone, is a crystalline substance.

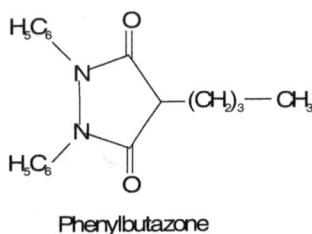

Phenylbutazone

It is used as a non-steroidal anti-inflammatory drug (NSAID) for the treatment of chronic pain, including the symptoms of arthritis. Its use is limited by its side-effect of suppressing the production of white blood cells by bone marrow. Phenylbutazone has an active metabolite, oxyphenbutazone (Tandearil®, Oxalid®), which was marketed separately, but has also been taken off the market. It has the same potency and toxicity as phenylbutazone itself.

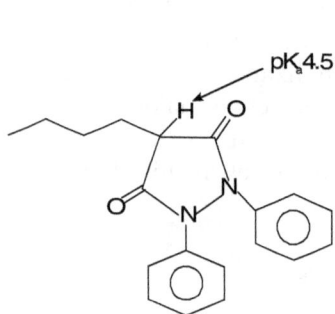

Phenylbutazone (Azolid, Butazolidin)   Oxyphenbutazone (Tandearil, Oxalid)

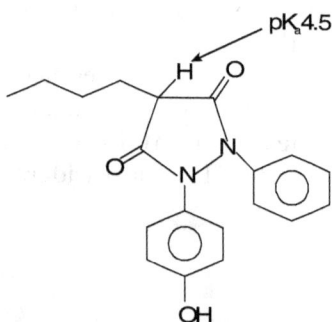

Sulphinpyrazone   Another related analogue sulphinpyrazone (Anturane) has a strongly electron-withdrawing (-Is) phenylsulphoxide group on the alkyl side chain at position 4, and thus the pK_a of the enolic hydrogen is lower (2.8). For this reason, it is much less active as an anti-inflammatory agent, but is an excellent uricosuric used in the treatment of gout. It retains the same side effects of phenylbutazone.

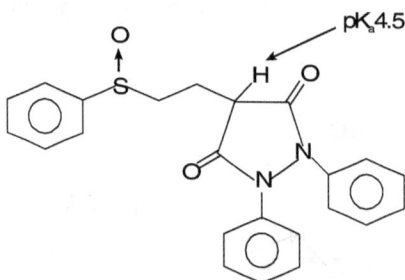

Sulphinpyrazone (Anturane)

Acetanilide

Aniline derivatives have strong antipyretic action but their easy absorption and action on RBCs lead to serious side effects. Attempts have been made to discover new aniline derivatives with strong antipyretic action and less or no side effects. Acetanilide the acetyl derivative of aniline is the cheapest antipyretic. This possesses all the harmful effects of aniline. Its antipyretic action is accompanied by the slow liberation of aniline. Refluxing aniline is a mixture of acetic anhydride and glacial acetic acid produces acetanilide. When the acetic reaction mixture after refluxing is poured into cold water, acetanilide solidifies out.

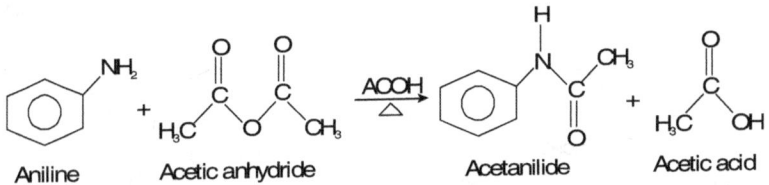

Figure 7.5   Synthesis of acetanilide

Derivatives of *p*-aminophenol

Acetanilide is known to undergo oxidation in the body to *p*-aminophenol. This led to the discovery of many *p*-aminophenols, which includes, phenacetin, *p*-acetamol, etc.

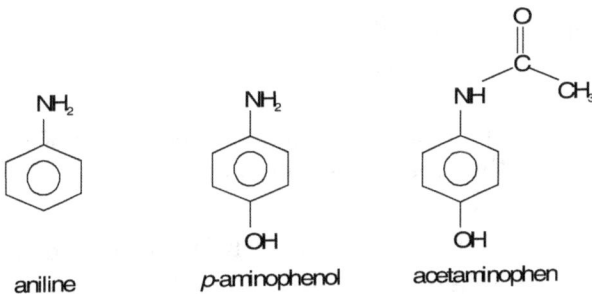

Phenacetin   There are a number of acetaminophen (*p*-acetamol) analogues, which have been synthesized, but none are superior to acetaminophen itself. Acetaminophen (Tylenol) is good for relieving pain and fever. It is safe and less irritating to the stomach. The ethyl ether derivative phenacetin used to be a component of many over-the-counter preparations. However, the drug is more toxic than acetaminophen, and is immediately O-dealkylated to acetaminophen. The acetaminophen analogue phenetsal (*p*-acetaminophenyl salicylate, Salophen®) is a true salol, since it is metabolized to acetaminophen and salicylic acid.

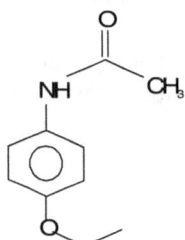

Phenacetin

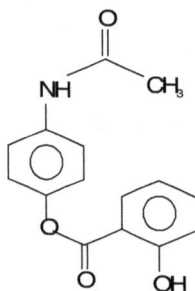

Phenetsal

The conversion of acetaminophenol into phenacetin is an example of the Williamson's ether synthesis using phenol as the alcohol. The net reaction is shown below:

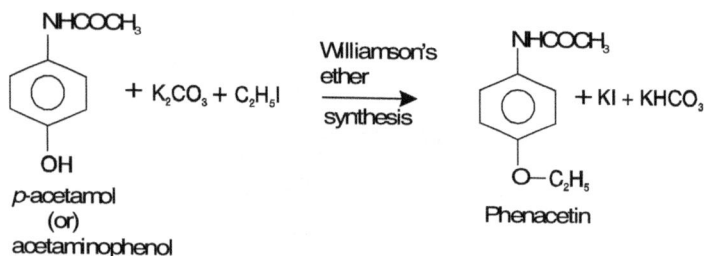

It is a mild antipyretic and has similar effect as that of aspirin. It is more toxic than *p*-acetamol. Prolonged use leads to renal damage.

p-acetamol   Aminophenols are relatively less toxic than aniline. The para isomer has the least toxicity, and significant antipyretic action.

Reduction of *p*-nitrophenol and acetylating the resulting *p*-aminophenol using acetic anhydride and glacial acetic acid yields *p*-acetamol.

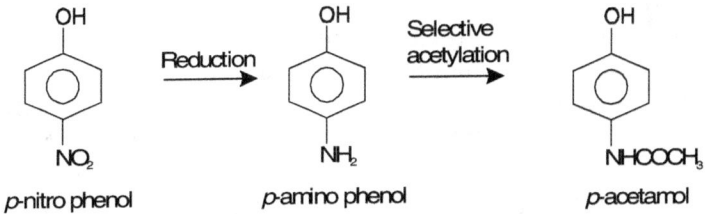

| p-nitro phenol | p-amino phenol | p-acetamol |

The crude product can be purified by recrystallization from water : ethanol mixture (1 : 1) or any other suitable solvents. It is used in arthritic and rheumatic conditions, common headache, myalgias and neuralgias. It is particularly useful in aspirin-sensitive cases.

Salicylates

Salicylates were first discovered when the observation was made that chewing willow bark could relieve pain. Some time later, the active constituent of willow bark, a glycoside of glucose and salicyl alcohol (salicin) was found to be the active species. Later, methyl salicylate was found to have similar activity. Acetylsalicylic acid (aspirin) was first synthesized in 1898 and introduced as a pain reliever in 1899, at which time it was used in doses of 650 mg every 4 h.

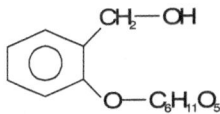

Salicin
(glycoside of glucose
and Salicylalcohol)

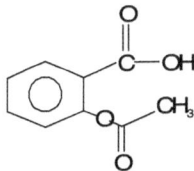

Acetylsalicylic acid
(Aspirin)

Aspirin acts on the hypothalamus, increases the peripheral blood flow, and sweating, therefore reduces the temperature of the feverish body. Although most of its analgesic properties are due to relief of inflammation, it also has a low degree of true analgesia. It is uricosuric in higher doses. Aspirin requires an acidic environment such as the stomach for optimal absorption, but is poorly soluble under these conditions. Some aspirin products contain buffering agents, which raise the pH locally and aid in dissolution of the drug.

Most carboxyl-modified salicylates are salts, and various salt forms of this drug are available. The sodium salt is somewhat unstable, and is not often used. The choline salt is marketed as Trilisate®. The counter-irritant compound methyl salicylate, found in Oil of Wintergreen, is fairly toxic when taken internally, although it is used as a flavouring agent in small quantities. It is mainly used in topical preparations such as Ben-Gay®. Another salicylate ester, phenyl salicylate, is an example of a "salol", since it is cleaved in the gastrointestinal tract to salicylic acid and phenol, which acts as an antiseptic agent. Salols are generally covalently bonded combinations of two drugs that individually have undesirable qualities such as taste, gastrointestinal irritation, etc. When combined, the adverse effects of each drug is masked.

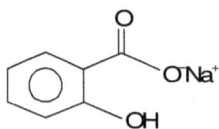

Sodium salicylate

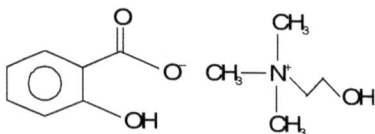

Choline salicylate (Trilisate®)

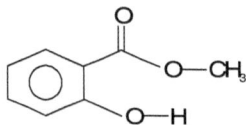

Methylsalicylate
(Oil of Wintergreen)

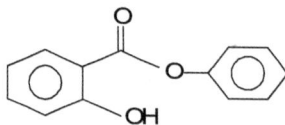

Phenyl salicylate
(a "salol")

Although there are many hydroxyl-modified salicylates that have been synthesized, the best member of this class is aspirin, which is the most successful drug ever produced. Other forms of salicylic acid such as sodium salicylate and choline salicylate, magnesium salicylate and salsalate are not effective NSAIDs, but they have a lower incidence of gastrointestinal effects, and they have negligible effects on platelet aggregation. Aspirin is a common choice for treatment of mild analgesia and inflammation, and is a reasonable antipyretic. Aspirin is also uricosuric in high doses. It is now commonly used to prevent recurrence of myocardial infarction in low chronic doses due to its ability to inhibit platelet aggregation.

Only two drugs in the salicylate class have been developed which are with nuclear substitutions. The first of these is flufenisal, which has twice the potency of aspirin, and a lower incidence of gastric irritation gastrointestinal. The second member diflunisal (Dolobid®) is three times more potent than aspirin. It is the more clinically useful drug.

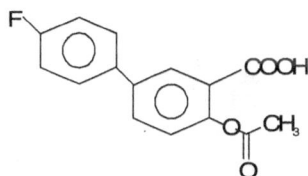

Flufenisal

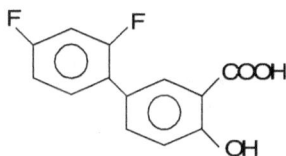

Diflunisal (Dolobid®)

Salicin was the first compound belonging to this class. It was employed as a substitute for quinine as a febrifuge. Salicin does not irritate the stomach as acetyl salicylic acid. Salicin is synthesized from salicylaldelyde. This on reaction with tetra-O-acetyl - α-bromo glucose produces an *o*-glycoside. This product on further reduction of aldehydic group, followed by hydrolysis to remove the protecting acetyl groups in glucose yields salicin.

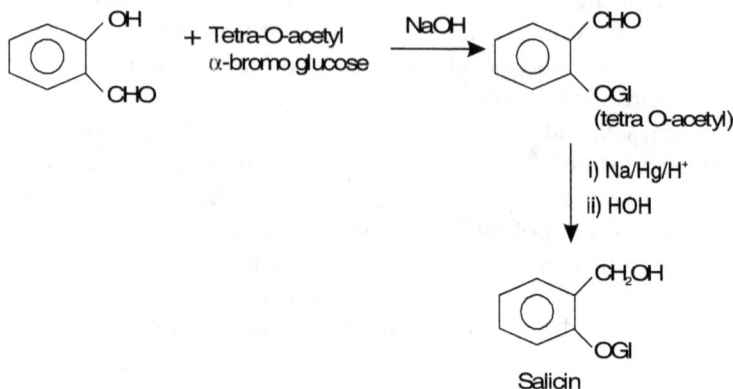

(tetra O-acetyl)

i) Na/Hg/H⁺
ii) HOH

Salicin

Salicylic Acid

It is prepared by Kolbe's synthesis.

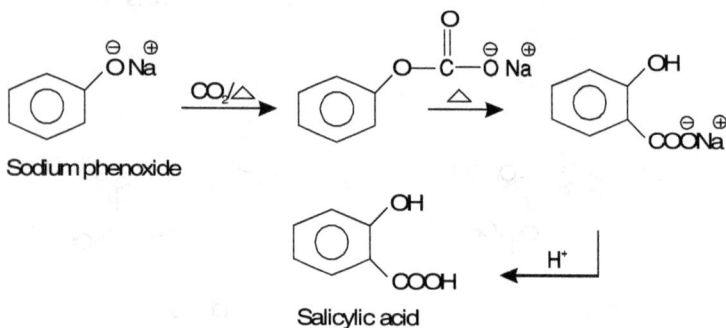

Sodium phenoxide

Salicylic acid

It is used to relieve headache, fever and rheumatic fever. This produces gastric disturbances.

The other derivatives of salicylic acid include:

Aspirin   This is also known as acetyl salicylic acid. It is prepared by the acetylation reaction of salicylic acid.

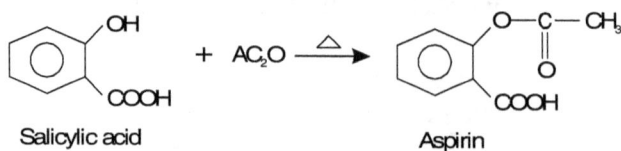

Salicylic acid   Aspirin

It is an antipyretic, analgesic, anti-rheumatic drug. A part of the aspirin that dissolves in the intestinal track and hydrolyses partially and the ester circulated in the bloodstream is then completely hydrolysed. The salicylic acid released in the bloodstream interferes with the nerve impulses, which are transmitted to hypothalamus thus increasing the threshold of pain.

Soluble aspirins   Calcium acetyl salicylate and aluminium aspirin are the two important members of soluble aspirin. Soluble aspirin is synthesized as shown.

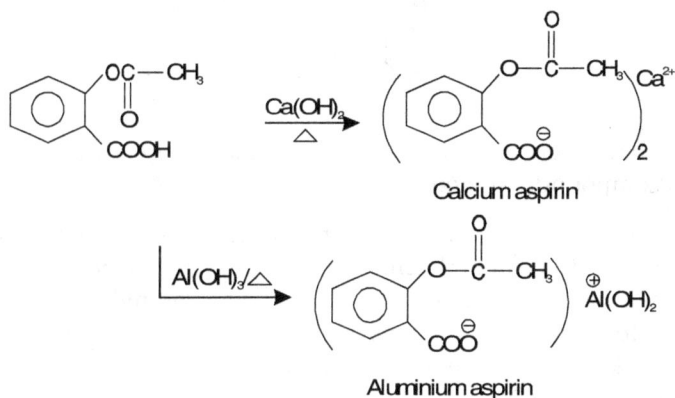

Calcium aspirin

Aluminium aspirin

They are recommended for people suffering from stomach ulcers.

Salol   It is prepared from salicylic acid.

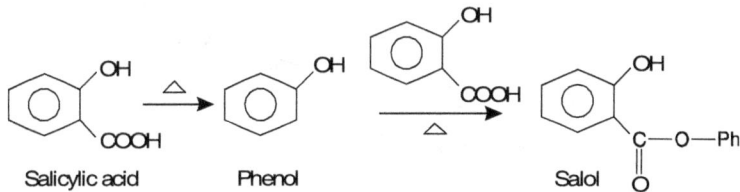

Salicylic acid   Phenol   Salol

Salol is used as an antipyretic as well as internal and external antiseptic. When it is administered, it passes unhydrolysed in the stomach, but slowly hydrolyses in the intestine into phenol and salicylic acid. The rate of hydrolysis is slow thus the concentration of the products released is low. Therefore it is better than the other salicylic acid derivatives.

N-aryl and Indolyl Compounds

Certain *N*-aryl and indolyl compounds are known for their antiseptic properties. This includes *N*-aryl anthranilic acid and aryl acetic acid derivatives. The *N*-aryl anthranilic acids are superior to aspirin, e.g. ponstel.

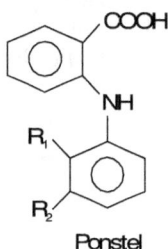

Ponstel

The aryl acetic acid derivatives include ibuprofen and the indolyl compounds—indomethacin.

Indole-3-acetic acid derivatives The representative analogue in this series is Indomethacin (Indocin), (*p*-chloro benzoyl-5-methoxy-2-methyl indole-3-acetic acid). The methylene group adjacent to the carboxyl moiety is termed α, and the carboxyl group has a pK_a of 4.5. Indomethacin, developed in 1965, is one of the more potent NSAIDs, and is about 2–3 times more potent than phenylbutazone. It is a potent anti-inflammatory agent, analgesic in rheumatic arthritis, and a good antipyretic. Its gastric effect is almost equal to aspirin, but it also causes headache, vertigo and blood dyscrasias in some patients.

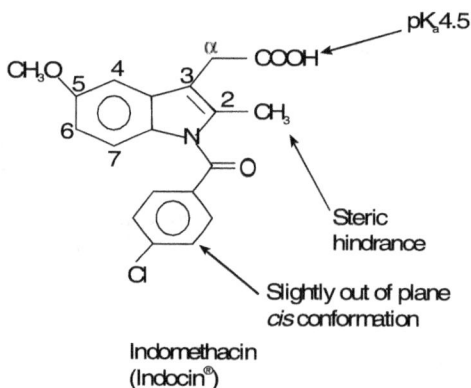

Indomethacin
(Indocin®)

The other compound in this class is Sulindac (Clinoril®), an aryl acetic acid with a pK_a of 4.7. In this, the nitrogen in the indole ring of indomethacin has been replaced by a double bond, thus is called as indene isostere, which has the same electronic character as the lone pair of the indole nitrogen. The electron withdrawing *p*-methylsuphoxide increases potency and the solubility of the drug.

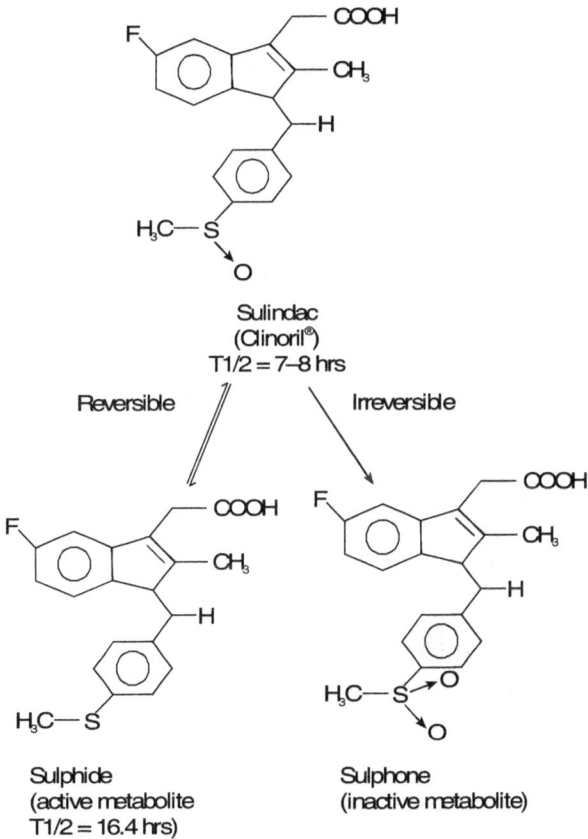

Sulindac
(Clinoril®)
T1/2 = 7–8 hrs

Reversible   Irreversible

Sulphide
(active metabolite
T1/2 = 16.4 hrs)

Sulphone
(inactive metabolite)

Sulindac may be considered a pro-drug due to its unusual metabolism.

The structures of other aryl acetic and aryl propionic acid derivatives are given below.

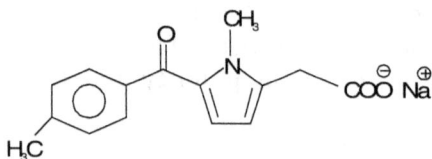

Tolmetin sodium
(Tolectin®)

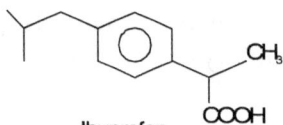

Ibuprofen
(Motrin, Relafen, Nuprin, Advil)

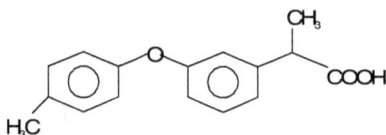

Fenoprofen
(Nalfon)

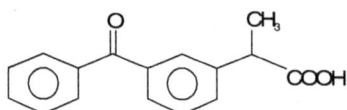

Ketoprofen
(Orudis)

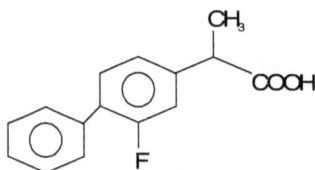

Flurbiprofen
(Ansaid®)

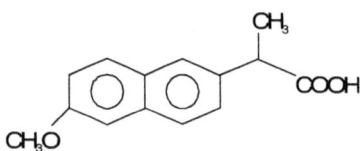

Naproxen
(Naprosyn®, Anaprox®)

There are a number of analogues in this class that have been marketed as non-steroidal anti-inflammatory agents. These agents generally have three structural features of indomethacin that allow them to retain activity: an acidic carboxyl group, and out-of-plane phenyl group and a flat, nitrogen-containing ring system.

Tolmetin sodium (Tolectin®) is about equal to ibuprofen as an NSAID, but with low incidence of gastrointestinal effects.

Ibuprofen is marketed under a number of trade names, and is widely used both over-the-counter and as an anti-inflammatory drug for many decades. It is an aryl propionic acid, and as such has an alpha methyl group, imparting stereochemistry to the molecule. In general, the S(+) isomer of any given NSAID is the active form, and the R(–) isomer is inactive. However, ibuprofen is administered as a racemate. Fortuitously, the R(–) isomer of most NSAIDs is converted to the corresponding S(+) isomer *in vivo*, so that the stereochemistry is not a concern in the synthesis and production of the drug. Ibuprofen is prescribed for mild to moderate pain and inflammation, and for dysmenorrhoea.

The S(+) isomer of naproxen is marketed as "Naprosyn®", while the R(–) isomer is marketed as "Anaprox®". A number of more recently marketed NSAIDs includes diclofenac. It is used extensively worldwide, and contains structural features of the arylalkanoic acid and anthranilic acid. Diclofenac undergoes extensive metabolism *in vivo*, by hydroxylation of the aromatic ring bearing the chlorines at the 3, 4 and 5 positions.

- Nabumetone (Relafen®) was introduced in the US in 1992. This drug does not contain acidic functional group, and must be converted by oxidation to the corresponding aryl acetic acid known as 6MNA. Since the non-acidic form is administered orally, it causes less gastric erosion. It has a very long half-life, therefore is dosed once a day.

- The pyranocarboxylic acid derivative etodolac (Lodine®) is a representative of a new class of NSAID.

- Two additional recent agents are Ketorolac (Toradol®) and Oxaprozin (Daypro®). Ketorolac is especially useful, since it can be used parenterally.

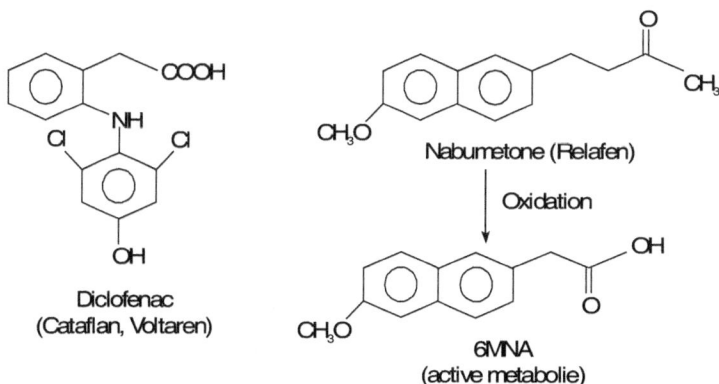

Diclofenac
(Cataflan, Voltaren)

Nabumetone (Relafen)

Oxidation

6MNA
(active metabolie)

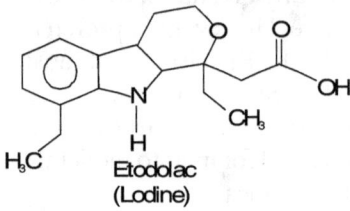

Etodolac
(Lodine)

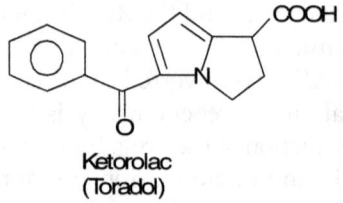

Ketorolac
(Toradol)

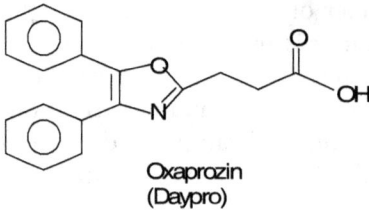

Oxaprozin
(Daypro)

8

TRANQUILIZERS, SEDATIVES AND HYPNOTICS

Tranquilizers are drugs used in the treatment of mental disorders. They produce a specific improvement in the mood and behaviour of the patient suffering from mental disorder. They are also called as *psychedelic drugs*. They exert a unique type of control on the CNS. They do not possess hypnotic action but predispose an individual to sleep from which he is readily arosen.

An exact demarcation cannot be drawn between hypnotics and sedatives since their action varies with dosages. Thus, a drug at lower concentration is a sedative but at higher concentration, it is a hypnotic. A hypnotic drug is the one which produces sleep resembling natural sleep. A sedative is a drug that reduces excitement. Both these drugs are CNS depressants. As already stated, the difference is only quantitative. It is important to mention that a few drugs are exceptional in exerting only one type of effect, e.g. potassium bromide is a good sedative and has no hypnotic action. Similarly, thiopental sodium is known for its use as a sedative.

CHARACTERISTICS OF A TRANQUILIZER

The following are the characteristics of an ideal tranquilizer:

- It should not produce any toxic effects.
- It must not possess any undesirable side effects.
- It should not impart unconsciousness.
- It should be effective.

Thus, it can be said that tranquilizers in general reduce mental tension, relieve anxiety, and result in a more calm outlook without producing any sedation or hypnosis and do not alter the level of consciousness. They are normally used when other hypnotics, sedatives and other central nervous system relaxants fail or are not applicable.

CLASSIFICATION OF TRANQUILIZERS

Tranquilizers are classified into two major groups:

1. Anti-psychotics or major tranquilizers
2. Anti-anxiety agents or minor tranquilizers

Anti-psychotics

The major tranquilizers reduce the agitation and disturbed behaviour, which are often associated with delusion and hallucination in schizophrenia. The second group members introduce calm effect in the anxiety state, which is associated with neurotic personality, situational crisis, or physical disease.

A few important members from both the groups are discussed.

In the field of tranquilizers, two powerful drugs were introduced simultaneously namely reserpine and chlorpromazine.

Reserpine, an indole alkaloid obtained from *Rauwolfia* alkaloid is a potent drug used in the treatment of mental disturbances. This is in use for nearly five centuries. Various *Rauwolfia* alkaloids and their synthetic analogues are mainly used in schizophrenia. These are important in the individuals who cannot tolerate phenothiazines which is one of the members belonging to the major tranquilizers.

The *Rauwolfia* alkaloids possess certain undesirable side effects such as lethargy, nasal congestion, nausea and vomiting, and sometimes severe depression. Later on it was found that tertiary amines having trimethoxy benzoyl group exhibit reserpine-like activity.

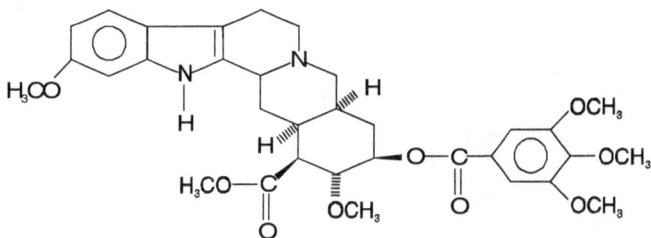

Reserpine

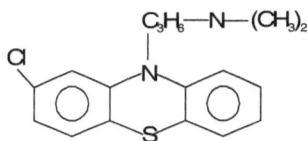

Chlorpromazine

Other important members belonging to the major tranquilizers are:

- Phenothiazine derivative, e.g. chlorpromazine
- Butyrophenones, e.g. haldol
- Diphenyl butyl piperidines
- Indole derivatives

Anti-anxiety Agents

Members belonging to the minor tranquilizers are:

- Benzodiazepines, e.g. librium, valium, diazepam, calmpose, etc. All the benzodiazepines have barbituric–hypnotic effect, except that they produce less sedation.
- Diphenyl methanes, e.g. piperadol, hydroxyzine, benactyzine. They are anti-anxiety agents without causing drowsiness, used under the conditions of anxiety and emotional tension.
- Meprobamate

PSYCHEDELIC DRUGS

Psychedelic drugs are also called as hallucinogens, psychodysleptics, or psychotogenic drugs. These substances produce mental changes resembling that of psychotic state—psychosis—a state characterized by maladaptive behaviour, in which an individual reacts inappropriately to his environment. They are chiefly used by people seeking a new experience or an escape. This is used by psychiatrists in supervised therapeutic treatment to help the patient to gain insight and improved mental ability. These drugs produce depersonalization, changes in mood, and a variety of effects on memory and behaviour.

The psychedelic drugs are classified into

- Those containing indole ring, e.g. LSD.
- Those without indole ring, e.g. *Cannabis indica* and *Cannabis sativa*.

LSD is one of the most powerful hallucinogenic drugs known. It was invented in 1938 by the Swiss chemist, Albert Hoffman, who was interested in developing medicines from compounds in ergot, a fungus that attacks rye. Although LSD is purely synthetic, clues to its biological activity can be found by tracing the history of the fungus from which it is derived.

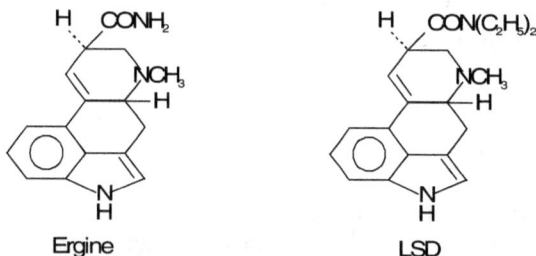

Ergine   LSD

A. Hofmann (1979) found out that the morning glory seeds contain a lysergic acid alkaloid called Ergine (*d*-lysergic acid amide), better known as "natural" LSD. The more potent synthetic LSD is *d*-lysergic acid diethylamide.

The structure of LSD is very similar to other hallucinogenic drugs such as Mescaline and Psilocybin, all of which contain a substituted indole ring or a related structure. It is a potent hallucinogen. The dosage varies from 100–400 mg. It causes slight clouding of consciousness, disturbances in perception, hallucination and loss of personal identity. It is a habit-forming drug.

The mechanism by which LSD causes such profound effects on the human perception has not been established. LSD stimulates centres of the sympathetic nervous system in the midbrain, which leads to pupillary dilation, increase in body temperature, and rise in the blood sugar level. LSD also has a serotonin-blocking effect. Serotonin is a hormone-like substance, naturally occurring in various organs of warm-blooded animals. Concentrated in the midbrain, it plays an important role in the propagation of impulses in certain nerves and therefore in the biochemistry of psychic functions. LSD also influences neurophysiological functions that are connected with Dopamine, which is another naturally occurring hormone-like substance. Most of the brain centres receptive to dopamine become activated by LSD, while the others are depressed. Because of its hallucinatory properties, LSD was widely adopted by the Hippy culture of the 1960s, who claimed that it led to higher states of consciousness and helped them search for religious enlightenment. Although LSD is relatively non-toxic and non-addictive, various governments around the world outlawed it after a number of fatal accidents were reported. Such accidents involved, for example, people under the influence of LSD jumping to their deaths off high buildings thinking they could fly. Research in the '60s and '70s showed that there was also a considerable psychological risk with the drug and that high doses, especially in inappropriate settings, often caused panic reactions. For individuals who have a low threshold for psychosis, a bad LSD trip could be the triggering event for the onset of full-blown psychosis.

Charas (Hashish)

It is the resinous product from the exudates of the *Cannabis* female plant. It is one of the oldest herbal remedies. *Cannabis* is a hemp plant. The active principle is tetrahydrocannabinol. *Bhang* is obtained from the dried leaves and flowering shoots. *Ganja* is from the resinous mass of small leaves and bracts of inflorescence (*Marijuana* is any plant part or extract containing the active principle). When it is smoked, the tetrahydrocannabinol is rapidly absorbed, and effects appear immediately, which lasts for 2–3 hours. When given orally, the action is slow. It causes drowsiness, dreamy state, and feeling of well-being, excitement and inner joy. It also possesses mild analgesic, muscle relaxant, sedative and hypnotic actions. When compared with other social drugs it does not pose great risks. A new synthetic derivative of *Cannabis* is used in cancer therapy.

9

ANAESTHETICS

The drugs which produce loss of sensation, in other words, insensibility to the vital functions, of all types of cells mainly those of nervous system are termed as *anaesthetics*. The effect produced by an anaesthetic is reversible. Therefore, the individuals return to the normal state as soon as the concentration of the anaesthetics is decreased. Thus, it is used to produce temporary insensibility to pain or sensation in the whole body or a particular organ, which has to be treated surgically.

The basic requirements of an anaesthetic include the following:

- It should be safe and pleasant.
- It should produce sufficient relaxation.
- It should have prompt recovery.

The conditions for an ideal anaesthetic include:

- inertness
- potent and non-inflammable
- rapid and smooth induction
- non-irritant to mucous membrane
- should also produce analgesia and muscular relaxation

- must not alter blood pressure and respiration
- non-toxic
- should not produce nausea and vomiting
- economical
- stable to light and heat

CLASSIFICATION OF ANAESTHETICS

According to the mode of action, anaesthetics are classified into two types.

1. General or central anaesthetics—produce insensibility to the entire system. Thus they act on the central nervous system and lead to unconsciousness.
2. Local anaesthetics—make the part under treatment insensitive.

GENERAL ANAESTHETICS

The general anaesthetics are further classified into:

1. Volatile or gaseous anaesthetics   They are administered by inhalation, e.g. ether, vinyl ether, methoxy fluranes, halo hydrocarbons, nitrous oxide, cyclopropane, etc.
2. Non-volatile or fixed anaesthetics   They are administered intravenously, e.g. thiopental sodium, methohexitone, propandid, etc.

The action of volatile anaesthetics are divided into four stages.

1. Preanaesthesia—the induction stage
2. Delirium or dream stage
3. Surgical anaesthesia
4. Stage of medullary paralysis

Volatile—Gaseous Anaesthetics

Nitrous oxide   This is commonly called as the laughing gas. It was the first anaesthetic prepared by heating ammonium nitrate to 200°C. It is then purified and liquefied. It is used as a mixture of 85% N_2O and 15% O_2. The induction period is short and unpleasant; recovery is almost immediate as soon as the mask is removed.

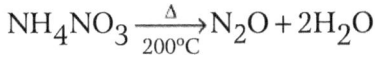

$$NH_4NO_3 \xrightarrow[200°C]{\Delta} N_2O + 2H_2O$$

The principal uses of laughing gas are

   i. to induce anaesthesia to be followed by other anaesthetics

  ii. for shorter dental surgeries

 iii. to supplement anaesthesia with thiopental sodium

 iv. to provide a state of unconsciousness during surgery under local anaesthesia

Ether   It was the first liquid anaesthetic used. It is prepared by the Williamson's synthesis.

Williamson's synthesis   Diethyl ether is prepared from ethyl halide.

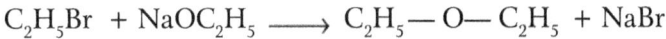

$$C_2H_5Br + NaOC_2H_5 \longrightarrow C_2H_5 — O— C_2H_5 + NaBr$$

Ethylbromide   Diethylether

It contains 96–98% $C_2H_5 — O — C_2H_5$ and the remaining water. Diethyl ether is commonly associated with impurities like aldehyde and peroxides. The latter compounds are known to delay the onset of anaesthetic effect. To maintain its anaesthetic effect stabilizers such as hydroquinones, sodium pyrogallate, etc. are added. Diethyl ether is a safe anaesthetic acting on the CNS and is inexpensive and stable.

The disadvantages are that it boils at lower temperature and is thus difficult to administer in tropical temperatures. Its vapours are inflammable and irritating to the mucous membrane. Its blood solubility is high and induction period is too short.

Vinyl ether/Divinyl ether/Vinethene   Vinyl ether is unstable to light and air. It decomposes to formaldehyde. In combination with nitrous oxide it is used as induction anaesthesia. It is more potent than ether, has speedy recovery, and is used only in minor surgeries.

$$\text{(structure: divinyl ether with central oxygen)}$$

$$2Cl\!-\!CH_2\!-\!CH_2\!-\!OH \xrightarrow{H_2SO_4} Cl\!-\!CH_2\!-\!CH_2\!-\!O\!-\!CH_2\!-\!CH_2\!-\!Cl$$

$$H_2C\!=\!CH\!-\!O\!-\!CH\!=\!CH_2 \xleftarrow{Et\,OH/KOH}$$

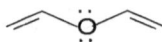

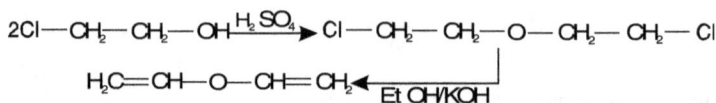

The disadvantages are that it affects liver on repeated administration, is inflammable and is less volatile.

Methoxy flurane   This is commonly called as penthrane $CHCl_2CF_2OCH_3$. It is a liquid anaesthetic. It is more potent, non-inflammable in nature. It posseses analgesic and muscular relaxation properties.

The disadvantages are that it is less volatile, liberates fluoride ions during its prolonged stay in the body and renal failure is often observed.

Halogenated hydrocarbons   The anaesthetic property of halogenated hydrocarbons is increased as hydrogen is successively replaced by the halogens. The anaesthetic property increases in the order as

$$CH_4 < CHCl_3 < CH_2Cl_2 < CHCl_3 < CCl_4$$

Along this order the inflammable property decreases. However, these two advantages are reduced by its increased toxicity. The halogenated hydrocarbons include chloroform, ethyl chloride, halothane and trichloroethylene.

Chloroform   Chloroform is more potent than ether. It is toxic to liver and kidney. Sudden cardiac arrest is also observed during the induction phase. It is non-inflammable but decomposes to phosgene when exposed to air and light. A smaller impurity of ethyl alcohol is added to preserve it by decomposing phosgene to harmless products.

$$COCl_2 + 2C_2H_5OH \longrightarrow O\!=\!C(OC_2H_5)_2 + 2HCl$$

Ethyl chloride   It is volatile and induces rapid induction. It resembles chloroform in its side effects. It is used in minor surgeries, and sprayed on the skin as local anaesthetic. It evaporates quickly causing the tissues to freeze and paralyses temporarily the sensory nerve endings.

Trichloroethylene   This is commonly called as trilene. It is prepared by the decomposition of tetrachloroethane.

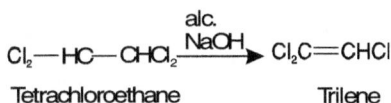

$$Cl_2-HC-CHCl_2 \xrightarrow{\substack{\text{alc.} \\ \text{NaOH}}} Cl_2C=CHCl$$

Tetrachloroethane   Trilene

This can also be prepared by the addition of controlled amount of chlorine and HCl to acetylene. It is a liquid, resembling the odour of chloroform. It is recommended for dental surgery. It is also used to induce analgesic state in the relief of pain in migraine.

The disadvantages include: it is not a good general anaesthetic, not suitable for administration into closed system; muscular relaxation is poor, disturbs cardiac rhythm, hepatotoxic and is less volatile.

Halothane   Particularly fluorothane is used as a general anaesthetic. It is more potent than ether.

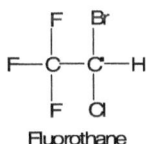

$$\underset{F}{\overset{F}{\underset{|}{\overset{|}{F-C-C-H}}}} \overset{Br}{\underset{Cl}{}}$$

Fluorothane

Anaesthetic capacity of fluorothane is the same as that of chloroform and it is less toxic. The solubility and vapour pressure are in reasonable range for its use. It is used in all types of surgeries and is the most widely used anaesthetic.

The disadvantages are: it leads to respiratory and cardiovascular depression, incomplete muscular relaxation, poor analgesia and is hepatotoxic.

Cyclopropane   Cyclopropane is a gaseous anaesthetic. It is prepared by the action of zinc on 1,3-dichloropropane. It is nearly an ideal anaesthetic, more potent, has no mentionable side effects. It is used widely as a low risk anaesthetic in aged patients and in all major surgeries. The disadvantages are: it has less abdominal relaxation, and is explosive with oxygen.

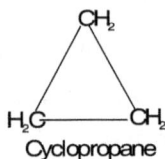

Cyclopropane

Non-volatile General Anaesthetics

Thiopental sodium   N-substituted barbiturates are the first known compounds administered as intravenous anaesthetics. The important drug under this group is thiopental sodium. It is a thiocarbonyl analogue of barbituric acid. It is prepared from thio urea and substituted malonic ester.

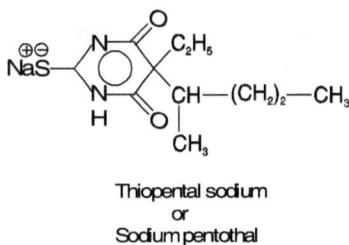

Thiopental sodium
or
Sodium pentothal

It is the most widely used, non-volatile general anaesthetic. It is unstable in solution, therefore is available as dry powder in sealed ampoules. It is dissolved in distilled water just prior to administration. It is used in many surgeries and to produce muscular relaxation in shock treatment. As its action is very short, it is not recommended for patients having respiratory depression and children of age below 10 years. The synthesis of thiopental sodium is as shown below.

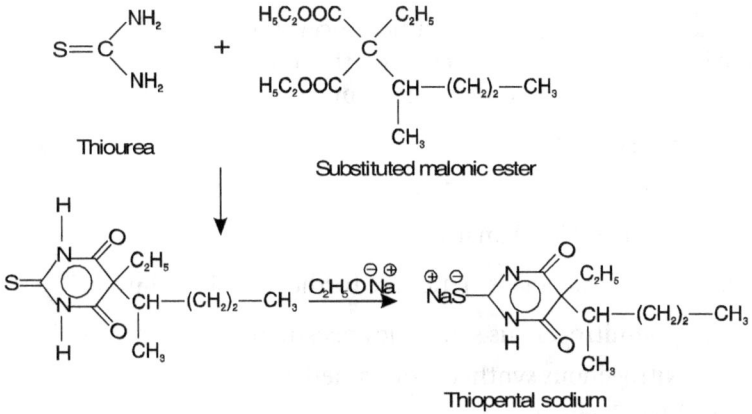

Thiourea

Substituted malonic ester

Thiopental sodium

Methohexitone   It is also a barbiturate. The sodium salt is known as brevital or brietal.

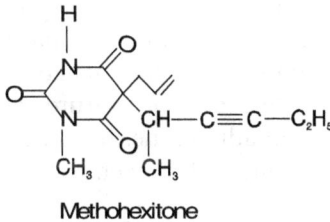

Methohexitone

Propandid   This is another short acting IV anaesthetic. Sweating, salivation are some of the undesirable side effects.

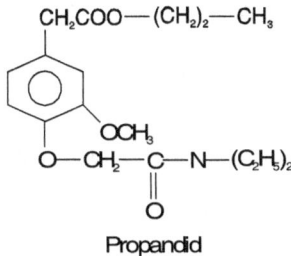

Propandid

LOCAL ANAESTHETICS

These are the agents in therapeutic concentration which block the nerve impulses reversibly. They block the sensation in the localized areas without leading to the state of unconsciousness.

They are classified into

1. Natural local anaesthetics
2. Synthetic local anaesthetics

Synthetic local anaesthetics are further classified into

  i. Non-nitrogenous synthetic anaesthetics
  ii. Nitrogenous synthetic anaesthetics
  iii. Miscellaneous

Natural Local Anaesthetics

Examples of natural local anaesthetics are cocaine and its derivatives. Cocaine is a benzoyl ester of ecognine and is closely related to the chemistry of atropine. In addition to its local anaesthetic action, it stimulates the central nervous system, dilates the pupil of eyes, raises the body temperature and blood pressure and activates the action of adrenaline. It produces euphoria and abolishes the sense of fatigue and hunger.

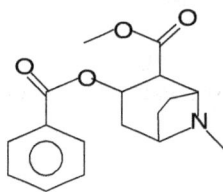

Cocaine

Activity Draw the structures of tropine and ecognine.

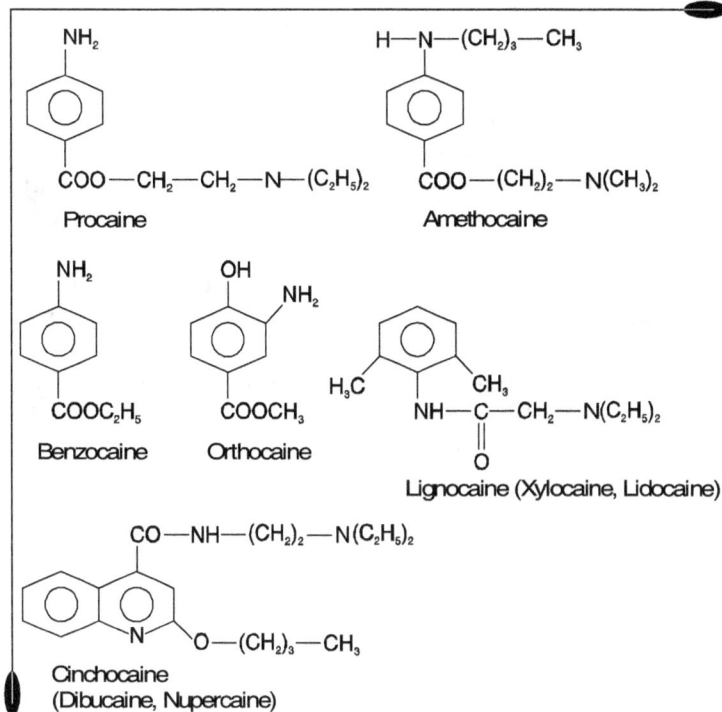

Procaine

Amethocaine

Benzocaine

Orthocaine

Lignocaine (Xylocaine, Lidocaine)

Cinchocaine
(Dibucaine, Nupercaine)

Activity Identify the common structural features of all the above structures and how can these be related to their activity? (Answer is discussed in the text in later portions of the chapter).

The concentration used to produce local anaesthetic effect is poisonous to many structures like leucocytes and tissue cells. It is a protoplasmic poison and drug causing addiction. Therefore it is used occasionally in surgery in the concentration of 5–10% solution. Allergic reactions to this drug are frequent.

Synthetic Local Anaesthetics

Nitrogenous drugs include

- Benzoic acid and para amino benzoic acid derivatives like: procaine, amethocaine, benzocaine and orthocaine.

- Derivatives of acetanilide like lidocaine.
- Quinoline derivatives like cinchocaine.

They possess various degrees of solubility in water and lipids. The nervous system is rich in lipids, thus lipoidal solubility of a drug is essential for its migration into neuronal fibre. Water solubility helps the drug migrate to the site of action from the site of administration. Hence, a drug with high lipid solubility and low water solubility will not be of much use, because of the difficulty in transportation through the aqueous phase surrounding the neuronal fibre. The potency of the drug is related to lipid solubility. Highly lipid-soluble drugs readily cross membranes, the higher lipid/water partition coefficient, the more potent and longer duration of action (DOA) of the drug, e.g. DOA of prilocaine-0.9, lignocaine-2.9, bupivicaine-28.

All the local anaesthetics have the following structural features in common:

- a hydrophilic amino group
- an intermediate chain
- a lipophilic aromatic moiety

The following structure is common to all local anaesthetic agents.

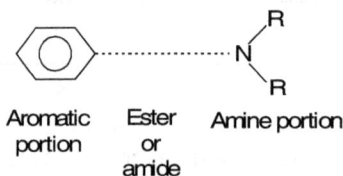

|Aromatic portion|Ester or amide|Amine portion|

Local anaesthetics can be structurally of two different types with respect to their intermediate chain, containing either an ester or an amide.

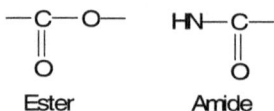

Ester   Amide

What determines the onset time of action?

- **lower pK$_a$** indicates better absorption into nerve tissue.
- higher pK$_a$ indicates more effective blockade within nerve.

The speed of onset of blockade of the local anaesthetic can be accelerated by alkalinization of the solution (by adding bicarbonate). By elevating tissue pH it raises the base–cation ratio (brings the pH and pK$_a$ closer together) and increases absorption of the local anaesthetic into the nerve tissue. Rare use of ester type local anaesthetics is in practice because esters are rapidly broken down and consequently tend to have a very short duration of action.

Orthocaine

3-amino-4-hydroxy-benzoic acid methyl ester (ortho form or orthocaine)

Orthocaine is used chiefly in the form of its hydrochloride as a local anaesthetic and antiarrhythmic agent. It is used topically and also for producing dental anaesthesia. Orthocaine/orthoform is prepared from *p*-hydroxy benzoic acid as shown by the following reaction.

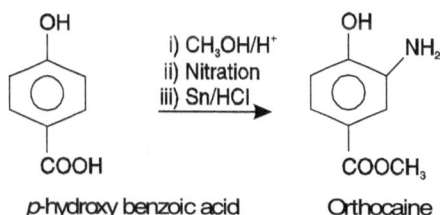

p-hydroxy benzoic acid   Orthocaine

Benzocaine   This is chemically 4-aminobenzoic acid ethyl ester, an anaesthetic commonly used as a topical pain reliever. It is the active ingredient in many over-the-counter analgesic ointments, including Anbesol and Orajel.

Fischer esterification is the process of forming an ester by refluxing a carboxylic acid and an alcohol in the presence of an acid (catalyst).

   p-amino benzoic acid (4-amino benzoic acid) on esterification in the presence of ethyl alcohol and acid catalyst yields benzocaine.

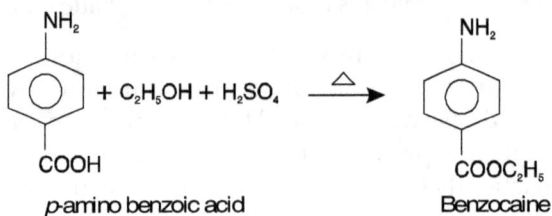

p-amino benzoic acid   Benzocaine

Procaine   This is chemically 2-(diethyl amino) ethyl *p*-amino benzoate. It is also called as novocaine and is the most widely used anaesthetic. It is applied along with penicillin G to prolong the action of the antibiotic.

   Nitro benzoylchloride reacts with 2-dimethyl amino ethanol followed by the reduction of the nitro group yielding procaine.

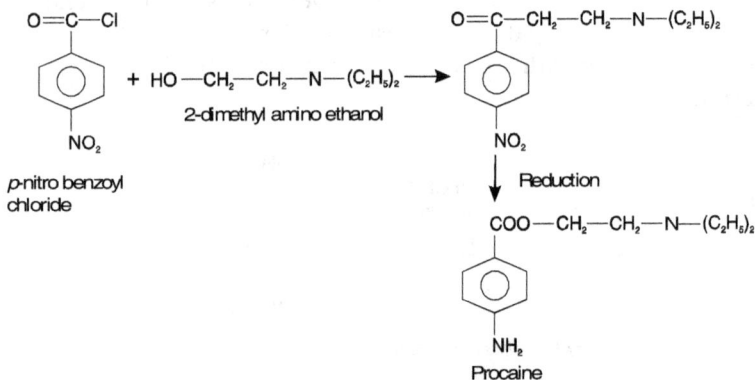

p-nitro benzoyl
chloride

Procaine

   Procaine hydrochloride is a local anaesthetic used primarily in dentistry.

   Procaine was first synthesized in 1905, and was the first injectable man-made local anaesthetic used. It was created by the German chemist Alfred Einhorn who gave the trade name Novocaine, from the Latin word *novus* (meaning "new") plus "caine" as in "cocaine". It was introduced into medical use by a surgeon Heinrich Braun.

Procaine is rarely used today since more effective (and hypoallergenic) alternatives such as lidocaine (xylocaine) exist. Prior to the discovery of procaine, cocaine was the most commonly used local anaesthetic. Procaine (like cocaine) has the advantage of constricting blood vessels which reduces bleeding, unlike other local anaesthetics like lidocaine, without the euphoric and addictive qualities of cocaine.

Procaine, an ester anaesthetic, is metabolized in the plasma by the enzyme pseudocholinesterase through hydrolysis into para-aminobenzoic acid (PABA), which is then excreted by the kidneys into the urine. Allergic reactions to procaine are usually not in response to procaine itself, but to PABA. About 1 in 3000 people have an atypical form of pseudocholinesterase, which does not hydrolyse ester anaesthetics such as procaine, resulting in a prolonged period of high levels of the anaesthetic in the blood, and increased toxicity.

Amethocaine It is an ester obtained by the reaction of 4-*N*-butyl amino benzoic acid and 2-dimethyl amino ethanol.

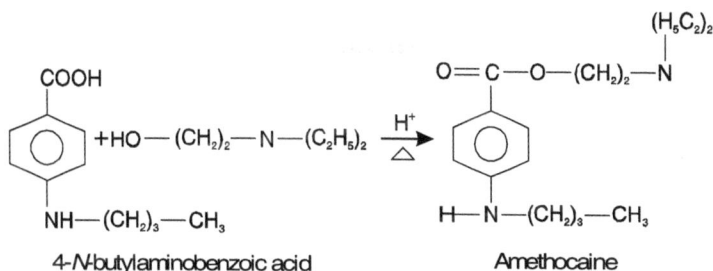

4-*N*-butylaminobenzoic acid   Amethocaine

It is used for topical anaesthesia in eye surgery or placement of IV cannulae. It should be used with care as it is an ester and is rather toxic as it is very slowly hydrolysed by pseudocholinesterase. Traditionally this drug was very popular as spinal anaesthesia. It is a long-acting and powerful local anaesthetic. It is mainly used as a surface anaesthetic.

Eucaines The piperidine derivatives are called as eucaines. These are also benzoic acid derivatives similar to orthocaine, cocaine, etc. and are known for their function as local anaesthetics.

β-Eucaine   This is one of the first synthetic local anaesthetic. It is superior to cocaine, because it is not a habit-forming drug. Its action is slow and less irritating.

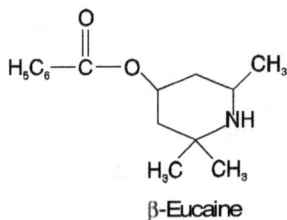

β-Eucaine

α-Eucaine   Structurally this is similar to *β*-Eucaine and can also be replaced by *β*-Eucaine for its painful irritant nature when injected.

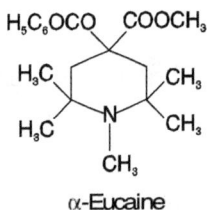

α-Eucaine

10

ANTINEOPLASTIC AND HYPOGLYCAEMIC DRUGS

CANCER

Cancer can broadly be described as the stage during which the cells lose their ability to regulate their own growth, or more specifically, replication. However, the loss of this regulation can be ascribed to changes in the cellular DNA, in the genetic make-up of the cell. Numerous causes of DNA damage are known and a good number of these causes can be traced back to a chemical origin. Indeed, enormous volume of knowledge is now available on the chemical causes of certain cancers, much based on "epidemiology" and the occupational links to various cancers. We will deal with just a few examples of these to illustrate how chemicals can cause DNA damage (Table 10.1).

Table 10.1   Etiology of common tumours

Cancer site	Approximate percentage of all malignancies	Major hypothesized cause
Esophagus	2	Smoking plus alcohol, diet and burnt products
Stomach	7	Preservatives and other aspects of diet
Large intestine	7	Intestinal flora and diet
Rectum	5	Beer
Pancreas	3	Smoking
Lung	19	Smoking and occupation
Skin	9	UV radiation and occupation
Prostate	4	Endogenous hormones
Bladder	5	Occupation diet and smoking
Brain	2	Drugs
Lymphoma	2	Viruses, benzene and irradiation
Leukemias	2	Viruses, benzene and irradiation
Female breast	10	Radiation, hormones and reproductive history
Uterine cervix	4	Viruses
Body of uterus	2	Exogenous hormones and reproductive history
Ovary	2	Reproductive history
Total	85	

Cancer cells may invade nearby tissues; they may spread through the bloodstream and lymphatic system to other parts of the body. This abnormal cell division leads to a newer growth-nodule or a tumour.

The tumour is of two types. They are:

1. Non-malignant otherwise called as benign tumour. This is more common and grows at a slower rate, remains restricted to a particular place thus does not spread to other parts of the body. It does not develop into a cancer.

2. Malignant also called as cancerous tumour. This initially begins as a smaller growth then rapidly grows and spreads to other parts of the body. The spreading manner resembles the movement of a crab. Due to the uncontrolled cell division, the tumour formed moves rapidly into the blood- stream and spreads to different parts of the body.

CAUSES FOR CANCER

As of now, it is not possible to assign or predict exact reasons for the cause, but it is possible to identify the substances or factors responsible for cancer.

Any constant irritation, taking place in any part of the body is prone to develop cancer. This includes the habit of smoking, which irritates the tissues of the lungs, chewing tobacco, an irritant to the mucous membrane of the mouth causing oral–buccal cancer, the broken edge of a teeth rubbing against the tongue causes tongue cancer. The other causes for cancer are: exposure to high energy and ionizing radiations such as X-ray and UV rays, etc. continuous handling or contamination of certain aromatic amines.

The damage to DNA by chemicals is now very clearly related to the chemical's ability to generate electrophiles or electrophilic behaviour. The reason for this is that nucleic acid components have a range of nucleophilic sites. Most damage occurs to the DNA/RNA purines themselves and the most likely sites of electrophilic attack are indicated below:

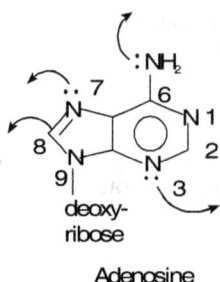

Adenosine

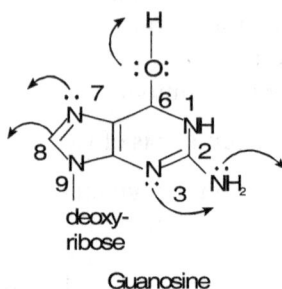

Guanosine

Many chemicals can damage DNA and they are grouped into two categories:

1. Chemicals that are metabolized to an electrophilic form
2. Direct-acting electrophilic substances

The range and quantity of biologically active compounds is enormous. The polycyclic aromatics and the aromatic amines constitute two classes of compounds whose discovery as cancer-causing agents had its origins in the workplace.

Polycyclic Aromatics

In the late 18th century, an abnormal prevalence of skin cancer was observed among chimney sweepers. The cause was shown to be particles of soot trapped in the clothing of sweepers. Soot is the product of incomplete combustion of organic materials and has been found to contain a multitude of polycyclic aromatic compounds (the most famous being *buckminsterfullerene*).

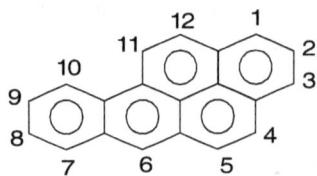

Benzo[α]pyrene

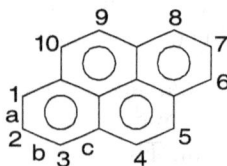

Pyrene

Later, shale oil and tars were shown to initiate a similar range of cancers. Soot, shale oil and petroleum oils are now known to contain a wide range of polycyclic aromatics, the most hazardous being benzo[α]pyrene which is also a constituent of tobacco smoke and thus a prime cause of lung cancer.

Many systematic studies of their carcinogenic activities have been carried out. Countless number of studies have investigated the activity and the activation of members of this class of organic compounds.

Aromatic Amines

Another class of substances that were first discovered to be potent cancer-causing agents through epidemiology is the aromatic amines. At the turn of the 20th century, a number of workers employed in the manufacture of aromatic amines for the dyestuff industry were diagnosed with bladder cancer. Systematic evaluation of the causes of deaths among workers in these industries over a 30-year period between 1921 and 1950 revealed a series of chemical culprits. While 1-naphthylamine appeared not to be a cause, its isomer, 2-naphthylamine, was a potent human carcinogen as were 4-biphenylamine, the closely related benzidine, and 2-acetylaminofluorene. The occupational hazards and the malignancy caused by some of the chemical compounds are briefly described in Table 10.2.

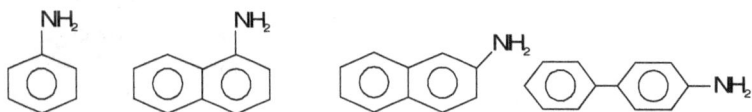

Aniline   1-Aminonaphthalene   2-Aminonaphthalene   4-Aminobiphenyl

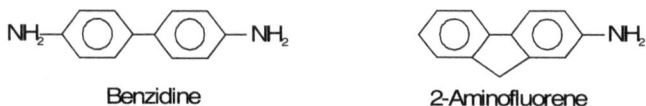

Benzidine   2-Aminofluorene

Table 10.2   Occupational hazards and malignancy of chemical compounds

Agents	Cancer sites	Latent intervals (years)	Occupation
X-rays	Bone marrow and skin	10–30	Medical and industrial
Radon gas, radium and uranium	Skin, lung, bone sarcoma and bone marrow	20–30	Medical and industrial chemists, dial painters and miners
UV radiation	Skin	Up to 70	Outdoor occupations
Polycyclic hydrocarbons in soot, tar, oil and resultant fumes and products of combustion	Lung, skin, larynx, bladder and nasal cavity	10–40	Furnaces and forges, foundries, shale oil workers, gas workers and retort men, chimney sweepers, lathe operators, textile workers, stokers, process workers and many others
Benzene	Bone marrow and lymph nodes	5–15	Process workers, painters, textile workers, explosive-using workers

(*Contd.*)

Table 10.2   (Continued)

Agents	Cancer sites	Latent intervals (years)	Occupations
1- and 2-naphthylamine, auramine and magenta, 4-biphenylamine and 4-biphenylamine and 4-nitrobiphenyl	Bladder	5–50	Dyestuff makers, rubber workers and other chemical plant process workers, shoe workers and printers
Mustard gas	Bronchial tree, lung and larynx	10–25	Production workers
Isopropyl alcohol	Nasal cavity	Over 10	Production workers
Vinyl chloride	Liver (angiosarcoma) and brain	5–40	Plastic manufacture
Chloroethers	Lung	At least 5	Chemical plant, process workers
Chloroprene	Skin, lung and liver	c10	Neoprene production
Arsenic	Skin, lung and liver	c10	Insecticide workers, miners and smelters, oil refiners
Chromium	Lung, nasal cavity and sinuses	10–25	Process and production workers, pigment workers

(Contd.)

Table 10.2   (Continued)

Agents	Cancer sites	Latent intervals (years)	Occupations
Cadmium	Lung, kidney and prostate	10–20	Battery workers, and smelters
Nickel	Lung and nasal sinuses	5–30	Smelters and process workers
Asbestos and similar fibres	Lung, pleura and peritoneum, larynx, stomach and large bowel	5–50	Miners, millers, manufacturers, users and demolition workers
Wood and leather particles	Nasal cavity	30–50	Wood and shoe workers
Unidentified organics	Lymph nodes	20–40	Chemists

Nitroaromatics

Many nitroaromatics are also carcinogenic compounds. 1-nitropyrene is a potent cancer-causing agent that was used originally as a component of photocopy toners and is generated in diesel engine exhaust.

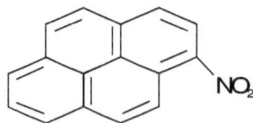

1-Nitropyrene

Many nitroaromatics are harmless on their own, but become potent carcinogens through metabolic activation. However, their mutagenic/carcinogenic properties often are in par with those of the aromatic amines. This is now understood in terms of the metabolic reactions of aromatic nitro compounds which undergo metabolic reduction to the aromatic amines with hydroxylamine as an intermediate. Thus for instance, 2-nitronaphthalene forms N-hydroxy-2-naphthylamine, on its way to 2-aminonaphthalene. The 2-aminonaphthalene would undergo metabolic oxidation back to N-hydroxy-2-naphthylamine.

Azo dyes

The azo dyes form another class of compounds that show carcinogenicity similar to aromatic amines. The azo dyes are formed by coupling of various diazonium salts with activated aromatic compounds.

Chloroethylene and Halogenated Olefins

Vinyl chloride or chloroethylene is a feedstock for the manufacture of polyvinylchloride (PVC) plastics. This class of compounds, like the aflatoxins relies upon metabolic activation for its mutagenic/ carcinogenic activity. The target is the double bond which undergoes epoxidation:

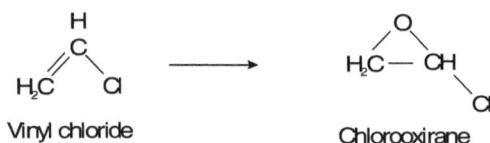

Vinyl chloride   Chlorooxirane

Diazomethane

Diazomethane is a commonly used methylating agent in organic synthesis. It is however known to be a human cancer-causing agent.

Sulphur and Nitrogen Mustards

Mustards are bis-alkylhalides and have the general structure:

A mustard   A sulphur   An aryl nitrogen
                   mustard   mustard

Sulphur mustards were used in the trenches during World War I. They inflicted enormous damage upon the lungs and mucous membranes of victims and they have since been proved to be carcinogenic. Both sulphur and nitrogen mustards react initially by an intramolecular SN2 reaction to form a cyclic sulphonium or an aziridinium ion and it is these ions that react with nucleophiles. They react at N7 of guanine.

A second reactive chloroethyl group allows a second alkylation reaction to take place and this can result in cross-linking of DNA. This is an important factor in making the nitrogen mustards and similar reagents useful as anticancer agents.

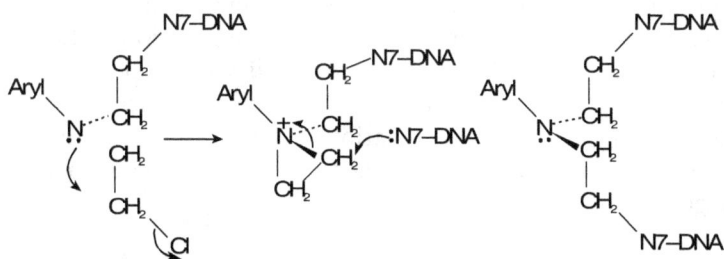

As seen above, we can understand that numerous chemicals damage DNA and a few result in mutations and cancer. This establishes the mechanisms of interaction of various chemical families with DNA. Chemicals that kill cells are called *cytotoxic* agents. Thus, of the many ways in which cells can be killed, modification of DNA is an important method. If the genetic code of DNA is altered significantly, either by several different mutation events, or by repetitive truncation of DNA in replication, eventually there is enough faulty translation for proteins or missing proteins through shortened DNA that the cell systems break down and the cell dies. Cells have an inordinate ability to repair DNA damage, and usually damage to cells is corrected before replication or transcription. However, cancer cells replicate much faster than normal ones and the repair mechanisms do not work as effectively. This forms the basis for using chemicals to kill cancer cells. Many compounds, similar to some of the cancer-causing agents described above, are also used in cancer therapy. If they harm the DNA in healthy cells, they can be expected to harm the DNA in cancer cells just as readily. However, since the damage to cancer cells has less chance of being detected and repaired by enzymes before the cell division and DNA replication, damaged genetic information is much more likely to be transmitted from one generation to the next. Compounded damage ultimately causes death of cancer cells.

Cancer chemotherapy works on this basis and it is therefore understandable that there is a good deal of healthy cell damage in patients undergoing cancer chemotherapy, particularly, damage to cell systems that naturally replicate rapidly. These include hair, stomach linings, mucous membranes and the liver. It is well known that cancer chemotherapy has severe side effects such as loss of hair. Alternatively, chemicals can be cytotoxic to cells if they cross-link DNA strands such that the separation of DNA into its two strands, an integral part of cell division, is prevented. If cross-linking is not repaired in time, death of cancer cells is assured. Changes in the conformation of DNA brought about by covalent or non-covalent bonding to DNA also leads to cell damage.

TREATMENT FOR CANCER

Cancer can be treated in many ways which is described in the following section.

Radiation Therapy

The cancer-affected part is exposed to high-energy radiations such as γ-radiation, radium radiation, cobalt isotopic radiation, etc.

Surgical Treatment

In this procedure, the affected part of the body is surgically removed. This procedure however depends upon the growth stages of the tumour.

Chemotherapy

When the above-mentioned methods fail to give the expected cure or are inapplicable or in order to supplement these methods, administration of certain chemical substances are followed; these preparations are known as **antineoplastic agents**.

Active research is in progress to discover drugs, which specifically interfere with the growth of cancer cells. Cell division by both normal and cancer cells progress through an orderly sequence of events called cell cycle. Drugs that inhibit the replication during a particular phase of the cycle are called cell-cycle specific. Methotrexate is a cell-cycle specific-antimetabolite. Mechlorathamine is a cell-cycle non-specific drug. It affects all phases of the cell cycle. Antineoplastic agents are capable of producing temporary regression of certain tumours, which are not curable by surgery or radiation.

The antineoplastic agents include:

1. Cytotoxic or alkylating agents
2. Antimetabolites
3. Plant-derived products
4. Hormones
5. Antibiotics
6. Miscellaneous

CYTOTOXIC OR ALKYLATING AGENTS AS ANTICANCER AGENTS

These agents act by modifying DNA causing cytotoxicity in cancer cells. There are many other types of interactions with DNA that inhibit DNA synthesis by non-reactive means. Wide ranges of cross-linking agents are in clinical use today. Their cytotoxic behaviour is attributable to their preventing DNA strands from separating through reactions on complimentary strands. Some of these reagents thus block DNA synthesis in the replication or transcription process.

Cytoxon

This has selective action among the cytotoxic agents. It is active *in vivo*. It is commonly administered orally. It is well absorbed in the intestinal tract. Cytoxon is converted into its active form before its action. Its exact metabolism is not known. It is also called as endoxan and cyclophosphamide.

Cyclophosphamide is a potent drug in the clinical treatment of cancer. It's unusual structure has its origin in the recognition that tumour cells contain high levels of an enzyme that hydrolyses phosphoramides. It was envisaged that hydrolysis would lead to formation of HN $(CH_2CH_2Cl)_2$ specifically in tumour cells. Although high activity was observed, the drug was metabolized differently forming a phosphoramide mustard. P–NH or the P–O bond cleavage is accompanied by the breakage of N–P bond. These cleavages are produced by cellular phosphatases or phosphamidases. The latter is present in larger amounts in cancer cells. Structurally, cytoxon is a nitrogen mustard. The latter are a group of compounds also called as alkylating agents as they transfer alkyl radical to a suitable receptor site. These alkylating agents in neutral or alkaline medium form highly reactive quaternary ammonium derivatives and they then can react with groups such as —NHR, —SH, —OH, present in the physiologically important molecules in the cells and render them unavailable for the normal metabolic activities. The alkylating agents being nucleophilic in nature react with the nucleic acids and alkylate them, thereby inhibiting the DNA synthesis. They also bring about a cross-linkage of DNA strands in the resting as well as in the dividing cells and interfere with the cell replication.

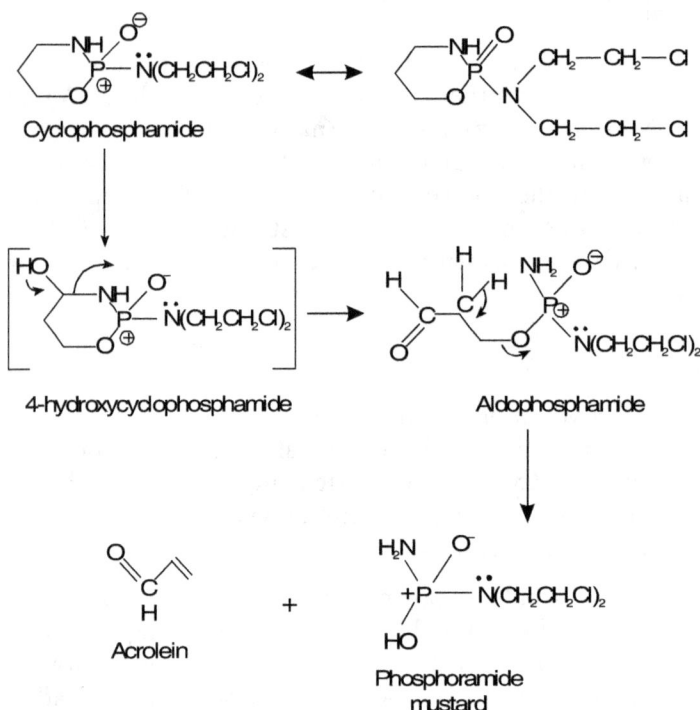

Cyclophosphamide

4-hydroxycyclophosphamide

Aldophosphamide

Acrolein

+

Phosphoramide mustard

Thiotepa   This is also a cytotoxic drug administered as injections.

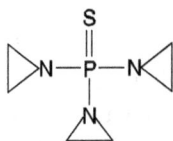

Thiotepa

Nitrogen mustards and related compounds

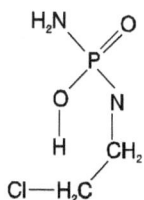

Phosphoramide mustard

Mechlorethamine was the first clinical cancer chemotherapeutic agent and it is still used today, as is chlorambucil, a 4-phenylbutyric acid derivative. Both function by causing interstrand cross-links.

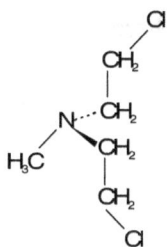

Mechlorethamine

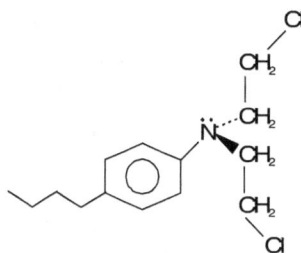

Chlorambucil

Nitrosourea Derivatives

The discovery that 1-methyl-1-nitrosourea was marginally active against leukemia and the discovery that mustards were successful cross-linking agents led to the development of hybrid drugs such as Carmustine and Lomustine.

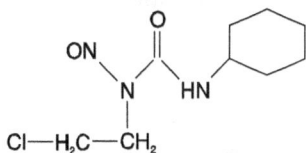

Lomustine

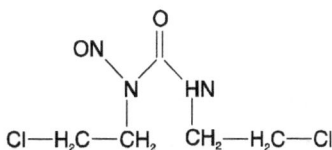

Carmustine

Cisplatin

Cisplatin is a tetra-coordinated platinum compound with unusual activity against testicular, ovarian, lung, bladder and cervical cancers. Various analogues have been made, but all contain two labile ligands (chlorides, mainly) and two inert ligands (NH_2 or amino groups). It acts by forming intrastrand bonds to two guanines.

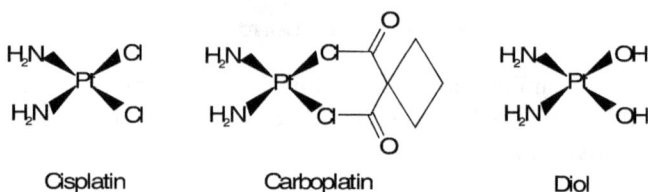

Cisplatin   Carboplatin   Diol

The chlorides are stable in the bloodstream where the chloride concentration is high but, once entering the cell, they are replaced by water molecules, which are converted to the diol, inter- and intrastrand cross-linking is formed by cisplatin as shown.

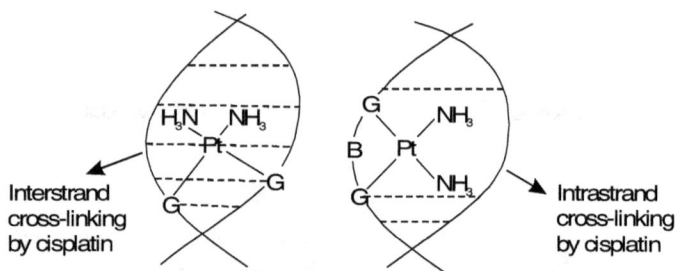

Interstrand cross-linking by cisplatin   Intrastrand cross-linking by cisplatin

The consequence of the intrastrand and interstrand cross-linking is a change of shape of DNA resulting in a block of DNA polymerase in the replication process. Transplatin, with the chlorides trans in the square planar complex, does not form intrastrand cross-links and is inactive.

Alkylsulphonates

Alkylsulphonates and polymethylene disulphonates were prepared and tested for their anti-cancer activity. Busulfan and clomesone are two examples.

$$CH_3SO_2O(CH_2)_4OSO_2CH_3 \qquad CH_3SO_2OCH_2SO_2CH_2CH_2Cl$$

Busulfan   Clomesone

Aziridines

Opening of three-membered rings is favourable particularly at acidic pHs. Thus, a number of poly aziridinyl systems are in use as anticancer agents. AZQ is a quinone, which is believed to undergo bioreduction to hydroquinone, which can protonate the aziridine nitrogens intramolecularly thereby enhancing their electrophilic character.

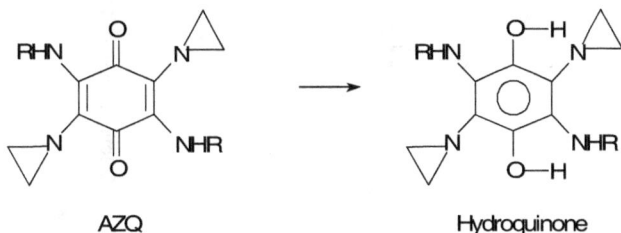

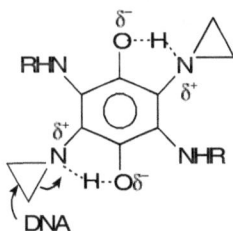

ANTIBIOTICS

Antibiotics with linear fused aromatic rings and flat structure interfere with the basic part of the nucleic acids. They interrupt the genetic coding, e.g. actinomycin, mithramycin, etc.

Actinomycin

Actinomycin was the first antibiotic reported to be able to halt cancer. It is a member of *actinomycin* group, derived from

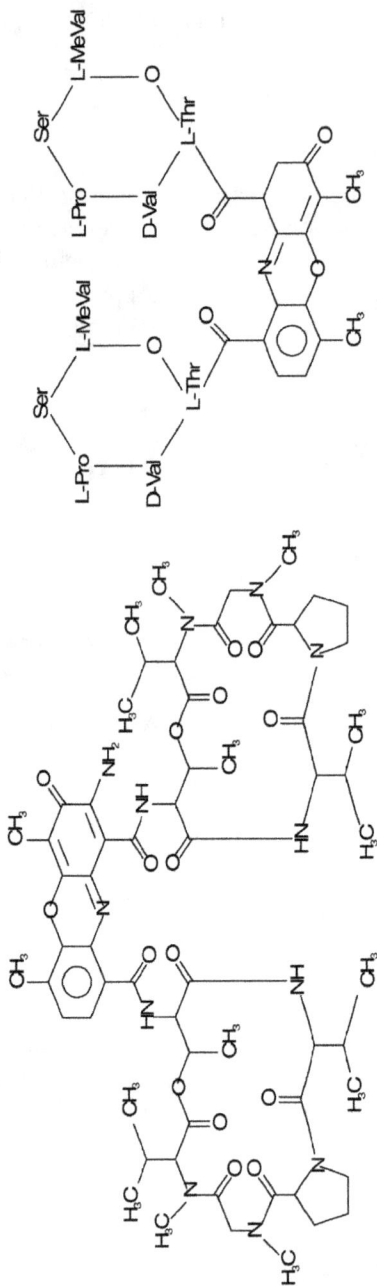

Actinomycin IV or Actinomycin C1 or Dactinomycin or Cosmegen

Streptomyces parvullus. It binds to DNA and inhibits RNA synthesis (transcription), with chain elongation more sensitive than initiation, termination or release. As a result of impaired mRNA production, protein synthesis also declines; however, it is not widely used to treat cancers since it is highly toxic to humans, interfering with the genetic material of cells. It is mainly used as an investigative tool in cell biology.

Mithramycin or Plicamycin

Mithramycin or Plicamycin belongs to the group of antineoplastics. It is one of the older chemotherapy drugs, which has been in use for decades in treating certain types of cancer. It is also used to treat hypercalcaemia or hypercalciuria (high concentration of calcium in the blood or urine) that may occur with some types of cancer. This also serves as a DNA-binding fluorescent dye and inhibits RNA synthesis. It is given via the intravenous route only.

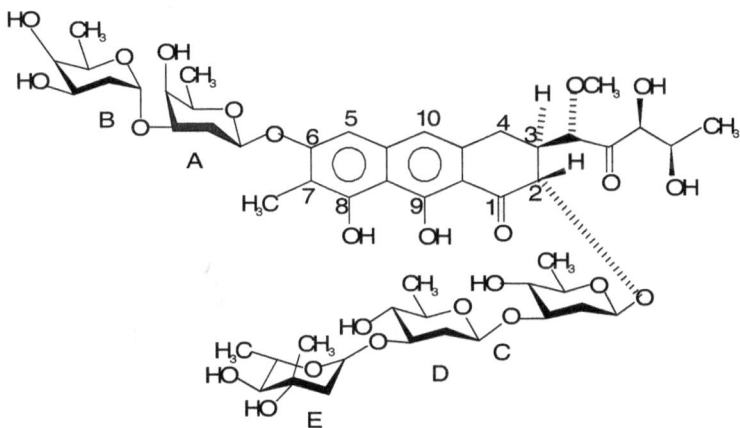

Mithramycin-Plicamycin

Mitomycin C

Naturally occurring substances made by microorganisms have been found to possess antitumour properties. Mitomycins are widely used as anticancer agents and possess activity similar to the other aziridine-containing synthetic drugs. Numerous members of the mitomycin family are now known but mitomycin C antibiotic is the representative. Not only can these substances alkylate by aziridine ring opening, but they also have a labile carbamate substituent that can be displaced by DNA nucleophiles. Thus, they are also known to be cross-linking agents.

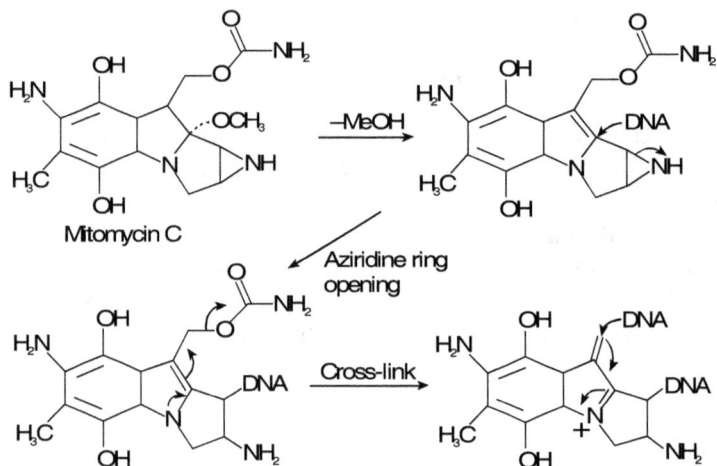

Mitomycin C

Aziridine ring opening

Cross-link

CC-1065 and Duocarmycin Antibiotics

CC-1065 is the name given to a very potent antibiotic isolated from *Streptomyces zelensis* in 1978. Duocarmycin A is a "cut-down" version of CC-1065 possessing remarkable potency. The three-unit structure of CC-1065, and its capacity to hydrogen bond through NH groups, allows it to bind strongly. The excellent binding ability

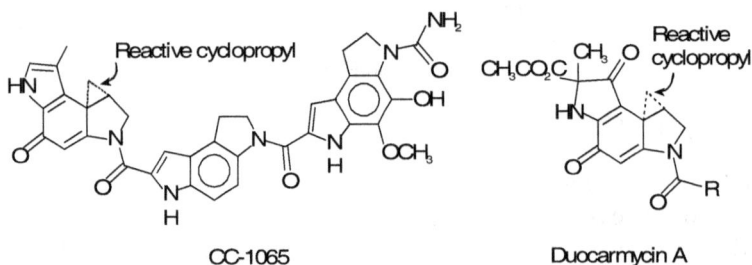

CC-1065

Duocarmycin A

of CC-1065 and other analogues combined with their alkylating ability make these and other analogues as attractive antitumour antibiotics. CC-1065 is apparently too toxic towards liver and bone marrow and other analogues may be better contenders for anticancer chemotherapy.

ANTIMETABOLITES

They are structural analogues of naturally occurring compounds that interfere with various metabolic processes and disrupt cell functions and proliferation. They incorporate themselves into the metabolic pathway and inhibit growth. Antimetabolites as antineoplastic agents are structurally similar to compounds that are found naturally in humans (vitamins, amino acids or precursors of DNA and RNA). They incorporate into either DNA or RNA (purine and pyrimidine nucleotides) and interfere with cellular function. They also inhibit catalytic action of enzymes necessary for macromolecular synthesis.

Examples of these include antagonists of purines (azathioprine, mercaptopurine and thioguanine) and antagonists of pyrimidine (fluorouracil and floxuridine). Cytarabine also has antiviral properties, and interferes with dihydrofolate reductase, which is necessary for the synthesis of tetrahydrofolate and subsequently for the synthesis of the folic acid needed for DNA formation.

As the antimetabolites act primarily upon cells undergoing synthesis of new DNA for formation of new cells, most of the toxicities associated with these drugs are seen in cells that are growing and dividing quickly. They are known to cause severe damage to the mucous membranes of the mouth and other parts of the gastrointestinal tract and also to produce skin disorders and hair loss. Anaemia can occur along with a decrease in the number of the white blood cells that are necessary to prevent infections. Methotrexate is used most often in the treatment of acute leukemia, breast cancer, lung cancer, and osteogenic sarcoma (osteosarcoma).

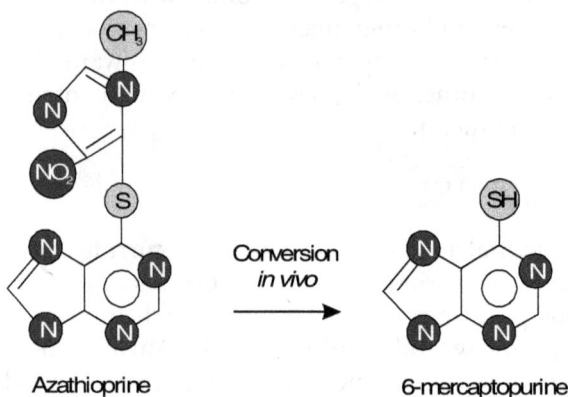

Azathioprine

Conversion
in vivo

6-mercaptopurine

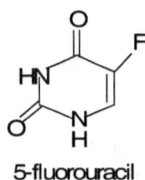

5-fluorouracil

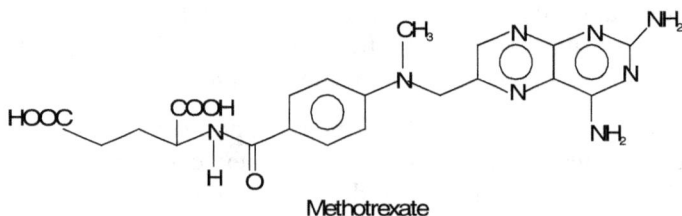

Methotrexate

PLANT PRODUCTS

Plant products like colchicine alkaloid is used in the treatment of leukemia. It prevents the mitotic cell division. *Vinca* alkaloids inhibit the RNA production.

STEROIDAL HORMONES

These are specific in their action. Cancer tissues are modified by these hormones. Alterations in the hormonal environment retard the growth of cancerous cells.

Hormonal agents that are of two types are used in the treatment of cancer:

Corticosteroid hormones   Corticosteroids are used in treating leukemia, multiple myeloma and lymphoma. They are also used to reduce swelling around tumours of the brain and spinal cord. They are used with other chemotherapy drugs in combination-chemotherapy. Examples of corticosteroids are prednisone and dexamethasone.

Sex hormones   The sex hormones change the action and production of female and male hormones. They are used to slow the growth of breast, uterine and prostate cancer, which grow in the presence of hormones. These drugs do not kill cells, as typical chemo drugs do, but they cut off the "food supply" to destroy cancers. Examples of sex hormones are tamoxifen and flutamide.

DIABETES MELLITUS

Diabetes mellitus has been known since ages and the sweetness of diabetic urine has been mentioned in Ayurveda by Sushruta. The word "diabetes" in Greek means "to flow through". The presence of sugar in the sample of urine was found to be responsible for the sweetness that was discovered in 1753. The substance that was responsible for diabetes was identified to be insulin. Decreased secretion of insulin enhances the blood sugar level leading to diabetes mellitus—hyperglycaemia. Insulin is secreted in pancreas in the islets of Langerhans. Hypoglycaemic drugs are those administered to decrease the blood sugar level by increasing the secretion of insulin. Insulin is a polypeptide, responsible for lowering the blood sugar level. It acts by:

- making the liver to convert more glucose into glycogen, which in turn is stored in liver.

- accelerating the oxidation of glucose by releasing more energy.

- reducing further addition of glucose to blood from other non-carbohydrate substances.

The main carbohydrates consumed by the human beings are starch, a polysaccharide and sugar, a disaccharide.

The ingested carbohydrates are digested successively by:

- Salivary enzymes
- Acidic medium in the stomach
- Pancreatic secretions in the stomach
- Intestinal juices

Most of the carbohydrates are broken down into various monosaccharides. They are absorbed by the walls of intestine at various rates. All these monosaccharides are to one another. Glucose is the major sugar to leave the liver, which is finally carried by the arterial blood to various cells. Various tissues require different amounts of glucose for their energy needs. The brain cells require only glucose while other tissues can utilize glucose and fatty acids for their metabolic functions. At the time of shortage of glucose, the skeletal muscles derive their maximum energy from free fatty acids.

In the absence of insulin, a larger amount of glucose is consumed by the tissues either for liberation of energy or converting glucose into glycogen. To meet this higher requirement, sugar level in blood should be high. Thus, the concentration of sugar in blood is more and this condition is called as hyperglycaemia. As insulin is responsible for the amount of sugar in blood, by administering insulin the sugar level in blood can be lowered.

During carbohydrate metabolism, the glucose molecules obtained, which are in excess, are stored as glycogen in muscle tissues and liver. Liver in turn can convert glycogen back to glucose. This is known as glycogenolysis. However, the muscle tissues do not do this conversion but instead convert the stored glycogen into lactate which is then carried to the liver and undergoes neoglucogenesis to provide glucose. (Neoglucogenesis is the process of obtaining glucose from sources other than carbohydrates). In diabetes, excess of neoglucogenesis takes place. On a healthy person, out of the total amount of glucose produced by the carbohydrate metabolism, nearly 3% is stored in liver and muscles as glycogen and 30% is converted into fatty acids. The rest is used for the production of energy aerobically and for the synthesis of amino acids.

Insulin is a polypeptide containing two amino acid chains, interlinked by —S—S— bonds. The number of amino acids vary between 21 and 30. The active centre of insulin is assumed to be the —S—S— bridges. Insulin synthesized in pancreas is stored as granules and this storage is helped by the micronutrient, zinc. This stored insulin is released into the bloodstream, which is stimulated specifically by glucose. Glucose is believed to act on a specific receptor on the cell wall of β-cells. This then mediates the release of stored insulin. The sensitivity of this mechanism depends on the prior intake of carbohydrate. It is tremendously decreased by the restriction of diet and short periods of fasting. A small rise in the concentration of glucose in the pancreatic artery leads to the release of insulin thus lowering the blood sugar level. Substrates other than glucose are also responsible for the release of insulin namely fatty acids of long chain, certain ketones, hormones and amino acids (*Refer Chapter 14 for structure of Human Insulin*).

TYPES OF DIABETES MELLITUS

Clinically there are two types of diabetes mellitus, namely type 1 and type 2.

Type 1 diabetes occurs when the pancreas cannot produce insulin, a hormone essential for moving glucose from the blood into cells. It is an autoimmune disorder, in which the body makes antibodies that attack the insulin-producing cells in the pancreas. The cause is complex and unclear, but may involve genetics, viruses, diet, chemicals and environmental factors. This was called juvenile diabetes because it is usually diagnosed in childhood or early adulthood. People with type 1 diabetes are supplied insulin by injection, pump or other methods.

Possible treatments include transplant of a pancreas or beta cells. Type 1 diabetes is labile. In cases of absolute deficiency of insulin, treatment by hypoglycaemic therapy is a failure. On such cases, the glycogen synthesis from glucose diminishes, proteins are converted into glucose at abnormally high rates and nitrogen residue occurs as larger amounts of urea.

Type 2 diabetes is the most common type of diabetes. It occurs when glucose builds up in the blood due to the body's inability to use insulin effectively. Insulin is a hormone that is essential to transport glucose, a sugar that is the body's primary fuel into the cells. This is called adult onset diabetes because it is usually diagnosed in adulthood. Excess fat, inactivity, and poor eating habits contribute to type 2 diabetes. Sometimes the disease can be prevented or controlled through diet and exercise, but some patients need insulin or other medications. Any group of hypoglycaemic drugs, such as tolbutamide, can be administered. They act on the beta cells of the pancreas and increase the secretion of insulin. However, all diabetics do not require insulin. They can be treated by controlled diet and proper exercises.

HYPOGLYCAEMIC DRUGS

Glucose-induced insulin release is enhanced by xanthines, sulphonyl urea, etc. Some of the antidiabetics are also orally administered. The basic requirement of oral hypoglycemic drugs include the following:

1. They should be effective.
2. They should be non-toxic and should correct the basic metabolic defects in addition to lowering the blood sugar level.

The following groups of chemicals are used in the oral administration.

1. Sulphonyl ureas
2. Biguanidines

The general structure of sulphonyl urea is

$$R_1 — Ar — SO_2 — NH — CO — NH — R_2$$

R_2 is any aliphatic group and is responsible for the lipophilic property. It is an alkyl group with 3–6 carbon atoms are more active.

When R_1 is methyl and R_2 is *N*-butyl, it reduces blood sugar level in adults effectively. When R_2 is aryl, it is toxic. When R_1 is Cl, R_2 is *N*-propyl, it cures stable diabetes and has longer duration of action.

Sulphonyl urea derivatives are a class of antidiabetic drugs that are used in the management of diabetes mellitus type 2 ("adult onset"). They act by increasing insulin release from the beta cells in the pancreas. These drugs are effective on patients having definite quantity of insulin.

Therefore absolute deficiency of insulin cannot be treated by these drugs. All sulphonyl ureas have a central phenyl ring with two branching chains, e.g. tolbutamide, chlorpropamide, tolazamide, etc.

The general structure of biguanidines is:

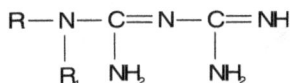

$$R - N - C = N - C = NH$$
$$\underset{\displaystyle R_1 \quad NH_2 \qquad NH_2}{|\qquad\quad|\qquad\quad\;\;}$$

Adverse Reactions of Insulin Therapy

Hypoglycaemia is a common adverse effect of insulin therapy. The common causes of hypoglycaemia include the following:

1. Too large a dose of insulin
2. Failure to eat
3. Violent physical exercise
4. Ingestion of alcohol

Renal failure is common among juvenile and aged diabetics. Patients of insulin therapy must be warned of the symptoms of hypoglycaemia which manifest as:

1. Palpitation
2. Profuse sweating
3. Headaches on waking up
4. Nightmares

In certain cases, hypoglycaemic attacks may be harmful to patients suffering from cardiac problems and often leads to neurological damage. The patients under hypoglycaemic drug therapy are warned of their missing meals, violent exercise, consumption of alcohol, etc. and they must be advised to carry a sample of sugar with them all the time, since oral glucose can relieve hypoglycemic symptoms if administered at early stages.

11

AIDS

HUMAN IMMUNODEFICIENCY VIRUS (HIV)

HIV stands for "human immunodeficiency virus", so adding the word "virus" to the acronym creates a redundancy. HIV is the name of the organism that is the cause of AIDS, not the name of the disease itself. A person may be HIV-positive (a result showing the person to be infected with the virus) without having yet developed AIDS (acquired immunodeficiency syndrome). *HIV is the cause, AIDS the result.*

HIV (Figure 11.1) presents a complex knot for scientists to unravel. After initial contact and attachment to a cell of the immune system (e.g. lymphocytes, monocytes), there is a cascade of intracellular events. The end product of these events is the production of massive numbers of new viral particles, death of the infected cells, and ultimate devastation of the immune system. However, the knot is being unravelled.

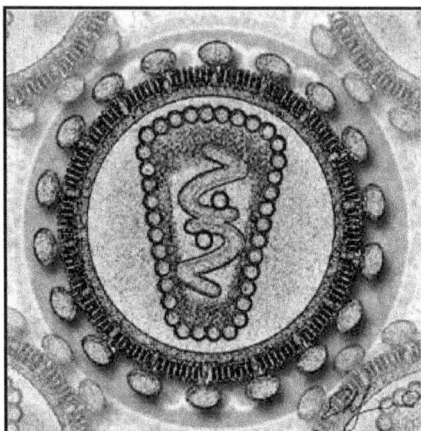

Figure 11.1   Human immunodeficiency virus

HIV INFECTION

The following section attempts to simplify HIV infection at the cellular level. The various stages involved in HIV infection (Figure 11.2 a and b) are as follows.

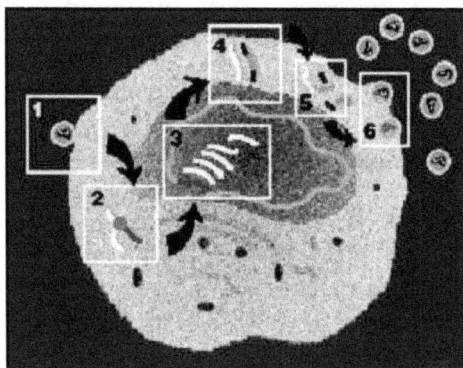

(a)

Figure 11.2 Simplified diagrams a and b showing stages involved in the HIV infection at the cellular level (*Continues*)

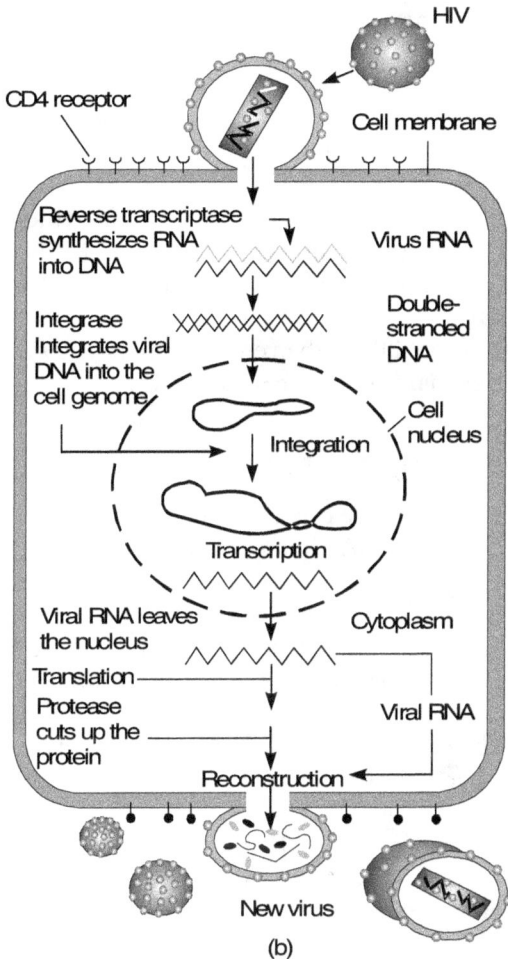

Figure 11.2 Simplified diagrams a and b showing stages involved in the HIV infection at the cellular level

Step 1 Entry of Virus and Attachment to the Lymphocyte Membrane

On the surface membrane of all living cells, there are complex protein structures called "receptors". A receptor is often compared to a lock into which a specific key or "ligand" will fit. There are at least two receptors on T lymphocytes to which the human immunodeficiency virus (HIV) attaches itself. The primary receptor is called "CD4", and a second receptor that loops through the cell membrane 7 times is critical for infection to occur. HIV infection of a lymphocyte requires attachment of the virus to the cell membrane through both of these "ligand-receptor" links. In cells whose "7-transmembrane receptor" is different, the HIV "key" no longer matches the lymphocyte "lock" and attachment is incomplete. Those cells may avoid infection by HIV. Tight attachment of the viral particle to receptors on the lymphocyte membrane enables fusion with the cell membrane. The viral contents, including viral RNA then empty into the cell's cytoplasm. Like other viruses that infect human cells, HIV commandeers the host's machinery to make multiple copies of itself.

Step 2   Reverse Transcription—From viral RNA to DNA

An enzyme that is part of the human immunodeficiency virus reads the sequence of viral RNA that has entered the host cell and transcribes the sequence into a complementary DNA sequence. The enzyme is called "reverse transcriptase". Without reverse transcriptase, the viral genome cannot be incorporated into the host cell, and cannot reproduce. Reverse transcriptase sometimes makes mistakes reading the DNA sequence. The result is that not all viruses produced in a single infected cell are alike. Instead, they end up with a variety of subtle molecular differences in their surface coat and enzymes. Vaccines, which induce the production of antibodies that recognize and bind to very specific viral surface molecules, are unlikely players in fighting HIV, because throughout the infection, HIV surface molecules keep continually changing.

AZT-like drugs inhibit reverse transcription. The first major class of drugs found useful in slowing HIV infections is collectively called "reverse transcriptase inhibitors". These include AZT, 3TC, D4T, DDC, and DDI that act by blocking the recoding of viral RNA into DNA. The chameleon-like nature of HIV, however, limits their continued effectiveness.

Step 3   Integration and Transcription of Viral DNA

Once the viral RNA has been reverse-transcribed into a strand of DNA, it can then be integrated (inserted) into the DNA of the lymphocyte. The virus has its own enzyme called "integrase" that facilitates incorporation of the viral DNA into the host cell's DNA. The integrated DNA is called as provirus.

As long as the lymphocyte is not activated or "turned-on", nothing happens to the viral DNA. However, if the lymphocyte is activated, transcription of the viral DNA begins, resulting in the production of multiple copies of viral RNA. This RNA codes for the production of the viral proteins and enzymes (translation) and will also be packaged later as new virus.

Step 4 Translation—from RNA to Viral Proteins

There are only nine genes in the HIV RNA. Those genes have the code necessary to produce structural proteins such as the viral envelope and core plus enzymes like reverse transcriptase, integrase, and a crucial enzyme called a protease. When the viral RNA is translated into a polypeptide sequence, the sequence is assembled in a long chain that includes several individual proteins (reverse transcriptase, protease, integrase). Before these enzymes become functional, they must be cut from the longer polypeptide chain.

Step 5 Viral Protease Cleaving Viral Proteins

Viral protease cuts the long chain into its individual enzyme components, which then facilitate the production of new viruses. Inhibitors of this viral protease can be used to fight HIV infection. By blocking the ability of the protease to cleave the viral polypeptide into functional enzymes , protease inhibitors interfere with continued infection. Mutations enable HIV to avoid treatments that involve only one drug, so there is growing use of multiple-drug therapies in which both a protease inhibitor and a reverse transcriptase inhibitor are combined.

Step 6 Assembly and Budding

Finally, viral RNA and associated proteins are packed and released from the lymphocyte surface, along with the viral surface proteins. These proteins will then bind to the receptors on other immune cells facilitating continued infection. Budding viruses are often exactly like the original particle that initially infected the host. In the case of HIV, however, the resulting viruses exhibit a range of variations, which makes treatment difficult.

Many people who are infected with HIV do not have symptoms for many years. One cannot rely on symptoms to know the HIV status.

The following may be warning signs of HIV infection:

- Severe weight loss
- Fevers
- Headaches
- Drenching night sweats
- Fatigue
- Severe diarrhoea
- Shortness of breath
- Difficulty in swallowing
- Non infective enlargement of lymph nodes
- Malignant tumours
- Widespread eruption of eczema, psoriasis, etc.

The symptoms can last for weeks or months at a time and do not go away without treatment. Since these symptoms are commonly seen in other diseases, these are not assumed to be HIV/AIDS-related until the individual gets tested.

A positive HIV test does not mean that a person has AIDS. A diagnosis of AIDS is made by a physician using certain clinical criteria. As with other diseases, early detection offers more options for preventive care and treatment.

HIV infection is transmitted by

- Using contaminated needles
- Transfusion of infected blood, blood products, semen, and transplantation of other infected organs
- Sexual intercourse

The ELISA is a fundamental tool of clinical immunology, and is used as an initial screen for HIV detection. Based on the principle of antibody–antibody interaction, this test allows easy visualization of results and can be completed without the additional concern of radioactive materials.

An HIV-ELISA, sometimes called an HIV enzyme immunoassay (EIA) is the first and basic test to determine if an individual is positive for a selected pathogen, such as HIV. The test is performed in a 8 cm × 12 cm plastic plate which contains an 8 ×12 matrix of 96 wells, each of which are about 1 cm high and 0.7 cm in diameter.

Figure 11.3 and 11.4 illustrate how an HIV-ELISA is performed.

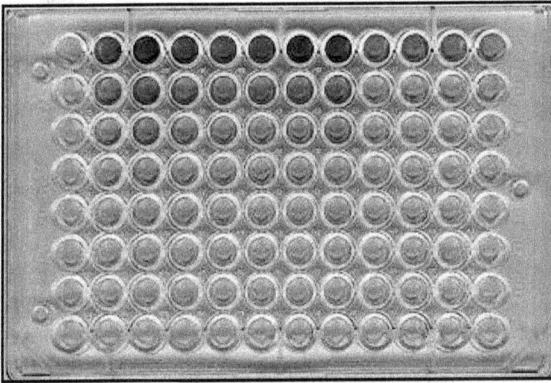

Figure 11.3   An ELISA plate

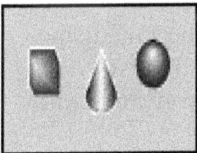

Partially purified, inactivated HIV antigens pre-coated onto an ELISA plate.

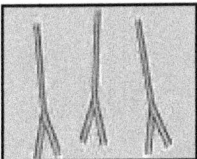

Patient serum, which contains antibodies. If the patient is HIV +ve, then this serum will contain antibodies to HIV, and those antibodies will bind to the HIV antigens on the plate.

Figure 11.4   ELISA method (*continues*)

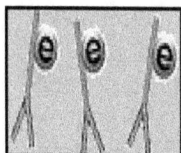

Anti-human immunoglobulin coupled to an enzyme. This is the second antibody, and it binds to human antibodies.

Chromogen or substrate, which changes colour when cleaved by the enzyme, attached to the second antibody.

Positive ELISA test

Negative ELISA test

Figure 11.4   ELISA method

Treatment

There is no cure for HIV infection. Prevention is the best medicine. However, with the development in the medicinal chemistry it is possible for prolonging the life of an HIV patient, and postponement of the onset of AIDS symptoms.

Some of the anti-HIV drugs are:

- AZT—Azidothymidine
- DDI—Dideoxyinosine
- DDC—Dideoxycytidine

As the cost of these anti-HIV drugs are high, efforts are continued to develop safer, effective, and economical drugs.

REVIEW QUESTIONS

1. Explain the following terms:
   - i.   Pharmacophore
   - ii. Antimetabolite
   - iii. Pharmacodynamics
   - iv. Pharmacokinetics
2. What is the name for the official code containing a list of drugs?
3. Name a few pharmacopoeias.
4. Define and differentiate the terms assay and potency.
5. What are the common modes of administration of drugs?
6. What is a vaccine?
7. Give an overall view on medicinal flora of India.
8. Write short notes on the medicinal value of the following plants:
   - i.   Neem
   - ii. Holy Basil
9. List the mineral and vitamin contents of mango.
10. Which fruit is termed as a nugget of vitamin C? Why?
11. What is the medicinal value of Malabar nut?
12. What is the medicinal role of *Phyllanthus amarus*?
13. Explain the importance of turmeric and thoothuvalai.
14. What does the word disease denote?
15. How are diseases caused?
16. What is the meaning of communicable diseases?
17. Mention a few modes of transmission of diseases.

18. List a few airborne diseases.
19. How does droplet contact occur?
20. What is oral transmission of disease?
21. For the following diseases/infections, identify their respective routes of transmission.

    i.   Rotavirus   ii.   Influenza
    iii. Warts   iv.   Dengue
    v.   Hepatitis A   vi.   Impetigo
    vii. HIV   viii. Giardiasis

22. Discuss the causative organism, symptoms, treatment and prevention for the following diseases:

    i.   Malaria
    ii. Cholera
    iii. Filariasis

23. What are the diseases caused by the following microorganisms?

    i.   *Vibrio cholerae*
    ii. *Plasmodium falciparum*
    iii. *Brugia timori*

24. Discuss vertical transmission of diseases.
25. Write a note on vector-borne diseases.
26. What does the term iatrogenic transmission mean?
27. Identify the diseases treated with the following medicines.

    i.   Mefloquine   ii.   Folic acid
    iii. Dried ferrous sulphate iv.   Doxycycline

28. What are the other names for folic acid?
29. What are the three structural moieties that constitute folic acid?
30. Why is jaundice not technically termed as a disease?
31. What is the origin of the word drug?
32. Enumerate a few requirements for an ideal drug.
33. Who laid the foundation for modern medicine?

34. What is the prediction of Paul Ehrlich about chemotherapeutic agents?

35. Define drug metabolism.

36. Explain the role of kidney in the metabolism of drugs.

37. Discuss phase-I reactions of metabolism of drugs.

38. State the contributions of Brodie in the studies of metabolism of drugs.

39. Show the reactions involved in the metabolism of sulphathiazole.

40. Explain the role of cytochromes in the metabolic processes.

41. What is cytochrome P450?

42. Referring to figure 4.4 delineate the physiological role of cytochrome P450.

43. Draw and explain the chemical reactions involved in the metabolism of halothane, pentobarbital, acetanilide, prontosil, chloramphenicol and codeine.

44. What are the products of metabolism of aspirin, procaine and DDT?

45. Aminopyrine is metabolized into monomethyl 4-aminopyrine. Which reaction is responsible for this?

46. Discuss conjugation reactions of drug metabolism.

47. Show the role of UDPGA by chemical reactions in the metabolism of phenols, amines and carboxylic acids.

48. How does the enzyme transferase help in

    i. The metabolism of sulphanilide

    ii. The conversion of glycine to hippuric acid

49. How is norepinephrine metabolized?

50. Discuss phase I and II transformations of paracetamol.

51. What does the term clearance refer to?

52. How are drugs classified?

53. What are sulpha drugs? Discuss their classification and applications.

54. Draw the structure of prontosil-S.

55. How are the following synthesized?
    i.   Sulphadiazine
    ii. Sulphathiazole
    iii. Prontosil

56. How do sulpha drugs function as antimetabolites?

57. Why are sulphonamides ineffective in the presence of pus and broken tissue?

58. Define antibiotics and how are they classified?

59. Explain the structural features of penicillin.

60. What are cephalosporins?

61. Mention the disadvantages of penicillin.

62. What are natural, synthetic and semi-synthetic penicillins?

63. Draw the structure of chloramphenicol with relevant stereochemical features.

64. Discuss the SAR of chloramphenicol.

65. What are tetracyclines?

66. What is the role of the sugar residue in a macrolide?

67. Give a few examples of macrolides, tetracyclines and semi-synthetic penicillin.

68. Define and differentiate the terms antiseptics and disinfectants.

69. What is Rideal–Walker coefficient?

70. Draw the structure of nitrofurazone and discuss its anti-infective property.

71. Explain the role of formaldehyde as an anti-infective agent.

72. How does the antiseptic property of phenols vary with respect to the substituents and position of substituents?

73. Prepare the following antiseptics suitably by chemical methods
    i.   Carvacrol   ii.   Thymol N-hexyl resorcinol

74. Identify the chief constituent of dettol and draw its structure.

75. What is the role of benzethonium chloride.

76. Discuss the mechanism of action of surfactants.

77. Draw the structures for cetyl pyridinium chloride and triton.

78. What are the applications of acridine dyes?

79. Name a few triphenyl methane dyes functioning as antiseptics.

80. Explain the role of organomercurials as anti-infective agents.

81. Give the chemical structures of thiomersal and mercurophen.

82. Discuss the importance of azo dyes as antiseptics.

83. What is the role of chlorhexidine?

84. Explain the role of iodine compounds as antiseptics.

85. What are iodophores?

86. What are the reactions involved in the germicidal action of molecular chlorine.

87. What are chloramines? Why are they preferred over other antiseptics?

88. Draw the structure of dequalinium chloride and account for its properties.

89. How useful are the aliphatic chloramines?

90. Define the terms analgesics and antipyretics.

91. Discuss the mechanism of action of analgesics and antipyretics.

92. What does NSAID denote?

93. What are narcotic analgesics? Give suitable examples.

94. Discuss the SAR of morphine.

95. What are the structural features essential for the pharmacological action of morphine?

96. What is heroin?

97. Draw the structure of nalorphine and account for its pharmacological activity.

98. Butarphanol resembles nalorphine structurally. How?

99. Name any two synthetic narcotic analgesics and give their preparation.

100. Name any three antipyrenes.

101. Draw the structure of novalgin.

102. Draw the structure of phenylbutazone and explain its applications.

103. Explain the role of aniline derivatives as antipyretics.

104. How are phenacetin and paracetamol prepared?

105. How do salicylates function as analgesics?

106. Discuss the steps involved in the preparation of salicin from salicylaldehyde.

107. What are soluble aspirins?

108. Draw the structures of flufenisal and diflunisal? Give their uses.

109. Salol is used as an internal antiseptic. Why?

110. Name a few aryl acetic acid derivatives used as anti-inflammatory drugs.

111. What is diclofenac?

112. What are tranquilizers? How are they classified?

113. Name two compounds functioning exclusively as sedatives.

114. Define the terms sedatives and hypnotics and give suitable examples.

115. What are antipsychotics?

116. Explain the term anti-anxiety drugs.

117. Identify the role of:

　　i.　Indole derivatives　　ii.　Librium

　　iii.　LSD　　iv.　Haldol

　　v.　Chlorpromazine　　vi.　*Cannabis sativa*

　　vii.　Meprobamate

118. What are hallucinogens?

119. Give a suitable classification of psychedelic drugs.

120. Draw the structure of LSD and discuss its physiological effect.

121. Write a note on charas.

122. What is the importance of serpentine alkaloids?

123. Name the active principle present in charas.

124. List the basic requirements of an anaesthetic.

125. Define the following terms with examples:
     i.   Anaesthetics
     ii. Local anaesthetics
     iii. General anaesthetics
126. Enumerate the conditions laid for an ideal anaesthetic.
127. On what basis are anaesthetics classified?
128. Discuss the anaesthetic effects of :
     i.   Nitrous oxide
     ii. Vinethene
     iii. Methoxy flurane
     iv. Trilene
129. Explain the anaesthetic property of halohydrocarbons.
130. Why are stabilizers added to the anaesthetic-grade diethyl ether?
131. What is the action of thiopental sodium?
132. Draw the chemical structures for:
     i.   Thiopental sodium
     ii. Propandid
     iii. Methohexitone
133. Draw the structure of the halothane that is optically active.
134. Discuss the anaesthetic action of cyclic propane.
135. Classify local anaesthetics.
136. To which class of anaesthetics does cocaine belong?
137. Explain the action of cocaine.
138. What are the basic structural features found in a local anaesthetic?
139. How vital are the structural features of a local anaesthetic?
140. Draw the general structure of a local anaesthetic.
141. State the factors deciding the onset of action of a local anaesthetic.
142. What was the first man-made local anaesthetic administered by injection?

143. How are the following prepared?
     i.   Orthocaine
     ii. Benzocaine
     iii. Procaine
144. How is procaine metabolized?
145. Draw the structure of amethocaine and mention its applications.
146. What are eucaines? Discuss their role as anaesthetics.
147. What is cancer?
148. How many types of tumours are known and what are they?
149. What are the factors that cause cancer?
150. Name a few aromatic compounds, which cause cancer.
151. Many nitro aromatics are potent carcinogens. Why?
152. Name a nitro aromatic which is a potent carcinogen.
153. Which methylating agent is identified to be a human cancer agent?
154. Illustrate the role of a mustard in bringing cross-linking of DNA. Write and discuss the reactions involved.
155. What are the commonly available treatments for cancer?
156. What are antineoplastics?
157. Discuss the role of cytotoxic drugs in the treatment of cancer.
158. Draw the structures of:
     i.   Thiotepa
     ii. Mechlorethamine
     iii. Chlorambucil
159. Name two nitrosoureas used in the treatment of cancer.
160. Explain the function of cisplatin.
161. What are busulfan and clomesone?
162. Draw the structure of AZQ.
163. Name the antibiotic that was first used in the treatment of cancer.
164. What are mitomycins?

165. Identify the amino acid residues constituting the peptide unit in dactinomycin.

166. Name the antibiotic, which, as a glycoside, is used in chemotherapy.

167. Explain the role of CC-1065 and duocarmycin as anticancer drugs.

168. Explain the function of antimetabolites as antitumour drugs.

169. Discuss the role of antimetabolites and steroidal hormones in the treatment of cancer.

170. What is diabetes mellitus?

171. How does insulin lower the sugar level?

172. State the different stages involved in the digestion of carbohydrates.

173. How many types of diabetes mellitus are known clinically?

174. Write a short note on type I diabetes mellitus.

175. Give a brief account of the structural features of insulin.

176. What is type II diabetes mellitus?

177. Define hypoglycaemia and hyperglycaemia.

178. What are hypoglycaemic drugs? Give examples.

179. Enumerate the basic requirements of hypoglycaemic drugs.

180. Discuss the role of sulphonyl ureas as hypoglycaemic drugs.

181. What are the adverse reactions of insulin therapy?

182. What does HIV stand for?

183. What is AIDS?

184. How is HIV infection transmitted?

185. What is reverse transcription?

186. List and explain the stages involved in HIV proliferation at the cellular level.

187. Give a few examples of reverse transcriptase inhibitors, which are used as anti-HIV drugs.

188. Name a few protease inhibitors used in the treatment of HIV infection.

189. What renders the treatment of HIV and AIDS difficult?

190. How does the HIV RNA get integrated into the DNA of the lymphocytes?

191. What are the symptoms of HIV?

192. What is the difference between a HIV-positive and an AIDS patient?

193. Name the fundamental tool used in clinical immunology.

194. How is HIV detected?

195. State the basic principle behind the ELISA method.

196. Describe how an HIV-ELISA is performed. List the different steps involved.

197. HIV infection or AIDS is often associated with psoriasis, malignant tumours, TB, etc. Give reasons.

PART II

BIOLOGICAL
CHEMISTRY

12

BLOOD

Blood serves as the principal transport medium of the body carrying oxygen, nutrients and chemical messages to the tissues, and waste products and synthesized metabolites away from them. The circulatory system provides access to all the cells of the body for the materials ingested or prepared elsewhere in the body. Blood courses through a complicated vascular network maintaining homeostasis with respect to temperature, oxidation–reduction potential and ionic concentration, throughout the body. The propulsive force for the system is being provided by the pumping action of the heart with assistance from the musculature of the arteries, and the hydrostatic and osmotic pressures generated. The presence of dispersed or dissolved nutrients, metabolites, hormones, etc. in the circulating blood, aids in counteracting infection and haemorrhage.

COMPOSITION OF BLOOD

Since blood has numerous functions, its composition is necessarily complex. The general composition includes:

1. A cellular fraction consisting of 45% of the total volume. The cellular composition includes erythrocytes, leucocytes and platelets.
2. A plasma fraction of 55%.

Plasma fraction consists of:

1. Non-diffusible constituents, including lipids, proteins and polypeptides, e.g. albumin, globulin, fibrinogen, enzymes, etc.
2. Diffusible constituents such as uric acid, glucose, electrolytes, e.g. Na^+, K^+, Ca^{2+}, vitamins, etc.

Blood plasma separated from blood cells without prior addition of any anticoagulants is known as *native*. If kept out of contact with air at 0–2°C it can maintain nearly the *in vivo* state. For the usual chemical analysis, plasma is obtained by the addition of an anticoagulant. Serum is the fluid obtained after the blood is allowed to clot. It is similar to plasma in composition but lacks fibrinogen and has very low concentration of other clotting factors. Serum is preferred to plasma for the determination of Ca^{2+}, albumin, globulin, etc.

Plasma Proteins

Plasma is a complex mixture containing a number of components, which differ in properties and function. The major components are albumin, globulin, fibrinogen and conjugate proteins.

Red Blood Corpuscles (RBCs)

The red cells or erythrocytes make up 45% by volume of blood. The rate at which the RBCs settle under force of gravity alone is referred to as erythrocyte sedimentation ratio (ESR). ESR is clinically an indication of the presence of tissue damage of various types, chronic infections, myocardial or pulmonary infections, etc.

RBCs are responsible for the opacity of blood. The opacity disappears if blood is diluted with water. When the fluid becomes transparent, the blood is said to be haemolysed. However, if the dilution is done with 0.9% NaCl solution, no haemolysis occurs.

This is due to the osmotic behaviour of RBCs. Each RBC is considered to be a mini osmometer. The water content of the cell depends upon the osmotic pressure of both the cell content and the surrounding medium. When the cell is in the medium of 0.9% NaCl solution, which has the same osmotic pressure as the cell constituents, the water content of the cell does not change, nor its size and therefore the medium is said to be isotonic with the cell. If the osmotic pressure of the medium is greater than the cell, the medium is then said to be hypertonic, then water is abstracted from the cell and the cell shrinks. When the osmotic pressure of the cell is greater, the solution is said to be hypotonic, then water enters from medium into the cell. This leads to increase in size of the cell. When the medium is sufficiently hypotonic, the swollen RBCs lose their ability to retain their contents and haemolysis results. The osmotic pressure at which haemolysis occurs is called as fragile point and its determination is of diagnostic value. Other agencies which bring about haemolysis are soaps, alkalies, chloroform, salts, certain drugs, etc.

Human RBCs are non-nucleated biconcave discs. The number of RBCs present in the human blood depends upon factors such as age, sex, altitude, exercise, etc. For a normal man, the count is 5,000,000 and a female it is 4,500,000. Increase in number is noticed on blood transfusion, higher altitude, strenuous physical exercise, etc. The increase is also noticed on loss of fluid from circulation such as due to starvation, after cold and hot bath, massages, etc.

The ability of RBCs to group together in masses is called as *clumping*. This occurs on a microscopic scale. The cell aggregates settle faster than the discrete cells. Clumping is a major factor in determining the ESR. When the clumping leads to the formation of visible clots, the process is known as *agglutination*.

Haemagglutination Haemagglutinins are polysaccharide complexes adsorbed on the surface of RBCs. The actual shape and the composition of the haemagglutinins are controlled by the genetic factors. The plasma serum contains a second factor, which is capable of reacting with certain haemagglutinins, causing alterations in the physico-chemical nature of RBCs and leading to their agglutination. Observation of agglutination is the basis for the establishment of blood types and is essential for blood transfusion.

BLOOD GROUPS

As already said blood consists of two principal components—the cellular and plasma fractions. In early days attempts on blood transfusion, in certain cases lead to the death of recipient for no determinable reason. Later on from the information and observation of clumping phenomenon of RBCs it was learnt that in certain cases this occurs at a macroscopic level leading to visible clumps only when mixed with the serum of certain specimens. This therefore, was attributed to the antigen–antibody reaction, which in turn is genetically determined.

Certain substances when injected into a living animal under appropriate conditions stimulate the animal to produce substances, which have the ability to react with the specific injected material. This injected substance is known as *antigen* and the specific substances produced are known as *antibodies*. Then the animal is said to be immunized. For a substance to function as antigen it must be a protein or a polysaccharide, and must be colloidal and foreign to the bloodstream of the injected animal. The antibodies are structurally and chemically modified serum proteins usually globulins, induced by the presence of the antigens in those organs in which the formation of serum proteins occurs. The presence of antibodies in the bloodstream of the immunized animal is recognized by their action on antigens. For example, if the antigen is RBC suspension, the blood serum of the immunized animal is the antiserum and this may acquire the property of haemolysing or agglutination of the added RBCs. If the antigen is a suspension of bacteria or any other cells, the antiserum will cause these cells to clump, i.e., agglutinate.

> Antigens are the agglutinogens.
> Antibodies are the agglutinins.

The antibodies are of two types: the acquired and natural antibodies. The acquired antibodies are produced only on the entry of a foreign antigenic substance into the bloodstream. Natural antibodies are the antibodies that are produced naturally in the bloodstream even in the absence of any antigen. The antibodies involved in the blood groups such as A, B, AB and O are the natural antibodies.

The differences in human blood are due to the presence or absence of certain protein molecules called antigens and antibodies. The antigens are located on the surface of the red blood cells and the antibodies are in the blood plasma. Individuals have different types and combinations of these molecules. The blood group to which one belongs depends on what he or she inherits from the parents. There are more than 20 genetically determined blood group systems known today, but the ABO and Rh systems are the most important ones used for blood transfusions. Not all blood groups are compatible with each other. Mixing incompatible blood groups leads to blood clumping or agglutination, which is dangerous for individuals. Nobel laureate Karl Landsteiner was involved in the differentiation of blood groups.

Human blood is grouped into four types A, B, AB and O. Each letter refers to a kind of antigen, or protein, on the surface of red blood cells—haemagglutinins. For example, the surface of red blood cells in Type A blood has antigens known as A-antigens.

Blood group A   Individuals belonging to the blood group A have A antigens on the surface of the red blood cells and B antibodies in the blood plasma.

Blood group B   If the individual belongs to the blood group B, then he/she will have B antigens on the surface of the red blood cells and A antibodies in the blood plasma.

Blood group AB   If one belongs to the blood group AB, he/she will have both A and B antigens on the surface of the red blood cells and no A or B antibodies at all in the blood plasma.

Blood group O   Persons belonging to the blood group O (null), will have neither A or B antigens on the surface of the red blood cells but will have both A and B antibodies in the blood plasma.

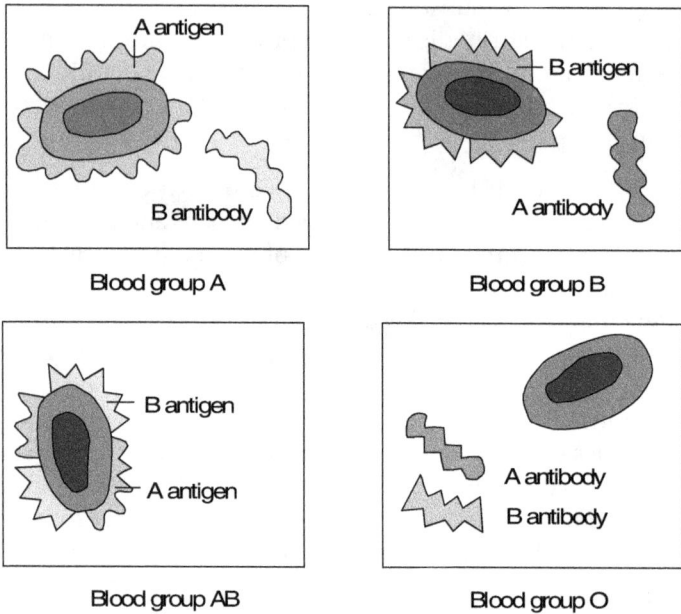

Figure 12.1 Representation of different blood groups according to the ABO system

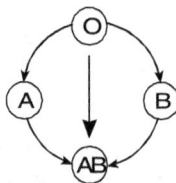

People with blood group O are called "universal donors" and people with blood group AB are called "universal receivers."

The ABO blood groups were the first to be discovered (in 1900) and are the most important in assuring safe blood transfusions.

Table 12.1 shows the four ABO phenotypes (blood groups) present in the human population and the genotypes that give rise to them.

Table 12.1　Blood groups with the corresponding genotypes

Blood group	Antigens on RBCs	Antibodies in serum	Genotypes
A	A	Anti-B	AA or AO
B	B	Anti-A	BB or BO
AB	A and B	Neither	AB
O	Neither	Anti-A and anti-B	OO

When red blood cells carrying one or both antigens are exposed to the corresponding antibodies, they agglutinate, i.e., clump together. People usually have antibodies against those red cell antigens that they lack.

Human red blood cells before (left) and after (right) adding serum containing anti-A antibodies are shown in Figure 12.2. The agglutination reaction reveals the presence of the A antigen on the surface of the cells.

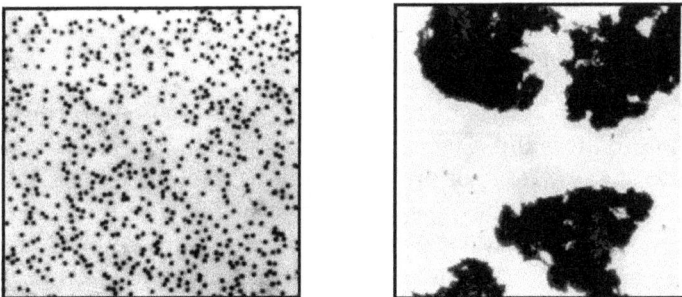

Figure 12.2 Human RBCs before (left) and after (right) adding serum containing anti-A antibodies

The critical principle to be followed is that transfused blood must not contain red cells that the recipient's antibodies can clump. Although theoretically it is possible to transfuse group O blood into any recipient, the antibodies in the donated plasma can damage the recipient's red cells. Thus all transfusions should be done with exactly-matched blood.

> Why do we have antibodies against red cell antigens that we lack? The answer for this question is that bacteria living in our intestine, and probably some foods, express epitopes similar to those on A and B. We synthesize antibodies against these if we do not have the corresponding epitopes; that is, if our immune system sees them as "foreign" rather than "self".

The Rh Factor

Each blood type is also grouped by its Rhesus factor, or Rh factor. Blood is either Rh positive (Rh+) or Rh negative (Rh–). About 85% of people have Rh+ blood. Rhesus refers to another type of antigen, or protein on the surface of red blood cells. The name "Rhesus" comes from Rhesus monkeys, in which the protein was discovered. Rh factor, the protein substance present in the red blood cells of most people is capable of inducing intense antigenic reactions. The Rh factor was discovered in 1940 by K. Landsteiner and A. S. Wiener, when they observed that an injection of blood from a rhesus monkey into rabbits caused an antigenic reaction in the serum component of rabbit blood. When blood from humans was tested with the rabbit serum, the red blood cells of 85% of the humans tested agglutinated (clumped together).

The red blood cells of the 85% (later found to be 85% of the white population and a larger percentage of blacks and Asians) contained the same factor present in rhesus monkey blood; such blood was typed Rh positive. The blood of the remaining 15% lacked the factor and was typed Rh negative (Figure 12.3). Under ordinary circumstances, the presence or lack of the Rh factor has no bearing on life or health. It is only when the two blood types are mingled in an Rh negative individual that the difficulty arises, since the Rh factor acts as an antigen in Rh negative persons, causing the production of antibodies. Besides the Rh factor, human red blood cells contain a large number of additional antigenic substances that has been classified into many blood group systems. However, the Rh system is the only one, besides the ABO system, that is of major importance in blood transfusions. If Rh positive blood is transfused into an Rh negative person, the latter will gradually develop antibodies called anti-Rh agglutinins, that attach to the Rh positive red blood cells, causing them to agglutinate. Destruction of the cells (haemolysis) eventually results. If the Rh negative recipient is given additional transfusions of Rh positive blood, the concentration of anti-Rh agglutinins may become high enough to cause a serious or fatal reaction. Therefore a person with Rh+ can receive blood from a person with Rh–.

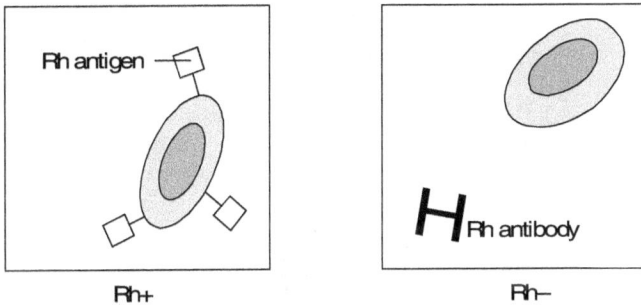

Figure 12.3   Representation of Rh factor in blood

Blood Group Notation

According to above blood grouping systems, one can belong to either of following 8 blood groups:

A Rh+	B Rh+	AB Rh+	O Rh+
A Rh–	B Rh–	AB Rh–	O Rh–

The same type of immune reaction occurs in the blood of an Rh negative mother who is carrying an Rh positive foetus. The probability of this situation occurring is high if the father is Rh positive. Some of the infant's blood may enter the mother's circulation, causing the formation of agglutinins against the foetal red blood cells. The first baby is usually not harmed. But, if the mother's agglutinins pass into the circulation of subsequent foetuses, they may destroy the foetal red blood cells, causing the severe haemolytic disease of newborns known as erythroblastosis foetalis and the haemolytic disease of the newborn, may be so severe as to kill the foetus or even the newborn infant. It is an example of an antibody-mediated cytotoxicity disorder. The infant's blood can also cross the placenta and attack the red cells of a subsequent Rh+ foetus. This destroys the red cells, producing anaemia and jaundice.

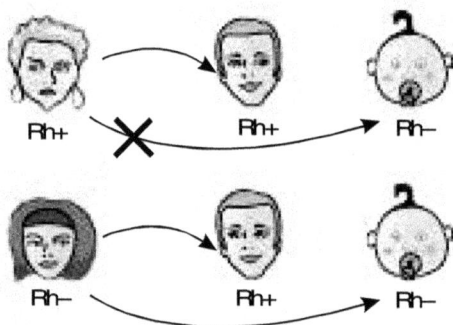

Figure 12.4   Rh incompatibility

Although certain other red cell antigens (in addition to Rh) sometimes cause problems for a foetus, an ABO incompatibility does not. This is because it turns out that most anti-A or anti-B antibodies are of the IgM class and these do not cross the placenta. In fact, an Rh–/type O mother carrying an Rh+/type A, B, or AB foetus is resistant to sensitization to the Rh antigen. Presumably, her anti-A and anti-B antibodies destroy any foetal cells that enter her blood before they can elicit anti-Rh antibodies in her. This phenomenon has led to an extremely effective preventive measure to avoid Rh sensitization. Shortly after each birth of an Rh+ baby, the mother is given an injection of anti-Rh antibodies. The preparation is called Rh immunoglobulin (RhIg) or Rhogam. These passively acquired antibodies destroy any foetal cells that got into her circulation before they can elicit an active immune response in her.

Blood Transfusions

Blood types become very important when a blood transfusion is necessary. In a blood transfusion, a patient must receive a blood type that is compatible with his or her own blood type, that is, the donated blood must be accepted by the patient's own blood (Table 12.2). If the blood types are not compatible, red blood cells will clump together, making clots that can block blood vessels and cause death. Type O– blood is considered the "universal donor" because it can be donated to people of any blood type. Type AB+ blood is considered the "universal recipient" because people with this type can receive any blood type.

Table 12.2   Compatibility within blood groups in the ABO system

Blood type	Who can receive this type
O+	O+, A+, B+, AB+
O–	All blood types
A+	A+, AB+
A–	A+, A–, AB+, AB–
B+	B+, AB+
B–	B+, B–, AB+, AB–
AB+	AB+
AB–	AB+, AB–

BLOOD PRESSURE

The volume of blood is more or less constant, and is greater than the total capacity of the blood vessels. Consequently, blood exerts a pressure on the walls of the vessels, and this is called as blood pressure. The blood pressure must be maintained constant by keeping constant water content and salts in the blood. A small volume of blood is always advantageous provided it is pumped rapidly through the vessels to ensure rapid and efficient transport. Heart and the arteries experience relatively a high blood pressure and this pressure depends upon the distance traversed and the metabolic state. When the ventricles contract, the pressure of blood in the blood vessel is highest and is called as systolic pressure; when the ventricles relax, no blood is forced into the arteries and will tend to become zero. Nevertheless, this is prevented by the elastic recoiling nature of the blood vessels. This pressure when the ventricles relax is known as diastolic pressure. The instrument sphygmomanometer is used to measure the blood pressure. This measures how high the blood pressure can push a column of mercury.

The values of systolic and diastolic pressures for adults are 120 mm Hg and 80 mm Hg respectively. The blood pressure is maintained within these limits by nervous and hormonal factors. The nerves take care of the diameter of the blood vessel. The hormones that are involved in the control of blood pressure are adrenaline and acetylcholine. Under emotional conditions, a larger amount of adrenaline is secreted, leading to constriction of blood vessels thereby increasing the blood pressure, while acetylcholine brings about expansion of blood vessels causing fall in the blood pressure. The blood pressure varies with respect to age and physiological state. In old people, the elasticity of the arteries decreases and the heart has to exert more force to pump blood with additional pressure. This results in increase in the pressure exerted on the arterial walls and the individual suffers from higher blood pressure.

Hypertension

Blood pressure is the force in the arteries when the heart beats—systolic pressure and when the heart is at rest—diastolic pressure. It is measured in millimetres of mercury (mm Hg).

Hypertension results in an adult when the blood pressure is 140 mm Hg of systolic pressure and 90 mm Hg of diastolic pressure or higher.

High blood pressure directly increases the risk of coronary heart disease leading to heart attack and stroke, especially when it is present along with other risk factors. High blood pressure can occur in children or adults, but it is more common among people over 35 years of age. It may run in families, but many people with a strong family history of high blood pressure never have it. People with diabetes mellitus, gout or kidney disease are more likely to have high blood pressure.

Hypertension is clinically divided into two types.

1. Primary or essential hypertension arising due to unknown reason.
2. Secondary or malignant hypertension due to hormonal variation or renal disorders.

The second one is more serious than the first one. The recommended blood pressure levels are shown in Table 12.3.

Table 12.3   Recommended blood pressure levels

Blood category	Pressure	
	Systolic (mm Hg)	Diastolic(mm Hg)
Normal	Less than 120	Less than 80
Prehypertension	120–139	80–89
Hypertension		
Stage 1	140–159	90–99
Stage 2	160 or higher	100 or higher

Hypotension

It is an abnormal condition where a person's blood pressure (the pressure of the blood against the walls of the blood vessels during and after each beat of the heart) is much lower than usual, which can cause symptoms such as dizziness or light-headedness. When the blood pressure is too low, there is inadequate blood flow to the heart, brain, and other vital organs.

A blood pressure level that is borderline low for one person may be normal for another. The most important factor is how the blood pressure changes from the normal condition. Most normal blood pressures fall in the range of 90/60 mm Hg to 130/80 mm Hg, but a significant change, even as little as 20 mm Hg, can cause problems for some people.

Low blood pressure is commonly caused by drugs such as the following:

- Medications used for surgery
- Anti-anxiety agents
- Antihypertensive agents
- Diuretics
- Cardiac drugs
- Some antidepressants
- Narcotic analgesics
- Alcohol

Other causes of low blood pressure include the following:

- Dehydration
- Heart failure
- Heart attack
- Changes in heart rhythm (arrhythmias)
- Anaphylaxis (a life-threatening allergic response)
- Shock (from severe infection, stroke, anaphylaxis, major trauma, or heart attack)
- Advanced diabetes

Another common type of low blood pressure is orthostatic hypotension, which results from a sudden change in body position, usually from lying down to an upright position.

The variation of blood pressure according to the age is:

1. Up to 30 years 110–145 mm Hg of systolic/68–92 mm Hg of diastolic pressure.
2. Up to 45 years 110–155 mm Hg of systolic/70–96 mm Hg of diastolic pressure.
3. 60 yrs 115–170 mm Hg/70–100 mm Hg.

Hypertension is treated by drugs acting centrally and drugs acting on CNS and the vascular muscles. They bring the blood pressure to normal level. The effects of antihypertensive/hypotensive drugs are only temporary. Reserpine, the indole alkaloid is widely used in treating the mild to moderate hypertension. Reserpine functions both as a hypotensive agent and as a mild tranquilizer. It is used in treating situation tension, and as an anti-anxiety drug. It is very effective with minimum side effects. Administering higher concentrations leads to mental depression, nasal block and gastric disturbances. Methyldopa, sodium nitroprusside and clonidine are some of the synthetic hypotensive drugs. They act in one of the following ways:

- Drugs acting on CNS, e.g. methyldopa, clonidine.
- Vasodilators, e.g. sodium nitroprusside, etc.
- Beta blockers reducing both systolic and diastolic pressure, e.g. chlorpromazine, phenoxybenzamine, etc.

13

DIGESTION AND ABSORPTION

INTRODUCTION

In higher animals most of the food is ingested in the form that is not directly available to the various organs and needs to be digested before absorption by the digestive tract. Digestion is the process by which the ingested food is broken down into smaller parts so that the body can use them to build and nourish cells and generate energy. The digestive system consists of a set of organs joined in the form of a long, twisting tube. Inside this tube, there is a lining called the mucosa. In the mouth, stomach and small intestine, the mucosa contains tiny glands that produce juices which help in digestion. The two other digestive organs namely, the liver, and the pancreas, also produce juices that reach the intestine through small tubes. In addition, other parts of organ systems such as nerves and blood also play a major role in the digestive system.

The first stage of digestion takes place in the mouth. Food is chewed and cut into smaller pieces which can be swallowed easily. This also mashes the food, which provides a larger surface area for the digestive enzymes and makes digestion more efficient. Saliva helps to lubricate the food for swallowing and the enzyme salivary amylase converts starch into glucose.

PROCESS OF DIGESTION

Digestion involves the mixing of food, its movement through the digestive tract, and chemical breakdown of the large molecules of food into smaller molecules. It begins in the mouth and is completed in the small intestine. The chemical process varies for different kinds of food.

Movement of food through the system requires:

1. A few seconds in the mouth.
2. A few seconds in the oesophagus.
3. Up to 3½ hours in the stomach.
4. A few minutes in the small intestine.
5. A few hours in the large intestine.

The muscular nature of the digestive tract through its contractile movement propels food and liquid. This also helps in mixing the food contents within each organ of the digestive system. This contractile movement of oesophagus, stomach and intestine is called as peristalsis. The action of peristalsis is like a wave moving through the muscle. The muscle of the organ produces a narrowing when it contracts and then propels the narrowed portion slowly down the length of the organ. These waves of narrowing push the food and fluid in front of them through each hollow organ. The first major muscle movement occurs when food or liquid is swallowed. Although the swallowing of food is by choice, once the swallowing begins, it becomes involuntary and proceeds under the control of the nerves.

The oesophagus is the organ into which the swallowed food is pushed. It connects the throat above, with the stomach below. At the junction of the oesophagus and stomach, there is a ring-like valve closing the passage between the two organs. However, as food approaches the closed ring, the surrounding muscles relax and allow the food to pass.

When the food enters the stomach, three mechanical tasks are performed.

1. The stomach stores the swallowed food and liquid. This allows the muscle of the upper part of the stomach to relax and accept large volumes of swallowed material.
2. The digestive juice produced by the stomach mixes with the food. The lower part of the stomach mixes these materials by its muscle action.
3. The stomach empties its contents slowly into the small intestine.

Several factors affect emptying of the stomach, including the nature of the food (mainly its fat and protein content) and the degree of muscle action of the emptying stomach and the next organ to receive the stomach contents (the small intestine). The food is digested in the small intestine and dissolved by the juices from the pancreas, liver and intestine, the contents of the intestine are then mixed and pushed forward for further digestion.

Finally, all the digested nutrients are absorbed through the intestinal walls. The waste products of this process include undigested parts of the food known as fibre, and older cells that have been shed from the mucosa. These materials are propelled into the colon, where they remain, usually for a day or two, until the faeces are expelled by a bowel movement.

Digestive enzymes are responsible for the chemical breakdown of the large food molecules into ones that are small enough to be absorbed into the blood. The major chemical process involved in digestion is hydrolysis. The three types of enzymes involved in this process are carbohydrases/carbases, proteases and lipases.

Carbohydrases hydrolytically break the bonds between complex sugar molecules or polysaccharides to make smaller molecules of sugar called disaccharides, e.g. maltose. This is further hydrolysed in the final stage of digestion to the monosaccharide, glucose, which is readily absorbed.

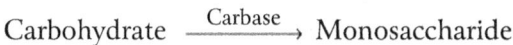

$$\text{Carbohydrate} \xrightarrow{\text{Carbase}} \text{Monosaccharide}$$

There are several different carbohydrase enzymes that break down complex polysaccharides, but the main one which helps in digestion is called amylase. This enzyme is found in saliva and also found in the digestive juices produced by the pancreas.

Proteases hydrolyse proteins into amino acids.

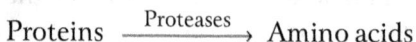

$$\text{Proteins} \xrightarrow{\text{Proteases}} \text{Amino acids}$$

Fats and oils are triglycerides of fatty acids. Lipases cleave hydrolytically the ester linkage in the triglycerides to liberate glycerol and fatty acids before they are absorbed in the small intestine.

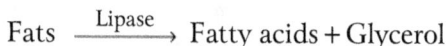

$$\text{Fats} \xrightarrow{\text{Lipase}} \text{Fatty acids} + \text{Glycerol}$$

The process of digestion takes place at three stages by the respective enzymes:

1. Digestion in the oral cavity
2. Digestion in stomach
3. Digestion in the small intestine

Digestion in the Oral Cavity

It is also known as salivary digestion. The salivary glands secrete saliva containing 99.5% water, a viscous glycoprotein, mucin, and enzyme ptyalin. Mucin acts as a lubricant making the food into bolus and helps in swallowing. The role of saliva is to provide a medium for the food particles to dissolve in which the enzyme hydrolase initiates the digestion. The breaking down of food into bolus increases the solubility and surface area for the enzyme attack.

The salivary enzyme ptyalin is an α-amylase and it catalyses the hydrolytic cleavage of α-glycosidic bonds of starch. pH of saliva is 6.8 and that of ptyalin is 6.6 and the conditions for the hydrolysis of starch in the mouth are suitable. However, the food remains in the oral cavity only for a shorter duration and soon moves down to acidic medium of the stomach. In the medium of lower pH, the activity of ptyalin ceases except in the interior part of the food, which is not immediately exposed to the acidic medium at that instance. However, by the churning action and peristaltic movement of the stomach, this is soon exposed to the digestive action in the stomach.

Digestion in the Stomach

The gastric secretions constitute 0.2–0.5% HCl providing a pH of ~1.0 and 97–99% water. The remaining percentage comprises of the enzymes pepsin, renin and lipase.

The digestive function of the stomach is in the initiation of protein digestion. Carbohydrates do not undergo further hydrolysis in the stomach. The lower pH and warmth of the stomach contents favour to some extent the hydrolysis of disaccharide sucrose, as it is labile in the acidic medium. However, the acidic medium of the stomach favours denaturation of proteins. By this, the tertiary structure of proteins is lost as a result of destruction of hydrogen bondings. This makes the polypeptide chains to unfold and become accessible to the action of proteolytic enzymes or proteases.

The lower pH also favours destruction of microorganisms entering the gastrointestinal tract. The flow of gastric juices into the stomach is under the influence of histamine. Histamine is released into the bloodstream when the smell and taste of food provides the necessary stimulus, which in turn increases the gastric flow. There are hormones which inhibit the gastric flow thus maintaining a delicate balance.

Pepsin is one of the proteases produced in the inactive form, zymogen, called as pepsinogen, which has a protective coat of polypeptide. The acidic medium of the stomach favours the conversion of pepsinogen into its active form pepsin. The liberated pepsin acts on the denatured proteins converting them into proteoses and then to peptones which are the large polypeptide molecules. Pepsin is an endoenzyme, because it hydrolyses the peptide bonds present in the interior part of the polypeptide structure.

The enzyme exopeptidase hydrolyses the C- or N-terminal peptide bonds. It is also specific for peptide bonds formed by the aromatic acids such as tyrosine or dicarboxylic amino acids like glutamate.

Renin is present in infants and prevents the rapid flow of milk from the stomach. It is absent in the adults.

Lipase brings about the hydrolysis of triglycerides of short and medium length fatty acids. The heat content of the stomach is important in liquifying the dietary lipids. The peristaltic contractions of stomach help in emulsifying the lipids. The role of lower pH is unimportant in this lipolytic process.

Digestion in the Small Intestine

Digestion is completed in the small intestine. This is done by the digestive juices secreted by pancreas, small intestine and liver.

These digestive juices are distinctly alkaline, to raise the pH of the acid contents of the food received from the stomach to a value nearing neutrality. The pH of the juice is 7.5–8 and even higher. The pancreatic juice is a watery fluid and contains some proteins and other organic and inorganic salts including Na^+, K^+, HCO_3^-, and Cl^- in large amounts and Ca^{2+}, Zn^{2+}, HPO_4^{2-}, SO_4^{2-} in small amounts. The pancreatic juice contains two zymogens namely trypsinogen and chymotrypsinogen. Trypsinogen is converted into its active form trypsin by an enzyme secreted by the intestinal wall. Trypsin in turn converts chymotrypsinogen into chymotrypsin.

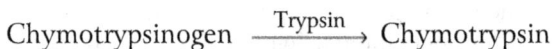

$$\text{Trypsinogen} \xrightarrow{\text{Enterokinase}} \text{Trypsin}$$

$$\text{Chymotrypsinogen} \xrightarrow{\text{Trypsin}} \text{Chymotrypsin}$$

Both trypsin and chymotrypsin are endopeptidases which hydrolyse the peptide bonds in the interior part of the protein molecule to smaller polypeptides. Trypsin is specific to basic amino acids; chymotrypsin is specific to uncharged amino acids containing aromatic residues. The enzyme elastase secreted by the pancreas, is broader in its action and attacks mainly the peptide bonds adjacent to small amino acid residues like glycine, alanine and serine.

The stomach contents, chyme, are intermittently introduced during digestion into the duodenum. The pancreatic and bile ducts open into the duodenum. The secretions of pancreas and bile being alkaline, neutralize the acidity of the chyme thereby making the medium alkaline. This change in pH advances the action of enzymes present in the pancreatic and intestinal juices. However, this is unfavourable for the action of pepsin. Carboxypeptidase, another pancreatic secretion, being an exoenzyme catalyses the further hydrolysis of smaller polypeptides by attacking the free C-terminal. The other secretions of pancreas include α-amylase and lipase. The action of pancreatic amylase is similar to that of ptyalin in hydrolysing starch and glycogen to maltose and maltotriose and a mixture of oligosaccharides and glucose. The pancreatic lipase along with the intestinal lipase catalyses the hydrolysis of triglycerides in a successive manner.

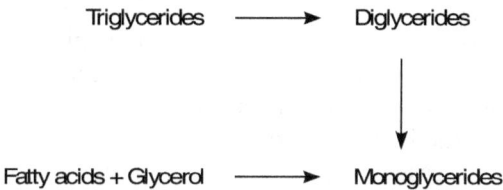

Triglycerides ⟶ Diglycerides

Fatty acids + Glycerol ⟶ Monoglycerides

Bile is produced by the liver and plays a vital role in the process of digestion of fats. The gall bladder, a sac-like structure attached to the bile duct, stores a definite amount of bile. During digestion, this rapidly supplies bile to the small intestine where it mixes with the pancreatic secretions. Bile contains no enzymes. The function of bile is that of an emulsifier, and facilitates the digestion and absorption of fatty acids and water-insoluble vitamins A, K, D, and E. When digestion is impaired, other foods are also poorly digested. This is because the food particles are covered with fat and prevent enzymes from acting on them. Under such conditions, the activity of intestinal bacteria causes putrefaction and production of gas.

The intestinal juice contains another exopeptidase, namely aminopeptidase. This catalyses the hydrolysis of smaller peptides at the free N-terminal. The intestinal enzymes sucrase, maltase and lactase hydrolyse the respective substrates which is evident from the respective names of the enzymes.

One of the unsolved puzzles of the digestive system is why the acid juice of the stomach does not dissolve the tissue of the stomach itself. Generally, the stomach mucosa is able to resist the acidic juice, although food and other tissues of the body cannot.

ABSORPTION

There is very little absorption from the stomach. The small intestine is the major organ, which absorbs about 99% of the ingested food.

Absorption of Carbohydrates

The products of carbohydrate digestion are absorbed into the bloodstream as monosaccharides mainly the hexoses (glucose, fructose, mannose, galactose) and pentoses. The two mechanisms by which absorption of monosaccharides takes place are active transport and simple diffusion. The former mechanism requires certain molecular configurational features. Because of the suitable configuration of the hydroxyl group on the second carbon and a pyranose ring, glucose and galactose are readily absorbed. This explains the slower rate of absorption of fructose, which occurs mainly by the diffusion process. As hydrolysis of polysaccharides, oligosaccharides, and disaccharides is rapid, their absorption is also faster and therefore their absorptive mechanism is quickly saturated. However, lactose which hydrolyses at a much lesser rate is absorbed slowly. Hence its absorptive mechanism does not get saturated rapidly unlike glucose and galactose.

Absorption of Fatty Acids

The fatty acids along with the partially hydrolysed di- and monoglycerides combine with the bile salts, emulsify and pass through the intestinal walls without undergoing further hydrolysis and finally get emptied into the bloodstream. Liver is the site of metabolism and resynthesis of triglycerides and phospholipids.

Absorption of Amino Acids and Proteins

Under normal conditions, the dietary proteins are completely digested to their constituent amino acids and readily absorbed from the intestine into the bloodstream. The natural L-aminoacids are actively transported, involving vitamin B6-pyridoxal phosphate as the coenzyme.

14

HORMONES AND THEIR PHYSIOLOGICAL EFFECTS

The word "hormone" is derived from the Greek word *horomao*, meaning excitation. A hormone is an organic substance produced by a system of ductless glands (the endocrine system) in small doses, and released into the bloodstream, being targeted to specific parts of the body. The hormones function as communication substances—they regulate and integrate various bodily functions. Hormones are directed to have a specific effect on the target areas and are indirect and slower and their influence is long-lasting. As a group, hormones do not resemble one another chemically, and their classification is therefore based on their physiological activity. Some of the important hormones are described below:

ADRENALINE

The first hormone discovered was adrenaline, also known as epinephrine. It is produced and secreted by the adrenal gland (and all its hormones are known as "stress hormones") and secreted as a direct reaction to stressful situations, such as by increasing heartbeat, blood pressure, sugar-level and muscle activity by dilation of bronchi and pupils, by vasoconstriction and sweating and by reducing the clotting time of blood. Blood is shunted from the skin and viscera to the skeletal muscles, coronary arteries, liver and brain. Besides its hormonal functions, adrenaline is also an excitatory neurotransmitter in the CNS (indirectly controlling its own production). It is involved in both neural and hormonal processes and its effects as a neurotransmitter are further reinforced by its hormonal function. It is the single most important hormone as regards to stress and playing a major role in the stress reaction.

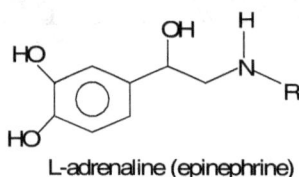

L-adrenaline (epinephrine)

L-adrenaline is chemically related to noradrenaline and to the family of adrenal medulla hormones. The hormone has influence on the storage and mobilization of glycogen and fatty acids and the corresponding metabolic pathways. Adrenaline has the opposite effect of insulin. It is a first messenger hormone and will be released when the glucose level in blood is low. Because of the binding to the β-adrenergic receptors, it triggers the adenylate cyclase cascade. This activating cascade effects the mobilization of glycogen (liver) and triacylglycerines (fat tissue) and a general increase of the metabolic rate. The resulting rise in blood sugar enables the fermentation of glucose in the muscles. Adrenaline furthermore reinforces these effects, because it increases the secretion of glucagon, a hormone with the same effects as adrenaline and decreases the release of insulin.

L-adrenaline has only a short lifetime because of its fast degradation. The oral intake of adrenaline has no effect. Therefore, it has to be administered parenterally. It is used as sympathomimeticum (drugs which support the beating of the heart), broncholyticum (drugs which relax the bronchial muscles) and antiasthmaticum (drugs against asthma). It is also used to staunch or prevent bleedings during surgery or in the case of inner organ bleeding. Because adrenaline leads to contraction of blood vessels, it is administered in combination with local anaesthetics.

In this combination, anaesthetics have a long-lasting effect and can be administered in smaller doses.

THYROXINE

The thyroid gland is a double-lobed structure located in the neck. Embedded in its rear surface are the four parathyroid glands.

The thyroid gland secretes three hormones namely:

1. Thyroxine or 3:5, 3':5' tetraiodothyronine (T_4) is the major hormone secreted by the thyroid gland. 99.5% of the secreted T_4 is protein-bound, principally to thyroxine-binding globulin (TBG). It is bound to a lesser extent with thyroxine-binding prealbumin (TBPA) and albumin. T_4 is involved in controlling the rate of metabolic processes in the body and influencing physical development.

2. 3:5,3' triiodothyronine (T_3) is 34 times more potent than T_4.

3. Calcitonin is totally unrelated to the other thyroid hormones, since it is involved in calcium homoeostasis.

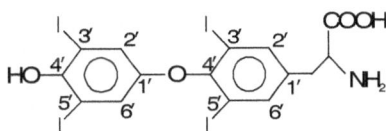

Thyroxine

Thyroxine abbreviated as T_4 is a hormone secreted by the thyroid gland and is a derivative of the amino acid tyrosine with four atoms of iodine. In the liver, one atom of iodine is removed from T_4 converting it into triiodothyronine (T_3). T_3 is the active hormone. The thyroid cells responsible for the synthesis of T_4 take up circulating iodine from the blood. It has many effects on the body. Among the most prominent of these are:

1. Increase in metabolic rate.

2. Increase in the rate and strength of the heartbeat.

T_4 and other thyroid hormones help to regulate growth and control the rate of chemical reactions—metabolism in the body.

Thyroxine increases the number and activity of mitochondria in cells by binding to the cell's DNA, thus increasing the basal metabolic rate. Administration of thyroid hormones, such as thyroxine, causes an increase in the rate of carbohydrate metabolism and a rise in the rate of protein synthesis and breakdown. This hormone, which excites the nervous system and leads to increased activity of the endocrine system, remains active in the body for more than a month. Thyroxine activity is controlled by thyrotropin, a substance released from the pituitary gland. Conversely, thyroxine regulates the effect of thyrotropin by feedback inhibition, i.e., high levels of thyroxine depress the rate of thyrotropin secretion.

Knowledge about the specific biochemical action of the thyroid hormones is less, nevertheless they are known to be involved in the stimulation of basal oxygen consumption and heat production. Synthetically prepared thyroxine is used clinically in the treatment of thyroid gland deficiency in adults and in the treatment of cretinism in children.

Thyroid Imbalance

The imbalance of the hormone causes hypothyroidism or hyperthyroidism

1. Hypothyroidism is caused by inadequate production of T_3.

2. Hyperthyroidism is caused by excessive secretion of thyroid hormones

High levels of thyroid hormones suppress the production of TSH through the negative-feedback mechanism. The resulting low level of TSH causes an increase in the number of bone-reabsorbing osteoblasts resulting in *osteoporosis*. T_4 (either total or free) along with thyroid stimulating hormone (TSH) is used as a biochemical indicator of thyroid function, which aids in the diagnosis and monitoring of either hyperthyroidism or hypothyroidism.

OXYTOCIN

It is a nanopeptide that is synthesized in hypothalamic neurons and transported down the axons of the posterior pituitary for secretion into blood. Oxytocin is also secreted within the brain and from a few other tissues, including the ovaries and testes. The most important stimulus for release of hypothalamic oxytocin is initiated by physical stimulation of the nipples or teats. Oxytocin acting within the brain plays a major role in establishing maternal behaviour.

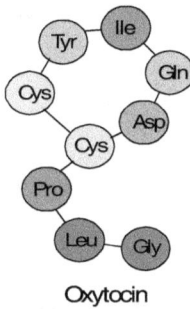

Oxytocin

Physiological Effects of Oxytocin

Earlier, oxytocin had the reputation of being an "uncomplicated" hormone, with only a few well-defined activities related to birth and lactation. As has been the case with so many hormones, further research has demonstrated many subtle but profound influences of this little peptide. The amount of oxytocin produced naturally, however, has little effect on uterine contractions and does not stimulate labour. When synthetic oxytocin is infused in larger amounts, it causes the smooth muscle in the wall of the uterus to contract and initiate the process of labour. Oxytocin's effect on uterine smooth muscle is dependent on the presence of oestrogen, and for that reason oxytocin has little effect on the uterus during the early stages of pregnancy.

INSULIN

It is a protein hormone. The first successful insulin preparations came from cows (and later from pigs). The pancreatic islets and the insulin protein contained within them were isolated from animals slaughtered for food.

A single insulin molecule consists of 2 polypeptide chains, A (21 amino acids) and B (30 amino acids). The structure of each chain can best be understood from figure shown. The two chains of insulin contain three disulphide bonds. Each disulphide bond is formed between the two cysteine amino acid residues. Four of these cysteine residues are found in chain A, and two are in chain B. Of the three disulphide bonds, two are formed between the two chains of the insulin molecule as interchain bonds. The third disulphide bond is an intra-chain bond, which holds part of chain A in a closed loop.

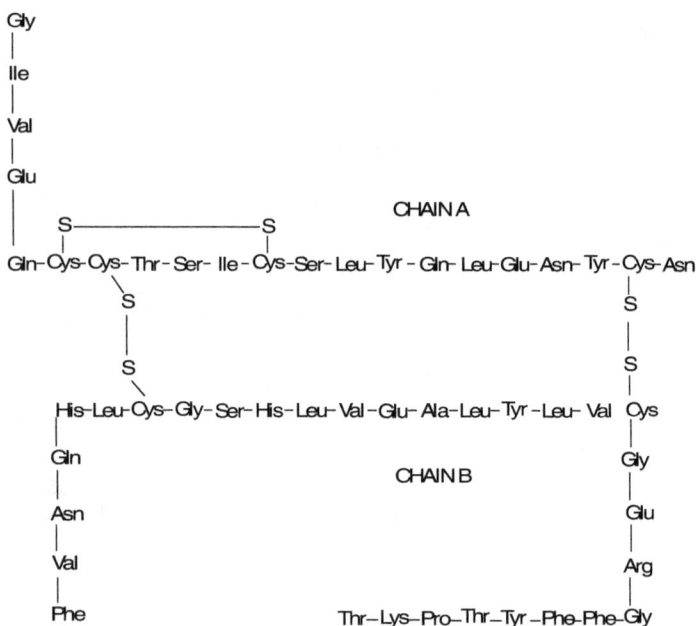

Human insulin

Proinsulin, the biological precursor of insulin is converted into insulin by the action of peptidases in the islets of Langerhans. The formed insulin is crystallized by the zinc ions and stored. On the receipt of certain signals triggered by the increase in blood sugar level, the stored insulin is released into the bloodstream, which is largely helped by the calcium ions. The concentration of insulin in the blood and tissues in human beings is estimated by certain special techniques such as radioimmunoassay, which showed it to be about 0.4 ng of insulin per millilitre. After a carbohydrate-rich meal, it rises to a maximum of 3–4 times the estimated levels. The normal human pancreas contains about 10 mg of insulin and the amount released into the blood daily is only about 1–2 mg. The release of insulin from the pancreas depends upon the concentration of glucose in the blood. When the glucose level increases above the normal levels of 80–120 mg per 100 ml the release of insulin is triggered. By this, the concentration of glucose declines in the blood. The half-life of insulin is only 3–4 minutes. Thus, release of insulin from pancreas is very responsive to fluctuations in glucose level of blood. The release of insulin is also sensitive to certain amino acids and certain other secretions by the stomach and small intestine.

Insulin acts to reduce extracellular (including blood plasma) levels of glucose by interacting in some way yet unknown with various cell membranes. In adipose (fatty) tissue, it facilitates the cellular uptake of glucose and its subsequent conversion to fatty acids, and inhibits the breakdown of fatty acids to simpler compounds. In muscle, it again facilitates the transport of glucose into cells and in addition stimulates its conversion to glycogen. It also increases protein synthesis in muscle. It also promotes protein synthesis from certain amino acids. In the liver, insulin facilitates glucose catabolism and its conversion to glycogen and inhibits its synthesis from simpler compounds.

Although the physiological effects of insulin are exerted by single protein molecules (monomers), the hormone is synthesized and stored in the pancreas as aggregates of six molecules (hexamers).

All therapeutic preparations of insulin are microcrystalline and hexameric.

SEX HORMONES

The sex hormones are steroidal in nature. They are highly specialized in their functions and do not produce general or systemic effects on metabolism. They are produced in the gonads.

Their activities appear to be controlled by the hormones secreted in the anterior lobe of the pituitary gland. It is for this reason that the sex hormones are also frequently referred to as *secondary hormones* and the hormones secreted in the anterior part of the pituitary are proteinous in nature and are called as primary hormones. The sex hormones are responsible for the sexual processes, and for the secondary sexual characteristics, that differentiate males from females.

The sex hormones are of three types

1. The androgens—the male hormones, e.g. testosterone, androsterone.

2. The oestrogens—the female/follicular hormones, e.g. oestradiol, oestrone, etc.

3. The gestogens—the corpus luteum hormones, e.g. progesterones.

Although the sex hormones act mainly on the sex accessory organs, some of them secreted by the adrenal cortex have pronounced effects on the metabolism of carbohydrates and proteins in many tissues.

15

MICRONUTRIENTS AND THEIR BIOLOGICAL ROLE

Good nutrition from eating a healthy diet is the foundation of all natural health-healing therapies. Micronutrients include vitamins and minerals. Bioavailability of nutrients is an important issue in nutrition. "Micronutrients" do not provide any energy to the body. Nutritionists call vitamins and minerals as micronutrients to distinguish them from those nutrients which are present in foods in much larger quantities such as proteins or lipids.

VITAMINS

Vitamins have many roles within the body and are essential for survival. However, they are required in only minute amounts. Most of the vitamins cannot be made by the body and must be provided through the diet. The three exceptions are vitamin D (produced by the action of sunlight on skin), vitamin K (produced by bacteria in the gut) and vitamin B or Niacin (obtained from the amino acid tryptophan). Anyone who eats a balanced and varied diet is unlikely to suffer from a vitamin deficiency. However, Asian and housebound elderly women living in northern countries are often deficient in vitamin D.

Vitamins are classified according to their solubility in fat or water. Fat-soluble vitamins (A, D, E and K) usually occur in foods that contain fats or oils. They are stable to heat and light, and are not easily destroyed by cooking. In general, fat-soluble vitamins are stored in fat deposits in the body, particularly in the liver. Water-soluble vitamins (B group and C) are not stored in the body. Because of their solubility, they are excreted in the urine, and must therefore be replenished frequently. Water-soluble vitamins are found in a variety of foods, but tend to be destroyed by poor storage or prolonged cooking.

Water-soluble Vitamins

Vitamin B   It is a complex of several vitamins. The name arises because it was once considered a single vitamin, much like vitamin C or vitamin D. Since later research has shown that it is in fact a complex of chemically distinct vitamins that happen to often coexist in the same foods, the name has gradually declined in use, being replaced by the generic term "the B vitamins", the vitamin B complex, or by the specific names of each vitamin.

Listed below are all the vitamins in the B-vitamin family. All the vitamins are important to maintain good health.

- Vitamin B_1 (Thiamine)
- Vitamin B_2 (Riboflavin)
- Vitamin B_3, also vitamin P or vitamin PP (Niacin)
- Vitamin B_5 (Pantothenic acid)
- Vitamin B_6 (Pyridoxine and Pyridoxamine)
- Vitamin B_7, also Vitamin H and Vitamin B-w (Biotin)
- Vitamin B_9, also Vitamin M and Vitamin B-c (Folic acid) (important for pregnancies)
- Vitamin B_{12} (Cyanocobalamin)

All the B vitamins are soluble in water. Most of the B vitamins have been recognized as coenzymes, and they all appear to be essential in facilitating the metabolic processes of all forms of animal life. The complex includes B_1 (thiamine), B_2 (riboflavin), niacin (nicotinic acid), B_6 (a group of related pyridines), B_{12} (cyanocobalamin), folic acid, pantothenic acid, and biotin.

Eating foods rich in these vitamins helps in protecting the heart. B-vitamins help lower homocysteine levels in the body. Homocysteine comes from the breakdown of protein in the body. High levels of homocysteine are related to many diseases of the heart. These vitamins can make the blood healthier by lowering homocysteine levels and make the vessels that carry blood to the heart stronger. Stronger vessels and a healthy heart can lower the chances of heart attacks or strokes.

Vitamin B₁ Other names for the water-soluble vitamin B_1 are thiamine, antiberiberi factor, aneurine and antineuritic factor. Thiamine pyrophosphate is the biochemically active ester of thiamine, involved in many important metabolic processes including:

- The decarboxylation of alpha-oxoglutaric acid in the citric acid cycle.
- The conversion of alanine to pyruvic acid, then to acetyl coenzyme A.

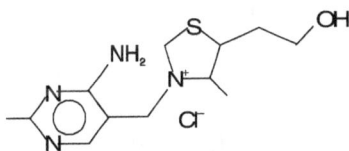

Vitamin B₁ (thiamine hydrochloride)

Although thiamine occurs widely in foods, it is generally present in small amounts. The principal source of thiamine is in dried brewer's yeast and there are other good sources including:

- Meat
  - Pork
  - Lamb
  - Poultry
- Whole grain cereals (the thiamine is present in the germ of the grain)
- Eggs
- Nuts
- Legumes

Thiamine is essential in the body for the release of energy from carbohydrate and fat, and is involved in conduction of nerve impulses. It is not stored well in the body (highest concentrations are in the brain, heart, liver and kidney) and so a continuous supply is required. Deficiency results in two diseases:

- **Beriberi** It causes disorders of the nervous and cardiovascular system:

  (Beriberi translated into English means "I can't, I can't")

  - **Dry Beriberi** It causes polyneuropathy with severe muscle wasting.

  - **Wet Beriberi** It is characterized by anorexia, muscle weakness, oedema, mental confusion and heart failure.

  - **Infantile Beriberi** It causes sudden appearance of vomiting, convulsions, abdominal distention and anorexia.

- **Wernicke–Korsakoff syndrome** It results from a combination of factors including:
  - Inadequate intake
  - Decreased absorption
  - Increased requirements

Vitamin B_1 is sensitive to heat, alkali, oxygen and radiation and considerable amounts of the vitamin can be lost during cooking.

Vitamin B_2   Vitamin B_2 or riboflavin is the second member of the B complex group. Earlier names for this water-soluble vitamin were lactoflavin, ovoflavin, hepatoflavin and verdoflavin, indicating the sources—milk, eggs, liver and plants—from which the vitamin was first isolated. Considerable loss of riboflavin may occur if foods are exposed to light. Thus, sun drying of foods destroys most of their riboflavin content. Ordinary cooking does not affect riboflavin, but cooking in a large quantity of water causes some of this vitamin to be drained out from the food. Sulpha drugs and alcohol can destroy vitamin B_2.

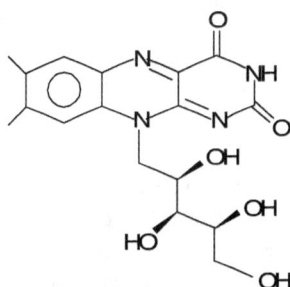

Riboflavin (Vitamin B₂)

Riboflavin widely occurs in nature and is present in all animal and plant cells. However, there are few sources, which are rich in riboflavin. The highest concentrations are in yeast and liver but the most common dietary sources (Table 15.1) include:

- Milk and milk products
- Egg whites
- Meats
- Green leafy vegetables
- Egg yolks

Table 15.1   Sources and the amount of B₂

Fruits	mcg	Milk and milk products	mcg
Ripe papayas	250	Skimmed milk powder	1640
Raisins	190	Whole milk powder	1360
Custard apples	170	Khoa	410
Black currants	140	Cow's milk	190
Apricots	130	Curd	160
Jackfruit	130	Buffalo milk	100

Riboflavin is essential for growth and general health. It functions as a part of a group of enzymes, which are involved in the metabolism of carbohydrates, fats and proteins. Vitamin B_2, or riboflavin is an intermediary during transfer of electrons in the cellular oxidation–reduction reactions, which generate energy from protein, carbohydrate and fat. The riboflavin coenzymes are also important for the transformation of vitamin B_6 and folic acid into their respective active forms, and for the conversion of tryptophan into niacin. It is involved in a number of chemical reactions in the body and is therefore essential for normal tissue maintenance. Riboflavin aids digestion and helps in the functioning of the nervous system. It prevents constipation, promotes healthy skin, nails and hair, and strengthens the mucous lining of the mouth, lips and tongue. Riboflavin also plays an important role in the health of the eyes and alleviates eyestrain. This vitamin is particularly helpful in counteracting the tendency towards glaucoma. An ample supply of vitamin B_2 provides vigour and helps to preserve the appearance and feeling of youth. Riboflavin is the precursor of flavoproteins:

- Flavin mononucleotide is produced in the mucosal cells of the intestine and is a coenzyme.

- Flavin adenine dinucleotide is a coenzyme synthesized in the liver.

Deficiency in riboflavin usually occurs with deficiencies of the other water-soluble vitamins. Specific symptoms of riboflavin deficiency include:

- Glossitis (magenta tongue)
- Angular stomatitis (fissures at the corner of the mouth)
- Itching
- Skin rash

Riboflavin is sensitive to light but is heat-stable. Riboflavin is often used as a colourant because of its bright yellow colour. The use of ethylene oxide in food sterilization can destroy riboflavin.

Niacin (B_3)   An earlier name for niacin was PP factor (pellagra-preventative factor). The designation vitamin B_3 also includes the amide form, nicotinamide or niacinamide. Niacin (nicotinic acid and the amide derivative nicotinamide) is one of the water-soluble B vitamins. In the blood, brain, kidney and liver it is converted to the coenzymes nicotinamide adenine dinucleotide (NAD) and nicotinamide adenine dinucleotide phosphate (NADP), both of them are involved in the generation of energy in cells. Tryptophan is an amino acid, which is a provitamin of niacin. Nicotinic acid and nicotinamide are both stable to light, heat, air and alkali.

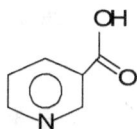

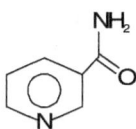

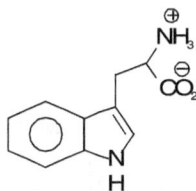

Nicotinic Acid   Nicotinamide   Tryptophan

Both forms of niacin widely occur in nature. Nicotinic acid is the predominant form in plants and nicotinamide in animals. The major dietary sources of niacin are:

• Yeast extract
• Liver
• Meats
• Oily fish
• Nuts
• Legumes

Other sources include:

• Green leafy vegetables
• Milk and milk products

Important dietary sources of tryptophan are:

• Meat
• Milk
• Eggs

Deficiency of niacin results in a disease called pellagra, the symptoms of which include:

• Dermatosis
• Dementia

- Diarrhoea
- Nervous disorders which can lead to paralysis of the extremities

Vitamin C (Ascorbic acid) Vitamin C, also known as ascorbic acid, L-ascorbic acid, dehydroascorbic acid, the antiscorbutic vitamin, L-xyloascorbic acid and L-threo-hex-2-uronic acidy-lactone, is an organic acid with antioxidant properties. Ascorbic acid and its sodium, potassium, and calcium salts are commonly used as antioxidant food additives. These compounds are water-soluble and thus cannot protect fats from oxidation. It is claimed as a cure for many diseases and problems—from cancer to the common cold. Yet, this miracle vitamin cannot be manufactured by the body, and needs to be ingested.

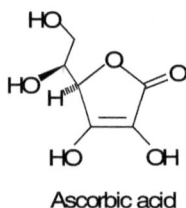

Ascorbic acid

2-oxo-L-threo-hexono-1,4-lactone-2,3-enediol

or

(R)-3,4-dihydroxy-5-((S)-1,2-dihydroxyethyl)furan-2(5H)-one

Good sources of vitamin C are green leafy vegetables, berries, citrus fruits, guavas, tomatoes, melons, papayas, etc. Some juices that are not normally a source of vitamin C have vitamin C added. Examples of these juices include apple and grape.

Functions of vitamin C It is important in forming collagen, a protein that gives structure to bones, cartilage, muscle, and blood vessels. Vitamin C also aids in the absorption of iron, and helps in maintaining capillaries, bones, and teeth. Ascorbic acid also promotes healthy cell development, proper calcium absorption, normal tissue growth and repair—such as healing of wounds and burns. It assists in the prevention of blood clotting and bruising, and strengthening the walls of the capillaries. Vitamin C is needed for healthy gums, to help protect against infection, and assisting in clearing up infections and is thought to enhance the immune system and help reduce cholesterol levels, high blood pressure and preventing atherosclerosis. Eating a variety of foods that contain vitamin C is the best way to get an adequate amount each day. Healthy individuals who eat a balanced diet rarely need supplements. Vitamin C being a biological reducing agent is also linked to prevention of degenerative diseases such as cataracts, certain cancers and cardiovascular diseases. Vitamin C can be lost from foods during preparation, cooking, or storage.

To prevent loss of vitamin C:

- Serve fruits and vegetables raw whenever possible.
- Steam, boil, or simmer foods in a very small amount of water, or microwave them for the shortest time.
- Cook potatoes with their skins. Be sure to wash the dirt off the outside of the potato.
- Refrigerate prepared juices and store them for no more than two to three days.
- Store cut raw fruits and vegetables in an airtight container and refrigerate. Do not soak or store in water.

Deficiency of vitamin C Scurvy is the only disease clinically treated with vitamin C. However, a shortage of vitamin C may result in "pinpoint" haemorrhages under the skin and a tendency to bruise easily, poor wound healing, soft and spongy bleeding gums and loose teeth, edema (water retention), weakness, a lack of energy, poor digestion, painful joints and bronchial infection and colds are also indicative of an under-supply.

Lipid-soluble Vitamins

Vitamin A (Retinol) Vitamin A is a fat-soluble, antioxidant vitamin, important in the process of vision and bone growth. It belongs to the family of chemical compounds known as retinoids. Vitamin A is required in the production of rhodopsin, the visual pigment used in low light levels. Maintaining vitamin A levels within a normal range is important, as either too little or too much of this vitamin lead to serious disease.

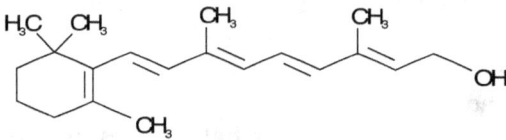

Vitamin A or retinol

Source   Vitamin A does not occur in plants, but many plants contain carotenoids such as beta-carotene that can be converted to vitamin A within the intestine and other tissues.

Some of the well-characterized effects of vitamin A include:

- *Vision*   Retinol is a necessary structural component of rhodopsin or visual purple, the light-sensitive pigment within the rod and cone cells of the retina. If inadequate quantities of vitamin A are present, vision is impaired.

- *Resistance to infectious disease*   In almost every infectious disease studied, vitamin A deficiency has been shown to increase the frequency and severity of disease.

- *Epithelial cell "integrity"*   Many epithelial cells appear to require vitamin A for proper differentiation and maintenance. Lack of vitamin A leads to dysfunction of many epithelia—the skin becomes keratinized and scaly, and mucus secretion is suppressed.

- *Bone remodelling*   Normal functioning of osteoblasts and osteoclasts is dependent upon vitamin A.

Vitamin A deficiency and excess states   Both too much and too little vitamin A are well known causes of disease in man and animals. Vitamin A deficiency usually results from malnutrition, but can also be due to abnormalities in intestinal absorption of retinol or carotenoids.

Some of the serious manifestations of vitamin A deficiency include:

- Blindness, while severe deficiency can result in severe dryness and opacity of the cornea (xerophthalmia).

- Increased risk of mortality from infectious disease. Supplementation with vitamin A has been shown to substantially reduce mortality from diseases such as measles and gastrointestinal infections.

- Abnormal function of many epithelial cells such as dry, scaly skin, inadequate secretion from mucosal surfaces, infertility, decreased synthesis of thyroid hormones.
- Abnormal bone growth.
- Disorders of the central nervous system and optic nerve.

Vitamin A excess states, while not as common as deficiency, also lead to disease. Vitamin A and most retinoids are highly toxic when taken in large amounts, and the most common cause of this disorder in both man and animals is excessive supplementation. In contrast, excessive intake of carotenoids are not reported to be toxic.

Teratogenic effects Both hypovitaminosis A and hypervitaminosis A are known to cause congenital defects in animals and likely to have deleterious effects in humans. Pregnant women are advised not to take excessive vitamin A supplements.

Vitamin D It is a fat-soluble seco-sterol hormone precursor that contributes to the maintenance of normal levels of calcium and phosphorus in the bloodstream. It is not precisely correct to describe it as a vitamin, but it might best be described as a conditional vitamin since human skin can manufacture it in under certain circumstances. It is also known as calciferol. Forms of vitamin D are the following.

- Vitamin D_1—molecular compound of ergocalciferol with lumisterol, 1:1
- Vitamin D_2—ergocalciferol or calciferol (made from ergosterol)
- Vitamin D_3—cholecalciferol (made from 7-dehydrocholesterol)
- Vitamin D_4—22,23-dihydroergocalciferol
- Vitamin D_5—sitocalciferol (made from 7-dehydrositosterol)

Vitamin D deficiency is known to cause several bone diseases, due to insufficient calcium or phosphate in the bones, which include the following.

- **Rickets**   A childhood disease characterized by failure of growth and deformity of long bones.

- **Osteoporosis**   A condition characterized by fragile bones.

- **Osteomalacia**   A bone-thinning disorder in adults that is characterized by proximal muscle weakness and bone fragility. Osteomalacia can only occur in a mature skeleton.

Vitamin E   It is a combination of α-tocopherol, β-tocopherol, γ-tocopherol, and δ-tocopherol. Vitamin E is a generic term used for a group of chemically similar compounds sharing the tocopherol and tocotrienol structures. Vitamin E is a fat-soluble vitamin present in many foods, especially certain fats and oils. It is one of a number of nutrients called antioxidants. Other well known antioxidants include, vitamin C and β-carotene. Antioxidants are nutrients that block some of the damage caused by toxic by-products released when the body transforms food into energy or fights off infection. The build up of these by-products over time is largely responsible for the aging process and can contribute to the development of various health conditions such as heart disease, cancer, and a host of inflammatory conditions like arthritis. Antioxidants provide some protection against these conditions and help reduce the damage to the body caused by toxic chemicals and pollutants.

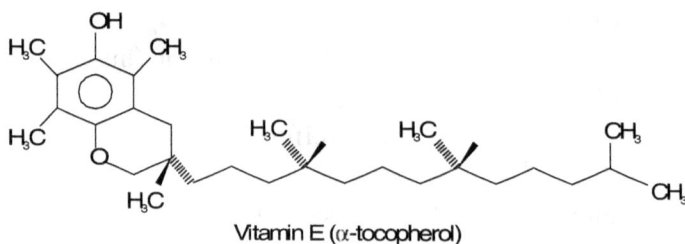

Vitamin E (α-tocopherol)

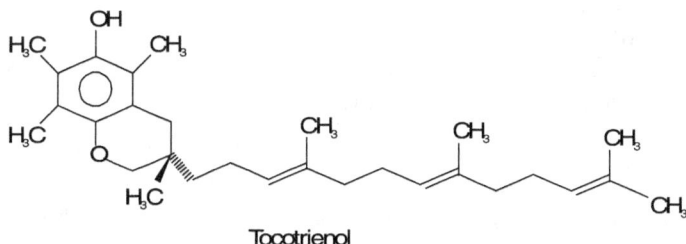

Tocotrienol

Sources

- Foods containing wheat are a good source of vitamin E. These foods vary in their content of vitamin E based on the particular source and processing involved. Wheat germ oil is the richest source of natural vitamin E. If the wheat product is processed to make other foods such as margarine, the content of vitamin E is reduced due to the methods involved in formulation and exposure to chemicals (acids and bases).

- Other dietary sources are nuts and seeds that are rich in oils, and vegetable oils derived from these sources.

Role of vitamin E   Vitamin E helps prevent arteries from clogging by blocking the conversion of cholesterol into the waxy fat deposits called plaque that stick to blood vessel walls. Vitamin E also thins the blood, allowing blood to flow more easily through arteries even when plaque is present. Studies in the last 10 years have reported beneficial results from use of vitamin E supplements as part of a prevention strategy for heart disease and other types of cardiovascular disease. There is some evidence for the use of supplemental vitamin E as a treatment for arterioscleroses.

Vitamin E deficiency can be seen in people unable to absorb fat properly. Such conditions include pancreatitis (inflammation of the pancreas), cystic fibrosis, and biliary diseases (illnesses of the gall bladder and biliary ducts). Symptoms of deficiency include muscle weakness, loss of muscle mass, abnormal eye movements, impaired vision, and unsteady gait. Eventually, kidney and liver function may be compromised. In addition, severe vitamin E deficiency can be associated with serial miscarriages and premature delivery in pregnant women. Deficiency in vitamin E is rare, but can arise during pregnancy (due to low dietary intake) and in newborn infants (due to feeding with formula milk that is low in vitamin E). This latter condition is generally associated with haemolytic anaemia in the newborn.

Vitamin K   It is a group name for a number of related compounds, which have in common a methylated naphthoquinone ring structure, and which vary in the aliphatic side chain attached at the 3-position. It is generally accepted that the naphthoquinone is the functional group, so the mechanism of action is similar for all K-vitamins.

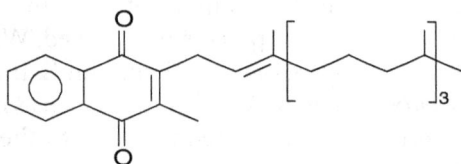

Vitamin K₁ (phylloquinone)

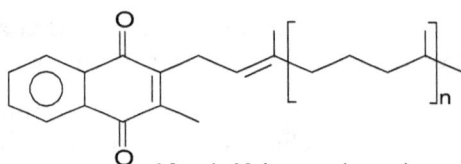

Vitamin K₂ (menaquinones)

Vitamin K denotes a group of 2-methylo-naphthoquinone derivatives. They are human vitamins and are needed for blood coagulation. Vitamin K_1 is found in plants. It is important in giving protection against osteoporosis. Vitamin K_2 (menaquinone, menatetrenone) is normally produced by bacteria in the intestines, and dietary deficiency is extremely rare unless the intestines are heavily damaged.

They play a key role in the regulation of three physiological processes namely

- blood coagulation
- bone metabolism
- vascular biology

Sources of vitamin K   Milk, eggs, liver and plants are good sources of vitamin K.

Vitamin K deficiency may occur by disturbed intestinal uptake (such as would occur in a bile duct obstruction).

MINERALS AND TRACE ELEMENTS

Like vitamins, minerals and trace elements are necessary for a variety of functions. Minerals (e.g. calcium, iron, phosphorus, magnesium, sodium, chloride and potassium) are required in relatively large amounts, whereas trace elements zinc, fluoride, selenium, iodine, copper, manganese, chromium and molybdenum are required in much smaller amounts.

Biological Role

Electrolytes of sodium and potassium ions play a major role in

- the maintenance of osmotic pressure
- the maintenance of electrical neutrality
- the production of energy
- the transmission of impulses

Calcium ions help in the formation of bones and clotting of blood.

Copper plays an important role in the formation of haemoglobin. It participates along with certain coenzymes in a number of enzymatic reactions. Enzymes requiring copper are, ascorbic acid oxidase, cytochrome oxidase, etc. The abnormal metabolism of copper is known as Wilson's disease. Liver is the chief source of copper.

Cobalt is a constituent of vitamin B_{12} and is an anti-anaemic agent.

Iron is important in the transport of oxygen as haemoglobin and is commonly referred to as anti-anaemic agent.

Iodine is essential for the synthesis of thyroid hormones.

Manganese is associated with the enzymes hydrolase and transferase used in the synthesis of glycoprotein.

Molybdenum is involved in the oxidation reactions.

Selenium is a synergistic antioxidant with vitamin E. It accompanies the enzyme peroxidase.

Zinc ions help in the storage of insulin as granules. In association with the enzyme carbonic anhydrase zinc ion is involved in the transport of carbon dioxide. Zinc ions act as cofactors of lactic dehydrogenase and alkaline phosphatases. It also maintains the concentration of vitamin A.

ANTIOXIDANTS

Some micronutrients are known as antioxidant nutrients, because of their role in defending the body against oxidative damage and the development of disease. Examples include, vitamins C and E, β-carotene, the yellow pigment present in some fruit and vegetables and the trace element, selenium. Everyday examples of oxidation include iron when it goes rusty and butter when it goes rancid. In our bodies, oxidative damage is an attack on a molecule by oxygen-containing free radicals, e.g. superoxide, hydroxyl and oxygen derivatives like ozone.

16

ENZYMES

INTRODUCTION

The study of enzymes began quite early when it was discovered that certain microorganisms were responsible for the formation of wines. Louis Pasteur was the first to demonstrate that yeasts can ferment glucose solution without being consumed or destroyed. This discovery then led to the identification of the catalyst, Ferment, present in the yeast cells which is responsible for fermentation. Later, the cell-free extracts of yeast that could ferment glucose was isolated. Since then, a lot of development has taken place to show that cells contain such chemical substances, that can catalyse a number of chemical reactions taking place in the cell. These were then named as *Enzymes*.

Having laid the foundation of enzymatic conversion of chemical substances in the living cells, several of them have been isolated and identified. Now it is established without any exceptions that all the enzymes are proteinous in nature, and have three-dimensional structure.

Enzymes are now collectively defined as the biological catalysts which are proteinous molecules formed by the living organism, that can catalyse a particular or a group of closely related biochemical reactions.

Hundreds of enzymes are now available in crystalline form; many of them contain a non-proteinous component called as the prosthetic group which may be bound to the enzymes, covalently or non-covalently. The enzyme in combination with the prosthetic group is referred as holoenzyme. If the prosthetic part is dissociated from the enzyme molecule, the catalytic activity of the enzyme is lost and is called as apoenzyme. Certain types of RNA can also serve as catalysts. These RNA molecules are called ribozymes.

Not all enzymes require the prosthetic part for their activity except those that get converted from an inactive form to an active form, which is depicted in the format shown below.

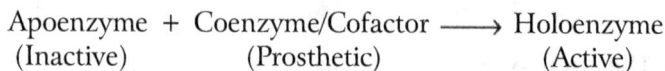

Apoenzyme + Coenzyme/Cofactor $\longrightarrow$ Holoenzyme
 (Inactive)   (Prosthetic)   (Active)

All the enzymes synthesized inside the cell can pass through the cell under certain conditions. There are many enzymes produced by the cell that carry out various metabolic functions within the cell and are called as endoenzymes. Certain enzymes liberated by the cell, catalyse reactions in the vicinity of the cell, and are called as exoenzymes. Enzymes show specificity towards substrates and this specificity is determined by the proteinous part of the enzyme.

Properties

The important properties of enzymes are as follows:

- Enzymes are present in the cell in low concentrations only.
- During the catalytic action, there occurs a reversible chemical change on the enzymes, thus inducing them to participate in several reactions repeatedly.
- The chemical equilibrium of the enzyme-catalysed reactions remain unchanged.

Nomenclature

The normal practice of naming an enzyme is based on the type of the reaction it catalyses and the substrate on which it acts, e.g. maltase, catalyses the hydrolysis of maltose. In 1972, the

International Union of Biochemistry (IUB) recommended not only the existing trivial names, but also the systemic names. These systemic names are on the basis of the reactions being catalysed.

Activity The student is advised to collect information about the recent nomenclature of enzymes based on number system.

CLASSIFICATION

There are approximately 3000 enzymes which have been characterized. These are grouped into six main classes according to the type of reaction catalysed. The various groups are defined as follows:

Oxido-reductases

These enzymes catalyse biological oxidation–reduction reactions involving the transfer of hydrogen atoms or electrons. The following are of particular importance in the design of enzyme electrodes. They are further divided into:

  i. Dehydrogenases, that catalyse removal of hydrogen from the substrate. This is done by certain cofactors such as nicotinamide adenine dinucleotide (NAD^+).

 ii. Oxidases, that catalyse hydrogen transfer from the substrate to molecular oxygen producing hydrogen peroxide as a by-product.

iii. Peroxidases, that catalyse oxidation of a substrate by hydrogen peroxide.

 iv. Oxygenases, that catalyse substrate oxidation by molecular oxygen. Thus molecular oxygen is incorporated into the substrate.

  v. Hydroxylases, that introduce hydroxyl groups.

 vi. Oxidative-deaminase, that catalyses oxidation of amino compounds with the elimination of ammonia.

Transferases

These enzymes effect transfer of group/groups between two substrates. They are further classified into:

- Amino transferases, which catalyse exchange of amino and keto groups between amino and keto acid.

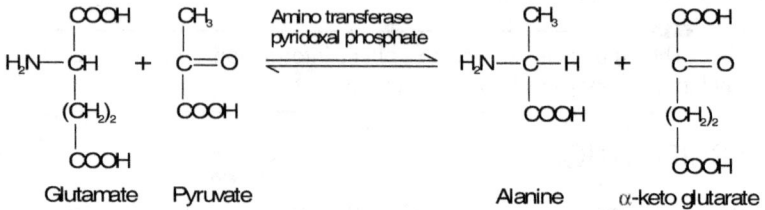

Glutamate   Pyruvate   Alanine   α-keto glutarate

- Kinases, which catalyse transfer of phosphate group, e.g. phosphorylation reactions.

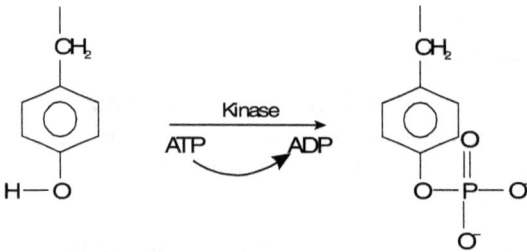

- Acyl transferases, which catalyse transfer of acyl groups, e.g. acetyl transfer reactions.

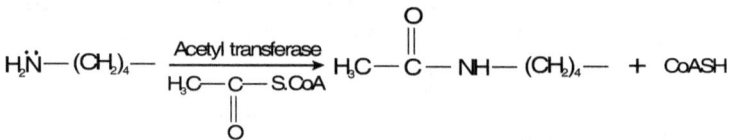

- Methylases which catalyse transfer of methyl group, e.g. methylation reactions.

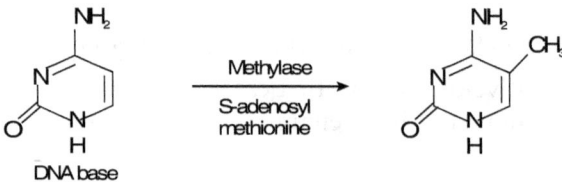

- Other enzymes like transketolases, transaldolases and transmethylases also belong to this group.

Hydrolases

These enzymes catalyse hydrolytic reactions. They are further classified based on the type of bond cleaved. They are:

- Peptidases catalyse hydrolysis of peptide bonds.
- Esterases cause hydrolysis of ester linkages.
- Glycosidases hydrolyse glycosidic bonds.
- Phosphatases hydrolyse phosphoric acid esters.
- Deamidases hydrolyse amides.
- Deaminases hydrolyse amines.
- Cellulases hydrolyse cellulose to glucose.
- Amylases hydrolyse starch to maltose.
- Proteases hydrolyse proteins to amino acids.
- Lipases hydrolyse fats to glycerol and fatty acids.
- Nucleases hydrolyse ribonucleic acid (RNA) and deoxyribonucleic acid (DNA) into smaller soluble molecules.

Lyases

These enzymes catalyse removal of groups non-hydrolytically from their substrates with the concomitant formation of double bonds or alternatively add new groups across the double bonds. They are further classified as:

- Decarboxylases
- Aldolases
- Dehydratases

Isomerases

Isomers catalyse geometric or structural changes within one molecule. According to the type of isomerism, they may be called racemases, epimerases, *cis-trans* isomerases, isomerases, tautomerases, mutases or cylcoisomerases, etc.

- Racemases catalyse racemization reactions. These enzymes catalyse inversion of the configuration around an asymmetric carbon in a substrate having one (racemase) or more (epimerase) centre(s) of asymmetry, as for example, hydroxyproline and ribulose phosphate respectively.

- Alanine racemase catalyses the interconversion of alanine enantiomers, and this represents the first step involved in bacterial cell wall biosynthesis.

- Mutases are also known as intramolecular transferases. They catalyse the transfer of groups within the molecule (intramolecularly).

- Epimerases catalyse changes to epimers, e.g. epimerase interconverts the stereoisomers ribulose 5-phosphate and xylulose 5-phosphate. This involves change in configuration at C3.

- Isomerase converts the ketose ribulose 5-phosphate to the aldose ribose 5-phosphate. This is an example of functional isomerism.

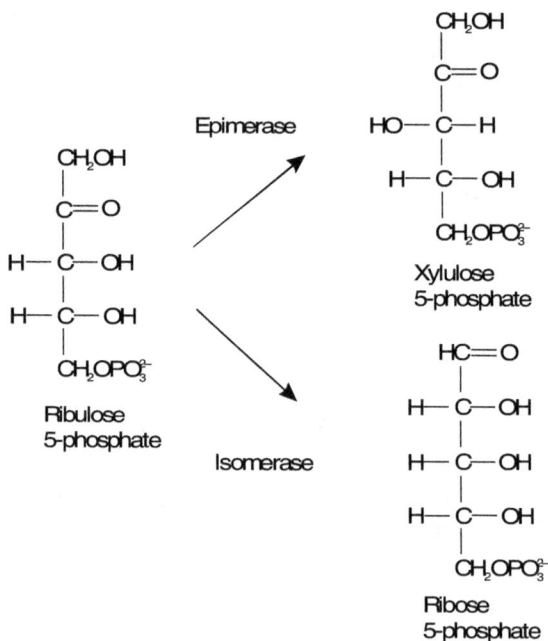

Activity Glucose isomerase catalyses the conversion of glucose into fructose.

The student is advised to draw the reaction and analyse this reaction.

• *cis-trans* isomerses (rotamases) bring about interconversion between the geometrical isomers (*cis* and *trans*).

For example, the role of rotamase and protein disulphide isomerase (PDI) are illustrated below.

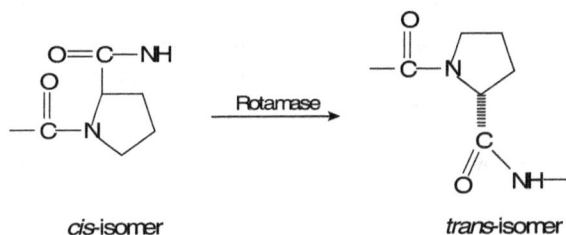

cis-isomer   *trans*-isomer

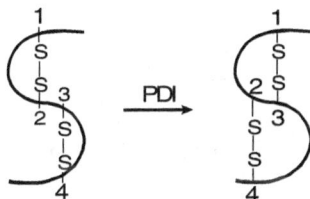

The reactions catalysed by these two enzymes can assist a peptide chain to fold into a correct three-dimensional structure.

Ligases

These enzymes catalyse joining of two molecules along with splitting of pyrophosphate linkage of ATP molecule without hydrolysis or oxidation. This reaction is usually accompanied by the consumption of high energy and it is therefore an ATP-dependent process. By this, formation of bonds between C—C, C—N, C—S results. These are synthesis reactions and the enzymes have been known for years as synthetases. Ligases take part in many of the steps involved in the synthesis of macromolecules such as proteins and many other compounds used as intermediates in nucleic acid biosynthesis.

When hydrogen peroxide is applied on a fresh wound, it produces a white "foam". Similarly, foaming is observed when a piece of raw liver is treated with hydrogen peroxide. Instantaneously the preparation foams up. This is a great demonstration.

The explanation for foaming is that in most of the cells of our body, especially liver cells, there is an enzyme called catalase-peroxidase which catalyses the following reaction:

$$2H_2O_2 \rightarrow 2H_2O + O_2$$

This reaction takes place very slowly at room temperature, in the presence of light but occurs more rapidly in the presence of the enzyme. Why do cells have these enzymes to break down hydrogen peroxide? In cells, these enzymes break down hydrogen peroxide produced by some of the cell's metabolic reactions such as the transfer of hydrogen from organic compounds like formaldehyde and ethyl alcohol to oxygen. In some parasites, this set of reactions serves to eliminate excess oxygen.

COFACTORS AND COENZYMES

Many enzymes require activating molecules in order to function as biocatalysts. For example, enzyme alcohol dehydrogenase requires organic molecules or various metals to function as biocatalysts. The biocatalysed reactions occur on certain sites of catalysts known as *active sites*. The activating groups occupy these active sites and trigger the catalytic action of the enzyme. The activating groups remain attached to the enzyme at the active site during the course of the reaction, perform the reaction, leave the active site to another locale, recharge, and return to their respective active site to continue the process. It is for this reason that these activating groups are also referred to as co-substrates. Many coenzymes are closely related to vitamins and some vitamins are known for their function as coenzymes. Vitamin is the main component of a coenzyme endowed with biocatalytic functions. Higher animals are unable to synthesize their own vitamins but bacteria can. Therefore, the nutritional requirements—vitamins for human beings—are derived from the diet or as supplements. These coenzymes are required only in small amounts as they are regenerated physiologically, thus massive doses of supplements are not essential.

The enzyme-activating moieties are either organic molecules or inorganic metal ions. If it is an organic molecule then the companion is often called a coenzyme and if inorganic, it is called cofactor. For instance, zinc is a cofactor for the enzyme alcohol dehydrogenase. Many vitamins of the dietary components are coenzymes. For example, the group of vitamins called the "B complex" are all important coenzymes.

- Vitamin B_2, riboflavin, is an important coenzyme in cellular respiration.

- Vitamin B_5, pantothenic acid, is called "Coenzyme A" in cellular respiration.

- Vitamin B_1, thiamine pyrophosphate, is a universal coenzyme. It is responsible for the decarboxylation of keto acids like pyruvic acid to the centrally important acetyl coenzyme A. The latter serves as the primary unit in the biosynthesis of several natural products.

The terms coenzyme and cofactor are used synonymously, but there is a subtle difference. The term "cofactor" also refers to either inorganic ions (part of essential micronutrients) or atoms or to organic molecules such as vitamins that work with enzymes and are not generally chemically altered during the reaction. In a few cases, there may be oxidation or reduction of such metal ions. These metal cofactors required in small amounts are generally available from the diet. They are bound tightly to the enzyme generally covalently and cannot be dissociated from the enzyme. Some of the important metal cofactors are described in Table 16.1.

The word "coenzyme" refers just to the organic molecules. Coenzymes may be covalently bound to the protein part—the apoenzyme and is dissociable from the enzyme. They generally act as carriers of specific functional groups. For this, they must exist in two forms; one form is converted to another during the catalytic action and the latter form is reconverted to the original form by another reaction. These two reactions may or may not follow each other.

Table 16.1   Metal cofactors

Cofactor	Enzyme(s)
Zn^{2+}T	Alcohol dehydrogenase, carbonic anhydrase, carboxypeptidases, some aminoacyl-tRNA synthetase
Mg^{2+}	Alcohol dehydrogenase, carbonic anhydrase, carboxypeptidases, some aminoacyl-tRNA synthetases
Mg^{2+}	Phosphohydrolases, phosphotransferases
Mn^{2+}	Arginase, phosphotransferases
Fe^{2+} or Fe^{3+}	Cytochromes, catalase, peroxidase, ferredoxin
Cu^{2+} or Cu^{1+}	Tyrosinase, cytochrome oxidase
K^+	Membrane ATPase
Na^+	Acetyl or other acyl group transfer, fatty acid synthesis and oxidation

A few coenzymes are nicotinamide nucleotides, e.g. NAD, NADP, biotin, ATP, etc.

Cellular enzymes are neither sporadic nor disorganized. All cells, including bacteria, have enzyme systems in which the enzymes work in an orderly sequence until a particular series of reactions has been completed. Many enzyme systems act in a kind of chain reaction; the product of one reaction becomes the substrate for the next reaction in the series and so on.

MECHANISM OF ENZYME CATALYSIS

All chemical reactions require some amount of energy to commence. This energy is called activation energy. The way enzymes operate is by effectively lowering the amount of activation energy required for a chemical reaction to start. In certain reactions enzymes might weaken a covalent bond within a substrate molecule. In other cases, this lowering of activation energy takes place because the enzyme holds the substrate molecules in a particular position that increases the chances of the molecules to react.

Energy hill diagrams help to visualize the effect of enzymes on activation energy. Figure 16.1 shows time on the horizontal axis and the amount of reactants energy involved in a chemical reaction on the vertical axis. The energy diagram clearly demonstrates that without the enzyme, much more activation energy is required for a chemical reaction to occur. Enzymes bind temporarily to one or more of the reactants of the reaction they catalyse. In doing so, they lower the amount of activation energy needed and thus speed up the reaction.

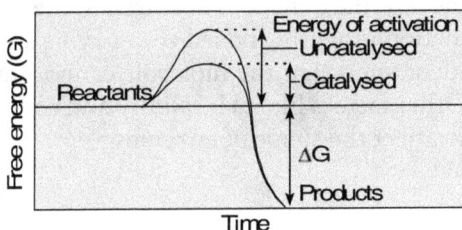

Figure 16.1   Enzymes and activation energy

A few enzyme-catalysed biochemical reactions include:

- The enzyme catalase, catalyses the decomposition of hydrogen peroxide into water and oxygen.

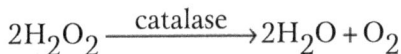

$$2H_2O_2 \xrightarrow{\text{catalase}} 2H_2O + O_2$$

One molecule of catalase can break 40 million molecules of hydrogen peroxide each second.

- Another enzyme carbonic anhydrase, found in red blood cells, catalyses the reaction

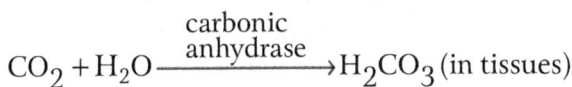

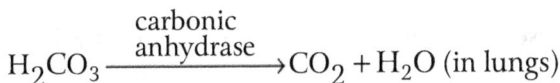

$$CO_2 + H_2O \xrightarrow{\text{carbonic anhydrase}} H_2CO_3 \text{ (in tissues)}$$

$$H_2CO_3 \xrightarrow{\text{carbonic anhydrase}} CO_2 + H_2O \text{ (in lungs)}$$

It enables red blood cells to transport carbon dioxide from the tissues to the lungs. One molecule of carbonic anhydrase can process 1 million molecules of CO_2 each second.

Factors Affecting Enzyme Activity

Enzymes operate best in a relatively narrow range of conditions. Many of the enzymes in our body work best at body temperature. At significantly lower temperatures, the substrate molecules do not possess enough kinetic energy for the reaction to proceed even in the presence of the enzyme. At temperatures significantly higher than normal, the enzyme will not work well because the kinetic energy from the molecules in the solution containing the enzyme is so high, that the enzyme's shape is distorted to the point that it is not able to function properly. Indeed the enzyme's structure is so disrupted or denatured that the molecule cannot return to its original shape. Other factors like pH, salinity and concentrations of other ions also affect the shape of enzymes.

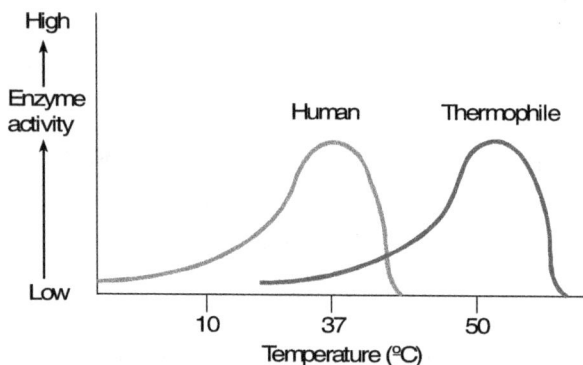

Figure 16.2   The enzymatic activity based on the effect of temperature

Effect of temperature and pH on enzyme activity   Figure 16.2 shows hypothetical relationships between temperature and enzyme activity for a human and for thermophilic (heat-loving) bacteria. Human enzymes work best at temperature (37°C) while the thermophilic bacteria's enzyme works best at a higher temperature. Some of these thermophilic organisms live even at the boiling point of water.

The effect of pH on the activity of enzyme is not surprising considering the importance of:

- Tertiary structure, i.e., shape in enzyme function.

- Non-covalent forces, e.g. ionic interactions and hydrogen bonds in determining the shape.

The enzyme protease pepsin works best at a pH of 1–2 (found in the stomach) while the protease trypsin is inactive at such a low pH but highly active at a pH of 8 (found in the small intestine as the bicarbonate of the pancreatic fluid neutralizes the arriving stomach contents).

Changes in pH alter the state of ionization of charged amino acids that may play a crucial role in substrate binding and/or the catalytic action itself. Hydrogen bonds are easily disrupted by increasing the temperature. This in turn, disrupts the shape of the enzyme thus diminishing its affinity for its substrate.

ENZYME ACTION

The operation of enzymes depends on the association between the enzyme and the reactant molecule or molecules participating in the reaction. This is shown in a simple fashion here.

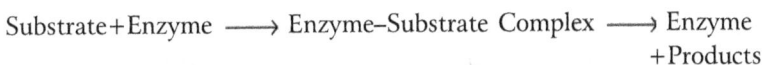

Substrate+Enzyme $\longrightarrow$ Enzyme–Substrate Complex $\longrightarrow$ Enzyme +Products

Enzyme–substrate complex The enzyme–substrate complex is a transition state when the substrates of a chemical reaction are bound to the enzyme. The forces that hold the enzyme and its substrate are non-covalent, an assortment of hydrogen bonds, ionic interactions and hydrophobic interactions.

Active site The area on the enzyme, where the substrate or substrates attach/bind to is called the active site. Enzymes are usually very large proteins and the active site is just a small region of the enzyme molecule. It was thought that the substrate and the active site had to have an exact fit for the enzyme to operate called the "lock and key" model of enzyme activity. The latest information on the model of how enzymes operate is called the "induced fit" model. The idea is that the enzyme's active site does not have the exact shape of the substrate, but the substrate brings about or induces a change in the shape of the active site. A good analogy is what happens when one tries to catch a softball. One need not hold the hand rigidly in the shape that best fits the ball, but can alter the shape of the hand to accommodate the ball.

Regulation of Enzyme Activity

Enzymes may have other binding sites, called control sites, which interact with other molecules. These molecules may improve or interfere with the enzyme's ability to recognize and bind to its substrate. Coenzymes may assist in the functioning of enzymes. They work with the enzyme and may serve as hydrogen or electron donors or acceptors. They may help to position the substrate molecules in ways that facilitate the initiation of reactions. Many enzymes are allosteric enzymes (allo = other; steric = site).

Often allosteric enzymes are located at metabolic branch points and are called branch-point enzymes. They can be inhibited by the presence of end product molecules—the molecules that are the end product of the reaction or of a series of related reactions. These end product inhibitors can "turn off" a whole metabolic pathway by preventing substrate molecules from entering the first reaction of a pathway. End product inhibition is an important mechanism for negative feedback control of metabolism. It provides a sensitive and responsive means for modulating the effective activity of an enzyme. Modulation of enzyme activity in response to the needs of the cell or the organism is a critical component of the regulation of metabolic activities at the cellular level.

INHIBITORS

Various substances interfere with the operation of enzymes. These substances are called inhibitors. The presence of inhibitors will change the shape of the enzyme, rendering it non-functional. When the inhibitor is removed, the once inactivated enzyme can resume its function. Organophosphates and certain other pesticides operate by inhibiting key enzymes in the nervous system. Both carbon monoxide and oxygen bind the same active site in the haemoglobin molecule and carbon monoxide binds so strongly, that it reduces the oxygen-carrying capacity of haemoglobin. Thus, when one substrate competes another for the same site it is called competitive inhibition.

Just as it is important for enzymes to catalyse biological reactions, so is the ability to control and regulate enzymatic activity. This is the role of small, specific molecules and ions known as **enzyme inhibitors**. Inhibitors are often molecules that are similar in shape to certain substrate molecules and can thus fit the active site of the enzyme that was intended to fit the substrate. Once the inhibitor occupies the active site, however, it does not act to catalyse the reaction as the enzyme would. Instead, it binds up the active site and does not allow any activity there; thus, the reaction is inhibited. Enzymes can also be inhibited, usually negatively, by drugs and toxic agents. Inhibition can even occur because of the enzymes producing too many product molecules. Enzyme inhibitors are classified as reversible/competitive or irreversible/noncompetitive.

Reversible/Competitive Inhibitors

The main characteristic of reversible inhibitors is the fact that they disassociate very quickly from the enzyme substrate after they attach to form an enzyme–inhibitor complex. In a competitive inhibition, one substrate competes with another for the same site. In this **competitive inhibition**, (Figure 16.3) a form of reversible inhibition, an enzyme is unable to form an enzyme–substrate (ES) complex and thus cannot complete the task of catalysing the reaction. Thus, the enzymes can form ES complexes or enzyme–inhibitor (EI) complexes, but not enzyme–substrate–inhibitor (ESI) complexes. In order to form an enzyme—inhibitor complex, many inhibitors take on a shape that is very similar to the substrates. They then bind to the enzyme at the active site, which prevents the substrate from binding at the site. A competitive inhibitor is any compound which closely resembles the chemical structure and molecular geometry of the substrate. The inhibitor competes for the same active site as the substrate molecule. The inhibitor may interact with the enzyme at the active site, but no reaction takes place. The inhibitor is "stuck" on the enzyme and prevents any substrate molecules from reacting with the enzyme. However, a competitive inhibition is usually reversible if sufficient substrate molecules are available to ultimately displace the inhibitor. Therefore, the amount of enzyme inhibition depends upon the inhibitor concentration, substrate concentration, and the relative affinities of the inhibitor and substrate for the active site.

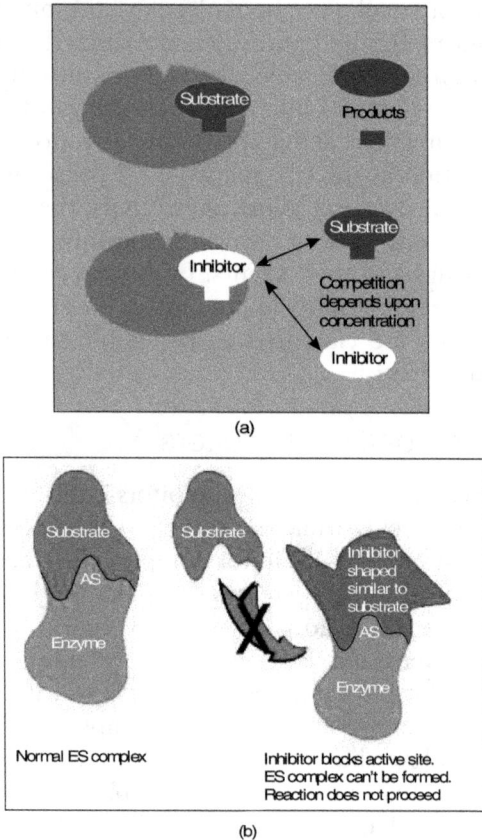

(a)

(b)

Figure 16.3   Mechanism of competitive inhibition

Since ESI complexes cannot be formed, the active site with the inhibitor attached essentially becomes useless. By adding a competitive inhibitor, the rate at which the catalyst can work is reduced by reducing the number of enzyme molecules bound to a substrate. One way to overcome competitive inhibition is to increase the concentration of the substrate.

> If the concentration of inhibitor is less than that of the substrate and the substrate has a higher affinity for the active site, is the enzyme inhibited hugely?
>
> **Answer**   The enzyme is inhibited only a little because the substrate wins the competition.
>
> If concentration of inhibitor is more than that of the substrate, is the enzyme inhibited by large amounts?
>
> **Answer**   The enzyme is inhibited largely because the inhibitor wins the competition.

Ethanol is metabolized in the body by oxidation to acetaldehyde, which is further oxidized to acetic acid by aldehyde oxidase enzymes. Normally, the second reaction is rapid so that acetaldehyde does not accumulate in the body. A drug, **disulfiram (Antabuse)** inhibits the aldehyde oxidase, which causes the accumulation of acetaldehyde with subsequent unpleasant side effects of nausea and vomiting. This drug is used to help people overcome the drinking habit.

Methanol poisoning occurs because methanol is oxidized to formaldehyde and formic acid, which attack the optic nerve causing blindness. Ethanol is given as an antidote for methanol poisoning because ethanol competitively inhibits the oxidation of methanol. Ethanol is oxidized in preference to methanol and consequently, the oxidation of methanol is slowed down so that the toxic by-products do not accumulate.

Irreversible/Noncompetitive Inhibitors

Another type of inhibition is known as **noncompetitive/ irreversible inhibition** (Figure 16.4). In this type of inhibition, the inhibitor binds to the enzyme at a site other than the active site. In doing so, the inhibitor changes the shape of the active site so that the substrate no longer fits it. As a result, substrate cannot come into contact with the active site and the enzymatic action cannot occur. *Noncompetitive inhibition cannot be overcome by increasing the concentration of the substrate.*

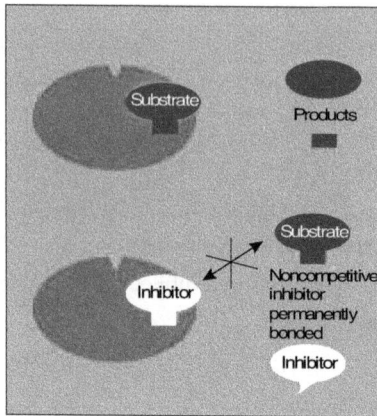

(a)

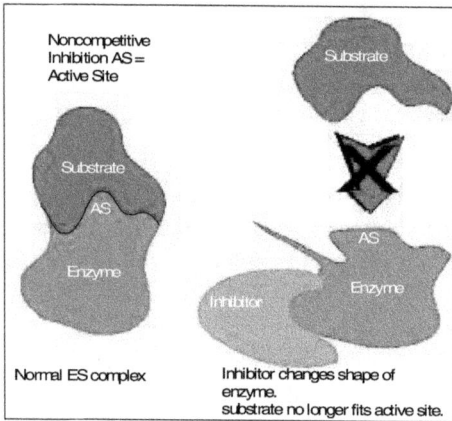

(b)

Figure 16.4   Mechanism of noncompetitive inhibition

An irreversible inhibitor is characterized by the fact that it dissociates very slowly from the enzyme. This is due to the fact that the inhibitor is tightly bound to the enzyme, by covalent or non-covalent bonds. Many toxic agents act as irreversible inhibitors when they enter the body. A noncompetitive inhibitor is a substance that forms strong covalent bonds with an enzyme. Consequently it cannot be displaced by the addition of excess substrate. Therefore, noncompetitive inhibition is irreversible. A noncompetitive inhibitor may be bonded at a spot, near, or remote from the active site.

In any case, the basic structure of the enzyme is modified to the degree that it ceases to work. Since many enzymes contain sulphydryl (–SH), alcohol, or acid groups as part of their active sites, any chemical, which can react with them, acts as a noncompetitive inhibitor. Heavy metals such as Ag^+, Hg^{2+}, Pb^{2+} have strong affinities for –SH groups. Nerve gases such as diisopropylfluorophosphate (DFP) inhibit the active site of acetylcholine esterase by reacting with the hydroxyl group of serine to make an ester. Oxalic and citric acid inhibit blood clotting by forming complexes with calcium ions, which act as the enzyme–metal ion activator.

Studies on inhibitors are useful for the following.

1. For mechanistic studies to learn about how enzymes interact with their substrates.

2. Understanding the role of inhibitors in enzyme regulation.

3. In the action of drugs if they inhibit aberrant biochemical reactions:

   - Penicillin, ampicillin, interfere with the synthesis of bacterial cell walls.

   - Methotrexate: anti-cancer drug that affects DNA metabolism in actively growing cells.

4. In understanding the role of biological toxins.

   - Arsenate mimics phosphate esters in enzyme reactions

   - Amino acid analogs are useful herbicides

Sometimes the product of a chemical reaction involving an enzyme attaches to a secondary site on the enzyme and inhibits the enzyme's ability to continue the reaction. This type of reversible inhibition is called feedback inhibition and it provides cells with a way to regulate the production of various compounds in the cell. Figure 16.5 shows a simple illustration of feedback inhibition for a metabolic pathway involving three compounds and two enzymes, E_1 and E_2. Therefore, inhibition mechanism provides another method for enzyme regulation. This is also called as positive feedback inhibition. It is an effective process to prevent any excess amount of products from being produced. A commonly known toxic agent that acts as an inhibitor is sarin nerve gas. This inhibits the enzyme acetylcholine esterase. Acetylcholine is important in maintaining nerve functioning and control.

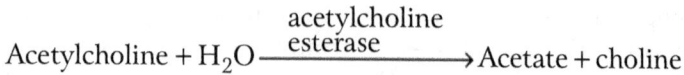

$$\text{Acetylcholine} + H_2O \xrightarrow{\text{acetylcholine esterase}} \text{Acetate} + \text{choline}$$

By inhibition acetylcholine is accumulated at the nerve endings causing blurred vision, profuse sweating, loss of motor function control also leading to paralysis.

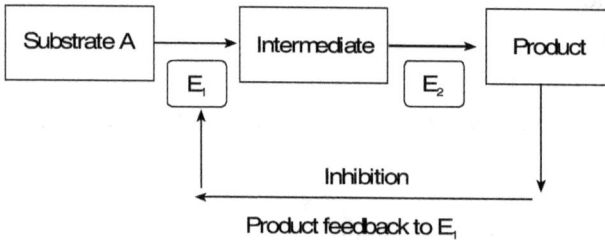

Figure 16.5   Feedback inhibition

Enzymes can also be regulated by covalent modifications such as phosphorylation. Phosphorylation may stimulate or inhibit enzyme activity, depending on the individual enzyme.

Making bonds lowers the energy content of a molecule, and breaking bonds raises it. Since reactions involve both bond breaking and bond making, a reaction's energy barrier is reduced if, in each step, the energy required to break one bond is supplied by making another. This is illustrated by the mechanism of a well-studied reaction, the hydrolysis of a peptide bond by the enzyme chymotrypsin shown in Figure 16.6.

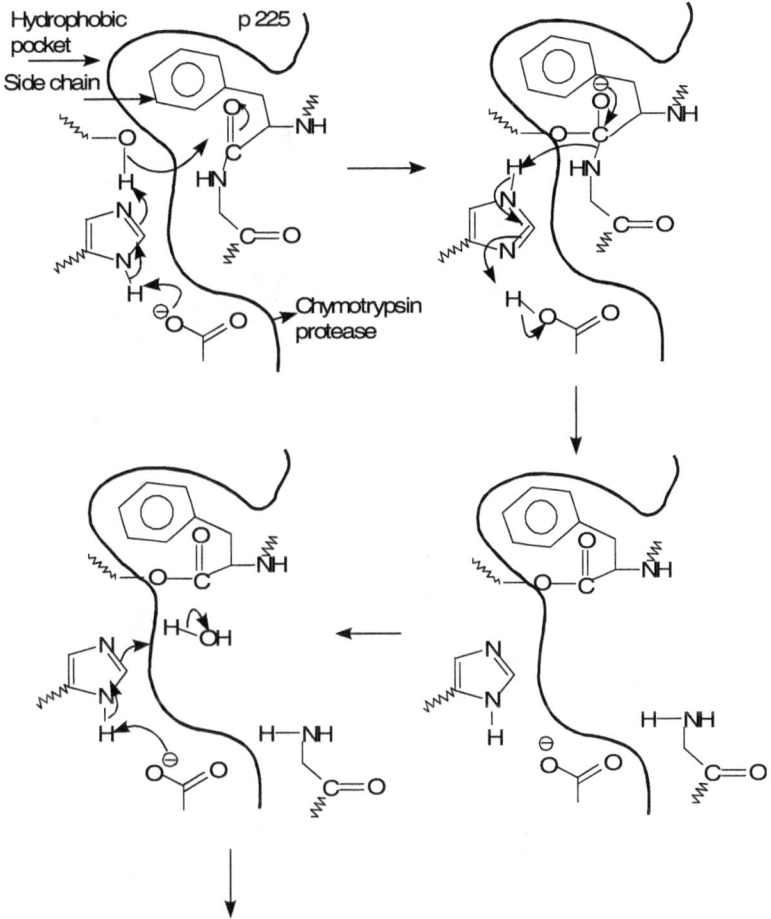

Figure 16.4   Hydrolysis of a peptide bond by chymotrypsin (*Continues*)

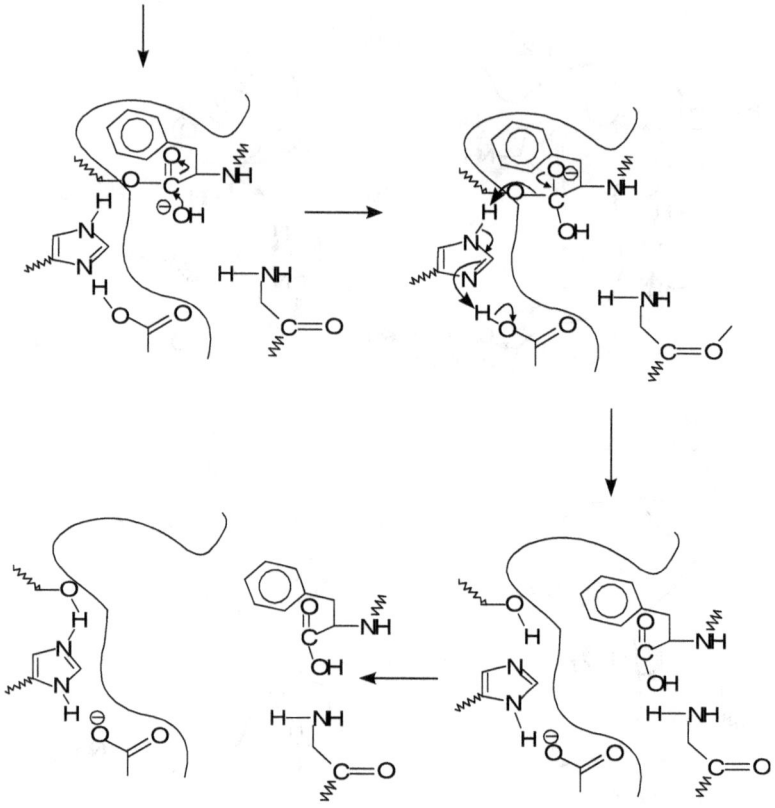

Figure 16.4   Hydrolysis of a peptide bond by chymotrypsin

REVIEW QUESTIONS

1. Blood serves as the principal transport system of the body. Explain.
2. State and explain the composition of blood.
3. What are the non-diffusible components of plasma?
4. What is serum?
5. Why are RBCs considered as the miniature osmometer?
6. What does ESR stand for?
7. What happens when a sample of blood is treated with 0.9% solution of NaCl?
8. What is fragile point and what is its importance?
9. What happens when RBCs are placed in a medium of:
     i. Higher osmotic pressure
     ii. Lower osmotic pressure
11. Do RBCs contain nucleus?
12. Mention the factors that affect the count of RBCs.
13. What is clumping?
14. Discuss the importance of clumping.
15. What is haemagglutination? Explain its significance.
16. Define the following terms
     i. Antigens
     ii. Antibodies
17. Explain the basis of blood grouping.
18. Discuss the four important groups of blood.
19. How is a sample of blood identified for its grouping?
20. In the following figure A, B, AB and O refer to various blood groups. Give the significance of the figure.

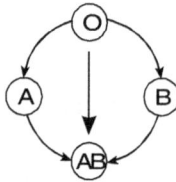

21. Tabulate the various blood groups along with the antigen and antibody.

22. What is Rh factor?

23. Enlighten the importance of Rh factor.

24. The figure given below represents a blood sample belonging to the group O. How can this be concluded on the basis of the figure?

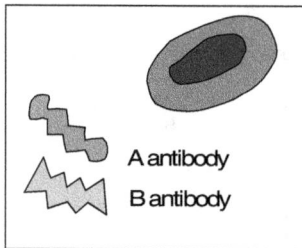

25. What is erythroblastosis foetalis?

26. Discuss blood transfusion.

27. Interpret the following diagram

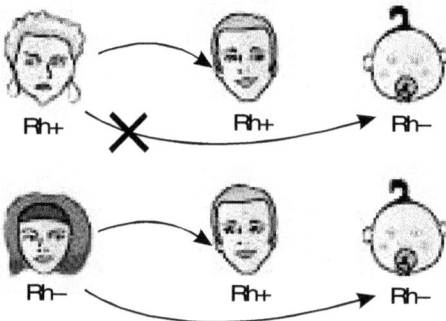

28. Define the following terms:
    i. Blood pressure
    ii. Hypertension
    iii. Hypotension
    iv. Systolic and diastolic pressure

29. What do the numbers 120/80 denote?

30. Name a few drugs that lower the blood pressure.

31. What are the drugs used in the treatment of hypertension?

32. How do the anti-hypertensive drugs act?

33. What is the role of:
    i. Sodium nitroprusside
    ii. Methyldopa
    iii. Reserpine

34. What is digestion?

35. Name the mechanical processes involved in the process of digestion of food after its entry into the stomach?

36. What are the enzymes involved in the process of digestion?

37. In how many stages does digestion occur?

38. Identify the products of the following reaction.
    i. Carbase
    ii. Carbohydrate
    iii. Proteases
    iv. Proteins
    v. Lipase
    vi. Fats

39. State the duration required for the movement of food through various digestive organs.

40. Enumerate the factors involved in emptying the stomach contents.

41. Discuss the essential features of oral digestion, and digestion in the stomach and small intestine.

42. What are the advantages of the low pH of gastric juice?
43. Explain the role of the following enzymes:
    i.   Pepsin
    ii. Exopeptidases
    iii. Renin
    iv. Chymotrypsin
    v.   Endopepdidases
44. How are lipids digested in the stomach?
45. Explain the digestion of proteins in the small intestine.
46. What is bile? Identify its role in digestion.
47. Write on absorption of:
    i.   Carbohydrates
    ii. Fatty acids
    iii. Proteins
48. What does the word hormone denote?
49. Draw the structure of the first discovered hormone and highlight its role.
50. What are the secretions of the thyroid gland?
51. Name the major thyroid hormone and draw its structure.
52. State the physiological functions of thyroid hormones.
53. What are the consequences of thyroid imbalance?
54. What are the causes of hyperthyroidism?
55. Give the sequence of amino acids present in oxytocin.
56. For the following deficiency diseases, identify the required vitamin/s:
    i.   Angular stomatitis
    ii. Beriberi
    iii. Glossitis
    iv. Rickets
    v.   Pellagra
    vi. Scurvy

57. How are vitamins classified?
58. Discuss the structure of insulin.
59. The steroidal hormones are known as secondary hormones. Why?
60. What are micronutrients? State their importance.
61. What are tocopherols?
62. What are the causes of vitamin E deficiency?
63. Where do the following vitamins occur?
    i.   Phylloquinine
    ii. Ascorbic acid
    iii. Riboflavin
64. Name the universal vitamin. Why is it called so?
65. State the importance of vitamin K.
66. Name the electrolytes that are found in our system. Mention their significance.
67. Delineate the role of zinc in our body.
68. Write notes on the biological role of antioxidants.
69. Give the names of certain elements and compounds functioning as antioxidants.
70. What is Wilson's disease?
71. Define enzymes and suggest a suitable classification for enzymes.
72. Give suitable illustrations for the function of the following enzymes.
    i.   Rotamase
    ii. Acetyl transferase
    iii. Kinases
73. What are cofactors and coenzymes?
74. Suggest the enzymes required for the following transformations taking place in a biological system.

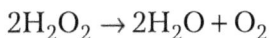

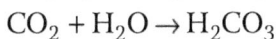

$$2H_2O_2 \rightarrow 2H_2O + O_2$$
$$CO_2 + H_2O \rightarrow H_2CO_3$$

75. Explain the factors affecting the action of enzymes.

76. Interpret the features of the following figure

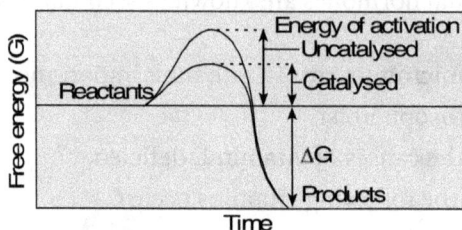

77. The enzyme pepsin works best at a pH of 1–2 but trypsin is inactive at such a low pH. Explain.

78. Interpret the following graph. What conclusions can be drawn about the enzymatic activity of enzymes in humans and thermophiles?

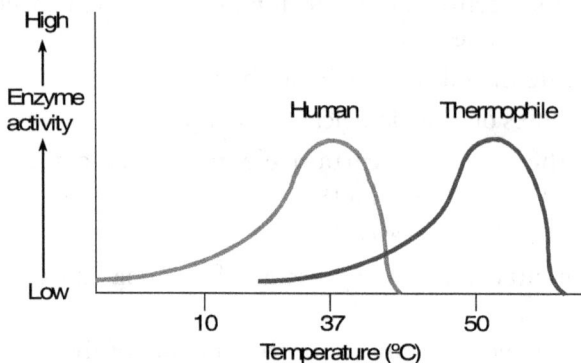

79. Write on enzyme–substrate (ES) complex.

80. What is an active site?

81. How is the activity of an enzyme regulated?

82. Discuss the role of enzyme inhibitors.

83. State the advantages of feedback inhibition.

PART III

INDUSTRIAL
CHEMISTRY

17

DAIRY CHEMISTRY

MILK AND MILK PRODUCTS

Milk can be described as

- an oil-in-water emulsion with the fat globules dispersed in the continuous serum phase.
- a colloidal suspension of casein micelles, globular proteins and lipoprotein particles.
- a solution of lactose, soluble proteins, minerals, vitamins and other components.

When viewing milk under a microscope at low magnification 5×, a uniform turbid liquid is observed. At 500× magnification, spherical droplets of fat, known as fat globules, can be seen (Figure 17.1a). At even higher magnification (50,000×), the casein micelles can be observed (Figure 17.1b). The main structural components of milk are fat globules and casein micelles.

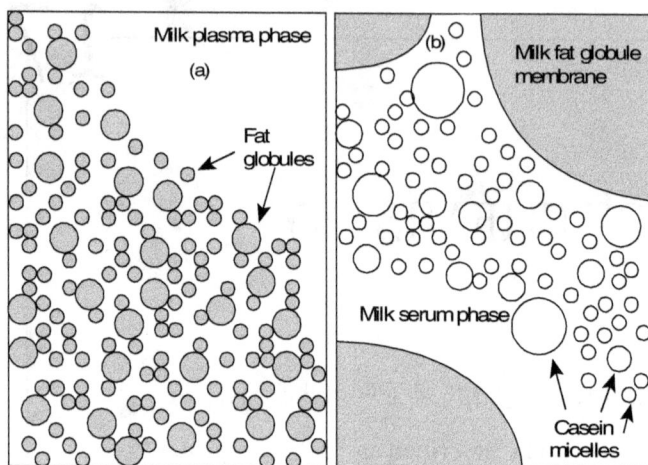

Figure 17.1 Drop of milk observed under a microscope   at
(a) 500 × magnification (b) 50,000 × magnification

COMPOSITION AND STRUCTURE OF MILK

The various components of milk include

- Milk lipids
- Milk proteins
- Caseins
- Whey proteins
- Enzymes
- Lactose
- Vitamins
- Minerals

Milk is a complex food. Many factors can affect the composition of milk such as breed variations, cow-to-cow variations and herd-to-herd variations including management and feed considerations, seasonal variations and geographic variations.

The composition of milk can be approximately given as:

Water 87.3% (range of 85.5%–88.7%)

Milk fat 3.9% (range of 2.4%–5.5%)

Solids-not-fat 8.8% (range of 7.9–10.0%):

Protein 3.25% (3–4% casein)

Lactose 4.6%

Minerals 0.65% (Ca, P, citrate, Mg, K, Na, Zn, Cl, Fe, Cu, sulphate, bicarbonate and many others)

Acids 0.18% (citrate, formate, acetate, lactate, oxalate)

Enzymes (peroxidase, catalase, phosphatase, lipase)

Gases (oxygen, nitrogen)

Vitamins (A, C, D, thiamine, riboflavin, etc.)

The following terms are used to describe milk fractions

Plasma = milk fat (skimmed milk)

Serum = plasma-casein micelles (whey)

Solids-not-fat (SNF) = proteins, lactose, minerals, acids, enzymes, vitamins

Total milk solids = Fat + SNF

MILK LIPIDS

Chemical Properties

The solid content of milk is of economic importance because milk is sold based on fat and solids-not-fat (SNF). Milk fatty acids originate either from microbial activity in the rumen, and transported to the secretory cells via the blood and lymph, or from synthesis in the secretory cells. The main milk lipids constitute a class of compounds called triglycerides, which comprises a glycerol backbone binding up to three different fatty acids. The fatty acids are composed of a hydrocarbon chain and a carboxyl group.

The major fatty acids found in milk are

Long chain

- C14-myristic acid (11%)
- C16-palmitic acid (26%)
- C18-stearic acid (10%)
- C18: 1-oleic acid (20%)

Short chain (11%)

- C4-butyric acid *
- C6-caproic acid
- C8-caprylic acid
- C10-capric acid

Lipid Structure

Triglycerides account for 98.3% of milk fat. The distribution of fatty acids on the triglyceride chain, while there are hundreds of different combinations, is not random. The fatty acid pattern is important for determining the physical properties of the lipids. In general, the SN1 position binds mostly long-chain fatty acids, and the SN3 position binds mostly short-chain and unsaturated fatty acids.

For example:

C4—97% in SN3

C6—84% in SN3

C18—58% in SN1

---

* Butyric fatty acid is specific for milk fat of ruminant animals and is responsible for the rancid flavour when it is cleaved from glycerol by lipase action.

Saturated fatty acids such as myristic, palmitic and stearic acids make up two-third of milk fatty acids. Oleic acid is the most abundant unsaturated fatty acid in milk with one double bond. While the *cis* form of geometric isomer is most commonly found in nature, approximately 5% of all unsaturated bonds are in the *trans* position as a result of rumen hydrogenation.

The small amounts of mono- and diglycerides, and free fatty acids in fresh milk may be a product of early lipolysis or simply incomplete synthesis. Other classes of lipids include phospholipids (0.8%), which are mainly associated with the fat globule membrane, and cholesterol (0.3%), which is mostly located in the fat globule core.

The positional specificity of fatty acid location on the glycerol molecule is designated as SN1, SN2 and SN3. The positional specificity is responsible for the plastic, spreadable nature of butter, for high digestibility of saturated fatty acids, etc.

Physical Properties

The physical properties of milk fat can be summarized as follows:

1. Density at 20°C is 915 kg m^{-3}.
2. Refractive index (589 nm) is 1.462 which decreases with increasing temperature.
3. Solubility of water in fat is 0.14% (w/w) at 20°C and increases with increasing temperature.
4. Thermal conductivity is about 0.17 Jm^{-1}s^{-1}K^{-1} at 20°C.
5. Specific heat at 40°C is about 2.1 kJ kg^{-1}K^{-1}.
6. Electrical conductivity is <10–12 ohm^{-1}cm^{-1}.
7. Dielectric constant is about 3.1.

At room temperature, the lipids are solid, therefore, referred to as "fat" as opposed to "oil" which is liquid at room temperature. The melting points of individual triglycerides range from –75°C for tributyric glycerol to 72°C for tristearin. However, the final melting point of milk fat is at 37°C because higher melting triglycerides dissolve in the liquid fat. This temperature is significant because 37°C is the body temperature of the cow and the milk would need to be liquid at this temperature. *Trans* unsaturation increases melting points. Chains with branches and odd numbered carbon atoms decrease melting points. Crystallization of milk fat largely determines the physical stability of the fat globule and the consistency of high-fat dairy products, but crystal behaviour is also complicated by the wide range of different triglycerides.

Fat Globules

More than 95% of the total milk lipid is in the form of a globule ranging in size from 0.1–15 µm in diameter. These liquid fat droplets are covered by a thin membrane, 8–10 nm in thickness, whose properties are completely different from both milk fat and plasma.

The native fat globule membrane (FGM) comprises the apical plasma membrane of the secretory cell, which continually envelops the lipid droplets as they pass into the lumen. The major components of the native FGM, therefore, are protein and phospholipids. The phospholipids are involved in the oxidation of milk. There may be some rearrangement of the membrane after release into the lumen as amphiphilic substances from the plasma adsorb onto the fat globule and parts of the membrane dissolve into either the globule core or the serum. The FGM decreases the lipid–serum interface to very low values, 1–2.5 mN/m, preventing the globules from immediate flocculation and coalescence, as well as protecting them from enzymatic action. It is well known that if raw milk or cream is left to stand, it will separate. Stokes' Law predicts that fat globules will convert to cream due to the differences in densities between the fat and plasma phases of milk. However, in cold raw milk, creaming takes place faster than what is predicted from this fact alone. IgM, an immunoglobulin in milk, forms a complex with lipoproteins. This complex, known as cryoglobulin, precipitates onto the fat globules and causes flocculation. This is known as cold agglutination. As fat globules cluster, the speed of rising increases and sweeps up the smaller globules with them. The cream layer forms very rapidly, within 20–30 min., in cold milk.

Homogenization of milk prevents this creaming by decreasing the diameter and size distribution of the fat globules, causing the speed of rise to be similar for the majority of globules. Homogenization causes the formation of a recombined membrane, which is much similar in density to the continuous phase. Recombined membranes are very different from native FGM. Processing steps such as homogenization decreases the average diameter of fat globule and significantly increases the surface area.

Some of the native FGM will remain adsorbed, but there is not enough of it to cover the entire newly created surface area. Immediately after disruption of the fat globule, the surface tension raises to a high level of 15 mN/m and amphiphilic molecules in the plasma quickly adsorb to the lipid droplet to lower this value. The adsorbed layers consist mainly of serum proteins and casein micelles.

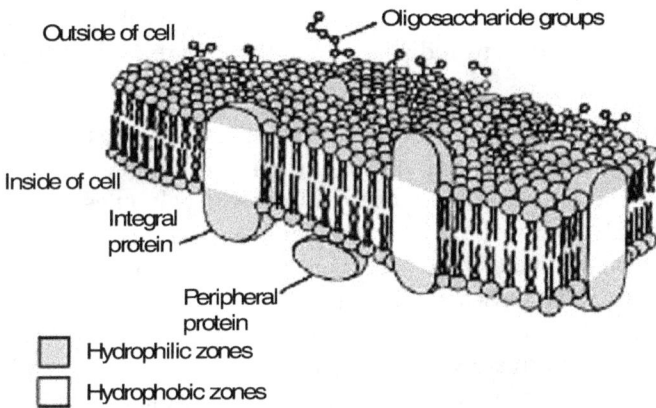

Figure 17.2 Structure of milk fat globule membranes using confocal scanning laser spectroscopy and atomic force microscopy

Figure 17.2 suggests that the phospholipid bilayer serves as the backbone of the membrane, which exists in a fluid phase. Peripheral membrane proteins are partially embedded or loosely attached to the bilayer. Transmembrane proteins extend through the lipid bilayer.

Fat destabilization While homogenization is the principal method for achieving stabilization of the fat emulsion in milk, fat destabilization is necessary for structure formation in butter, whipping cream and ice cream. Fat destabilization refers to the process of clustering and clumping (partial coalescence) of the fat globules, which leads to the development of a continuous internal fat network or matrix structure in the product. Fat destabilization (sometimes "fat agglomeration") is a general term that describes the summation of several different phenomena. These include the following.

Coalescence   It is an irreversible increase in the size of fat globules and a loss of identity of the coalescing globules.

Flocculation   It is a reversible (with minor energy input) agglomeration/clustering of fat globules with no loss of identity of the globules in the floc (the fat globules that flocculate); they can be easily redispersed if they are held together by weak forces, or they might be harder to redisperse and they share part of their interfacial layers.

Partial coalescence   It is an irreversible agglomeration/clustering of fat globules, held together by a combination of fat crystals and liquid fat, and retention of identity of individual globules as long as the crystal structure is maintained (i.e., temperature dependent, once the crystals melt, the cluster coalesces). They usually come together in a shear field, as in whipping, and it is envisioned that the crystals at the surface of the droplets are responsible for causing colliding globules to stick together, while the liquid fat partially flows between them and acts as the "cement". Partial coalescence dominates structure formation in whipped, aerated dairy emulsions, and it should be emphasized that crystals within the emulsion droplets are responsible for its occurrence.

Functional Properties

Like all fats, milk fat provides lubrication. It imparts a creamy feel in the mouth as opposed to a dry texture. Butter flavour is unique and is derived from low levels of short chain fatty acids. If too many short chain fatty acids are hydrolysed (separated) from the triglycerides, the product will taste rancid. Butterfat also acts as a reservoir for other flavours, especially in aged cheese. Fat globules produce a "shortening" effect in cheese by keeping the protein matrix extended to give a soft texture. Fat substitutes are designed to mimic the globular property of milk fat. The spreadable range of butterfat is 16–24°C. Unfortunately butter is not spreadable at refrigeration temperatures. Milk fat provides energy (1 g = 9 cal.), and nutrients (essential fatty acids, fat-soluble vitamins).

MILK PROTEINS

Fractionation of Milk Proteins

The nitrogen content of milk is distributed among caseins (76%), whey proteins (18%), and non-protein nitrogen (NPN) (6%). This does not include the minor proteins that are associated with the FGM. This nitrogen distribution can be determined by the Rowland fractionation method. In this method, precipitation at pH 4.6 separates caseins from whey nitrogen and precipitation with sodium acetate and acetic acid (pH 5.0) separates total proteins from whey and NPN. The concentration of proteins in milk is shown in Table 17.1.

Table 17.1   Concentration of proteins

	Grams/litre	% of total protein
Total protein	33	100
Total caseins	26	79.5
Alpha s1	10	30.6
Alpha s2	2.6	8.0
Beta	9.3	28.4
Kappa	3.3	10.1
Total whey proteins	6.3	19.3
Alpha lactalbumin	1.2	3.7
BSA	0.4	1.2
Beta lactoglobulin	3.2	9.8
Proteose peptone	0.8	2.4
Immunoglobulins	0.7	2.1

CASEINS

The casein content of milk represents about 80% of milk proteins. The major casein fractions are α-(s1) and α-(s2)-caseins, β-casein, and κ-casein. The distinguishing property of all caseins is their low solubility at pH 4.6. The common compositional factor is that caseins are conjugated proteins, mostly with phosphate group(s) esterified to serine residues. These phosphate groups are important to the structure of the casein micelle. Calcium binding by the individual caseins is proportional to the phosphate content.

The conformation of caseins is similar to that of denatured globular proteins. The high number of proline residues in caseins causes particular bending of the protein chain and inhibits the formation of close-packed, ordered secondary structures. Caseins contain no disulphide bonds. Also, the lack of tertiary structure accounts for the stability of caseins against heat denaturation because there is very little structure to unfold. Without a tertiary structure, there is considerable exposure of hydrophobic residues. This results in strong association reactions of the caseins and renders them insoluble in water. Within the group of caseins, there are several distinguishing features based on their charge distribution and sensitivity to calcium precipitation.

α-(s1)-Casein

Its molecular weight is 23,000, having 199 residues including 17 proline residues. Two hydrophobic regions containing all the proline residues are separated by a polar region which contains all but one of eight phosphate groups. It can be precipitated at very low levels of calcium.

α-(s2)-Casein

The molecular weight of β-casein is 25,000, having 207 residues including 10 prolines. It possesses concentrated negative charges near N-terminus and positive charges near C-terminus. It can also be precipitated at very low levels of calcium.

β-Casein

Its molecular weight is 24,000, having 209 residues including 35 prolines. They possess highly charged N-terminal region and a hydrophobic C-terminal region. They are amphiphilic proteins and act like a detergent molecule. Self-association is temperature-dependant and they form a large polymer at 20°C but not at 4°C. They are less sensitive to calcium precipitation.

κ-Casein

The molecular weight is 19,000, having 169 residues including 20 prolines. They are very resistant to calcium precipitation, stabilizing other caseins. Rennet cleavage at the Phe105-Met106 bond eliminates the stabilizing ability, leaving a hydrophobic portion, para κ-casein, and a hydrophilic portion called κ-casein glycomacropeptide (GMP), or more accurately, caseinomacropeptide (CMP).

Structure of Casein Micelle

Most, but not all, of the casein proteins exist in a colloidal particle known as the casein micelle. The biological function of it is to carry large amounts of highly insoluble $Ca_3(PO_4)_2$ to mammalian young ones in liquid form and to form a clot in the stomach for more efficient nutrition. Besides casein protein, calcium and phosphate, the micelle also contains citrate, minor ions, lipase, plasmin enzymes and entrapped milk serum. These micelles are rather porous structures, occupying about 4 mL/g and 6–12% of the total volume fraction of milk. The "casein sub-micelle" model has been the most accepted one for the past several years, and is illustrated and described, but there is no universal acceptance of this model, and there is research evidence to suggest that there is no defined sub-micellar structure to the micelle at all. It is thought that in the sub-micelle model, there are small aggregates of whole casein, containing 10 to 100 casein molecules, called sub-micelles. There are two different kinds of submicelles—with and without κ-casein. These sub-micelles contain a hydrophobic core and a hydrophilic coat, which partly comprises of the polar moieties of κ-casein. The hydrophilic CMP of the κ-casein exists as a flexible hair.

The open model suggests that there are denser and less dense regions within the micelle, but there is no well-defined structure. In this model, calcium phosphate nanoclusters bind caseins and provide for the differences in density within the casein micelle.

Colloidal calcium phosphate (CCP) acts as a binder between the hundreds or even thousands of sub-micelles that form the casein micelle. Binding may be covalent or electrostatic. Sub-micelles rich in κ-casein occupy a surface position, whereas those with less κ-casein are buried in the interior. The resulting hairy layer, at least 7 nm thick, acts to prohibit further aggregation of sub-micelles by steric repulsion. The casein micelles are not static; there are three dynamic equilibria between the micelle and its surroundings:

1. The free casein molecules and sub-micelles
2. The free sub-micelles and micelles
3. The dissolved colloidal calcium and phosphate

The following factors must be considered when assessing the stability of the casein micelle.

- *Role of Ca^{2+}* More than 90% of the calcium content of skimmed milk is associated in some way or other with the casein micelle. The removal of Ca^{2+} leads to reversible dissociation of β-casein without micellar disintegration. The addition of Ca^{2+} leads to aggregation.

- *Hydrogen bonding* This occurs between the individual caseins in the micelle but not much because there is no secondary structure in casein proteins.

- *Disulphide bonds* α-(s1) and β-caseins do not possess cysteine residues. Thus, if any S–S bond occurs within the micelle, they do not contribute to the stabilization.

- *Hydrophobic interactions* Caseins are among the most hydrophobic proteins and they play a definite role in the stability of the micelle. However, the hydrophobic interactions are temperature-sensitive.

- *Electrostatic interactions* Some of the subunit interactions may be the result of ionic bonding, but the overall micellar structure is very loose and open.

- *van der Waals forces* They do not play a prominent role in micellar stability.
- *Steric stabilization* As already noted, the hairy layer interferes with interparticle approach.

There are several other factors that affect the stability of the casein micelle system:

1. *Salt content* It affects the calcium activity in the serum and calcium phosphate content of the micelles.

2. *pH* Lowering the pH leads to dissolution of calcium phosphate until the isoelectric point (pH 4.6), all phosphate is dissolved and the caseins precipitate. At a temperature of 4°C, β-casein begins to dissociate from the micelle, and at 0°C, there is no micellar aggregation; freezing produces a precipitate called cryo-casein.

3. *Heat treatment* In this process whey proteins become adsorbed, altering the behaviour of the micelle. Dehydration by ethanol, for example, leads to aggregation of the micelles. When two or more of these factors are applied together, the effect can also be additive.

Casein Micelle Aggregation

Caseins are able to aggregate if the surface of the micelle is reactive. The Schmidt model further illustrates this. Although the casein micelle is fairly stable, aggregation can be induced in four major ways:

- Enzyme coagulation by chymosin or rennet or by other proteolytic enzymes as in cheese manufacturing
- Acid
- Heat
- Age gelation

Enzyme-coagulation Chymosin, or rennet, is most often used for enzyme coagulation. During the primary stage, rennet cleaves the Phe (105)-Met (106) linkage of κ-casein resulting in the formation of the soluble CMP, which diffuses away from the micelle and *para-κ*-casein and a distinctly hydrophobic peptide remains on the micelle. The patch or reactive site that is left on the micelles after enzymatic cleavage is necessary before aggregation of the *para*-casein micelles can begin. During the secondary stage, the micelles aggregate.

This is due to the loss of steric repulsion of the κ-casein as well as the loss of electrostatic repulsion due to the decrease in pH. As the pH approaches the isoelectric point (pH 4.6), the caseins aggregate. The casein micelles also have a strong tendency to aggregate because of hydrophobic interactions. Calcium assists coagulation by creating isoelectric conditions and by acting as a bridge between micelles. The temperature of coagulation is very important to both the primary and secondary stages. With an increase in temperature up to 40°C, the rate of the rennet reaction increases. During the secondary stage, increased temperatures increase the hydrophobic reaction. The tertiary stage of coagulation involves the rearrangement of micelles after a gel has formed. There is a loss of *para*-casein identity as the milk curd firms and syneresis begins.

Acid-coagulation   Acidification causes the casein micelles to destabilize or aggregate by decreasing their electric charge to that of the isoelectric point. At the same time, the acidity of the medium increases the solubility of minerals so that organic calcium and phosphorus contained in the micelle gradually become soluble in the aqueous phase. Casein micelles disintegrate and casein precipitates. Aggregation occurs as a result of entropically driven hydrophobic interactions.

Heat   At temperatures above the boiling point casein micelles will irreversibly aggregate. On heating, the buffer capacity of milk salts change, carbon dioxide is released, organic acids are produced, and tricalcium phosphate and casein phosphate may be precipitated with the release of hydrogen ions.

Age-gelation   Age-gelation is an aggregation phenomenon that affects shelf-stable, sterilized dairy products, such as concentrated milk and UHT milk products. After weeks to months of storage of these products, there is a sudden sharp increase in viscosity accompanied by visible gelation and irreversible aggregation of the micelles into long chains forming a three-dimensional network.

The actual cause and mechanism is not very clear, however, some theories exist.

Proteolytic breakdown of casein   Bacterial or native plasmin enzymes that are resistant to heat treatment may lead to the formation of a gel. Gelling also takes place through chemical reactions like polymerization of casein and whey proteins and through formation of κ-casein– β-lactoglobulin complexes.

WHEY PROTEINS

The proteins appearing in the supernatant of milk after precipitation at pH 4.6 are collectively called whey proteins. These globular proteins are more water-soluble than caseins and are subject to heat denaturation. Native whey proteins have good gelling and whipping properties. Denaturation increases their water-holding capacity.

The principle fractions are β-lactoglobulin, α-lactalbumin, bovine serum albumin (BSA) and immunoglobulins (Ig).

β-Lactoglobulins

Its molecular weight is 18,000, having 162 residues. This group, including eight genetic variants, comprises approximately half the total whey proteins. β-lactoglobulin has two internal disulphide bonds and one free thiol group. The conformation includes considerable secondary structure and exists naturally as a non-covalent linked dimer. At the isoelectric point (pH 3.5–5.2), the dimers are further associated to octamers, but at pH below 3.4, they are dissociated to monomers.

α-Lactalbumins

Its molecular weight is 14,000, having 123 residues. These proteins contain eight cysteine groups, all involved in internal disulphide bonds, and four tryptophan residues. α-lactalbumin has a highly ordered secondary structure, and a compact, spherical tertiary structure. Thermal denaturation and pH <4.0 results in the release of bound calcium.

ENZYMES

Enzymes are a group of proteins that have the ability to catalyse biochemical reactions and alter the speed of such reactions. The action of enzymes is very specific. Milk contains both indigenous and exogenous enzymes. Exogenous enzymes mainly consist of heat-stable enzymes produced by psychotropic bacteria: lipases and proteinases. Many indigenous enzymes have been isolated from milk. The most significant is the hydrolases group, namely lipoprotein lipase, plasmin and alkaline phosphatase.

Lipoprotein Lipase (LPL)

A lipase enzyme splits fats into glycerol and free fatty acids. This enzyme is found mainly in the plasma in association with casein micelles. The milk fat is protected from its action by the FGM. If the FGM has been damaged, or if certain cofactors are present, the LPL is able to attack the lipoproteins of the FGM. Lipolysis may be caused in this way.

Plasmin

Plasmin is a proteolytic enzyme; it splits proteins. Plasmin attacks both β-casein and α-(s2)-casein. It is heat-stable and is responsible for the development of bitterness in pasteurized milk and UHT processed milk. It may also play a role in the ripening and flavour development of certain cheeses, such as Swiss cheese.

Alkaline Phosphatase

Phosphatase enzymes are able to split specific phosphoric acid esters into phosphoric acid and the related alcohols. Unlike most milk enzymes, it has a pH of 9.8 and temperature optima differing from physiological values. The enzyme is destroyed by minimum pasteurization temperatures; therefore, a phosphatase test can be done to ensure proper pasteurization.

LACTOSE

It is the chief carbohydrate present in milk. Milk is the only substance that contains lactose. When milk is heated lactose reacts with protein, mainly casein and develops a brown colour. This reaction is called as Millard reaction. Reducing sugar reacts with the amino acid lysine and develops a brown colour (the brown colour in khoa, condensed milk and other dairy products is due to Millard reaction). By this reaction, the removal of amino acid lysine, leads to change in the quality of milk protein. The bacterial action of lactose produces lactic acid. Lactic acid is produced in the intestine by microbial action. This checks the growth of putrefactive bacteria and promotes the absorption of minerals particularly calcium. Lactose also increases the permeability of small intestine for calcium ions. Thus milk sugar for its controlled glycaemic effect is preferred as a source of carbohydrate. Lactose is a disaccharide, made up of glucose and galactose and comprises of 4.8–5.2% in milk, 52% in milk SNF, and 70% in whey solids. It is not as sweet as sucrose. When lactose is hydrolysed by β-d-galactosidase (lactase), an enzyme that splits these into monosaccharides, the result is increased sweetness, and depressed freezing point.

One of its most important functions is its utilization as a fermentation substrate. Lactic acid bacteria produce lactic acid from lactose, which is the beginning of many fermented dairy products. Because of their ability to metabolize lactose, they have a competitive advantage over many pathogenic and spoilage organisms. People suffering from lactose intolerance lack the lactase enzyme, hence they cannot digest lactose, or dairy products containing lactose. In addition to lactose, fresh milk contains other carbohydrates in smaller amounts, including glucose, galactose, and oligosaccharides. Lactose being relatively insoluble causes problems (discussed later in this chapter) in many dairy products like ice creams, and sweetened condensed milk.

Due to its role in nature, milk is in a liquid form. This may seem curious if one takes into consideration the fact that milk has less water than most fruits and vegetables. Water content of milk is dependent upon the synthesis of lactose. Without some water in the milk, milk would be a viscous secretion composed mostly of lipid and protein.

VITAMINS

Vitamins are organic substances essential for many life processes. Milk includes fat-soluble vitamins A, D, E, and K. Vitamin A is derived from retinol and β-carotene. Because milk is an important source of dietary vitamin A, fat-reduced products which have lost vitamin A along with the fat, are required to supplement the product with vitamin A.

Milk is also an important source of dietary water-soluble vitamins:

- B_1-thiamine
- B_2-riboflavin
- B_6-pyridoxine
- B_{12}-cyanocobalamin
- Niacin
- Pantothenic acid

There is also a small amount of vitamin C (ascorbic acid) present in raw milk but it is very heat-labile and is easily destroyed by pasteurization. Table 17.2 shows the vitamin content of fresh milk.

Table 17.2   The vitamin content of fresh milk

Vitamin	Contents per L
A (μg RE)	400
D (IU)	40
E (μg)	1000
K (μg)	50
B_1 (μg)	450
B_2 (μg)	1750
Niacin (μg)	900
B_6 (μg)	500
Pantothenic acid (μg)	3500
Biotin (μg)	35
Folic acid (μg)	55
B_{12} (μg)	4.5
C (mg)	20

MINERALS

All the 22 minerals essential to the human diet are present in milk. Table 17.3 shows the mineral content in fresh milk.

These include three families of salts

1. Sodium (Na), potassium (K) and chloride (Cl). These free ions are negatively correlated to lactose to maintain osmotic equilibrium of milk with blood.
2. Calcium (Ca), magnesium (Mg), inorganic phosphorous (P(i)) and citrate. This group consists of 2/3 Ca, 1/3 Mg, 1/2 P(i), and less than 1/10 citrate in colloidal (non-diffusible) form present in the casein micelle.
3. Diffusible salts of Ca, Mg, citrate, Ca^{2+}, and $HPO4^{2-}$. These salts are pH dependent and contribute to the overall acid–base equilibrium of milk.

Table 17.3   The mineral content of fresh milk

Mineral	Content per L
Sodium (mg)	350–900
Potassium (mg)	1100–1700
Chloride (mg)	900–1100
Calcium (mg)	1100–1300
Magnesium (mg)	90–140
Phosphorus (mg)	900–1000
Iron (µg)	300–600
Zinc (µg)	2000–6000
Copper (µg)	100–600
Manganese (µg)	20–50
Iodine (µg)	260
Fluoride (µg)	30–220
Selenium (µg)	5–67
Cobalt (µg)	0.5–1.3
Chromium (µg)	8–13

(*Contd.*)

Table 17.3   (Continued)

Mineral	Content per L
Molybdenum (μg)	18–120
Nickel (μg)	0–50
Silicon (μg)	750–7000
Vanadium (μg)	tr–310
Tin (μg)	40–500
Arsenic (μg)	20–60

Dairy foods are the major sources of calcium. This in addition to the significant amount of mineral content has the calcium phosphorus ratio as 1.2:1—the most favourable for the bone growth and development. The dairy products also contain other nutrients such as vitamin D and lactose favouring calcium absorption. Thus, calcium requirement cannot be met without the consumption of milk.

PROPERTIES OF MILK

Flavour and Aroma

The milk flavour is due to milk fat in the serum. The aroma of fresh milk is produced by the compounds such as acetone, acetaldehyde, dimethyl sulphide and short chain fatty acids.

DENSITY OF MILK

The density of milk (g/L) changes with temperature.

$$\text{Density of milk} = 1.003073 - 0.000179t - 0.000368F + 0.00374N$$

where,

t = temperature in degree Celsius,

F = percent fat and

N = percent non-fat solids.

Specific gravity is the relation between the mass of a given volume of any substance and that of an equal volume of water at the same temperature. Since 1 ml of water at 4°C weighs 1 g, the mass of any material expressed in g/ml and its specific gravity (both at 4°C) will have the same numerical value. The specific gravity of milk averages 1.032, i.e., at 4°C, 1 ml of milk weighs 1.032 g. Since the mass of a given volume of water at a given temperature is known, the volume of a given mass, or the mass of a given volume of milk, cream, skimmed milk, etc., can be calculated from its specific gravity. For example, one litre of water at 4°C has a mass of 1 kg, and since the average specific gravity of milk is 1.032, one litre of average milk will have a mass of 1.032 kg.

Density, the mass of a certain quantity of material divided by its volume, is dependent on the following:

- Temperature at the time of measurement
- Temperature history of the material
- Composition of the material (especially the fat content)
- Inclusion of air (a complication with more viscous products)

The density of milk and milk products is used for the following:

- To convert volume into mass and vice versa
- To estimate the solids content
- To calculate the physical properties (e.g. kinematic viscosity)

The density of milk normally varies between 1.028 and 1.034 kilograms per litre. The density depends on the content of water, fat and dried matter. If fat is removed from the milk, the density will go up. If the milk contains extraneous water, the density will go down. This fact can be used as an indicator for adulteration. The lactometer test is designed to detect the change in density of such adulterated milk. Carried out together with the Gerber butterfat test, it enables the milk processor to calculate the milk total solids (% TS) and solids-not-fat (SNF). In normal milk, SNF should not be below 8.5%.

Experiment to Measure the Density of Milk

The apparatus used to determine the density of milk is called lactometer.

Apparatus

Lactometer is a hydrometer (a device for measuring specific gravity) adapted to the normal range of the specific gravity of milk. It is usually calibrated to read in lactometer degrees (L) rather than specific gravity. The relationship between the two is:

$$(L/1000) + 1 = \text{specific gravity (sp.gr.)}$$

Thus, if L = 31, specific gravity = 1.031. The apparatus consists of a tall, wide, glass or plastic cylinder and a thermometer.

Procedure

1. Heat the sample of milk to 40°C and hold for 5 minutes. This is to get all the fat into a liquid state since crystalline fat has a very different density to liquid fat. After 5 minutes, cool the milk to 20°C.

2. Mix the milk sample thoroughly but gently. (Care is taken to avoid vigorous shaking, else air bubbles could be incorporated which affects the result.)

3. Fill the cylinder sufficiently with milk such that it overflows when the lactometer is inserted.

4. Holding the lactometer by the tip, lower it gently into the milk. Do not let go until it is almost in equilibrium.

5. Allow the lactometer to float freely until it reaches equilibrium. Then read the lactometer at the top of the meniscus. Immediately, read the temperature of the milk. This should be 20°C. If the temperature of the milk is between 17 and 24°C, the following correction factors are used to determine L:

Temperature (°C)	17	18	19	20	21	22	23	24
Correction	−0.7	−0.5	−0.3	−	+0.3	+0.5	0.8	1.1

For example, if the lactometer reading is 30.5 and the temperature is 23°C,

$$\text{Corrected lactometer} = L_c = 30.5 + 0.8 = 31.3$$

Calculations

All calculations always use L_c, the corrected lactometer reading. To calculate the specific gravity, divide the corrected lactometer reading by 1000 and add 1.

In the above example,

$$\text{Sp.gr.} = (31.3)/1000 + 1 = 1.0313$$

Determination of total solids (TS) and solids-not-fat (SNF) in milk

The total solids content of milk is the total amount of material dispersed in the aqueous phase and

$$\text{SNF} = \text{TS} - \%\ \text{fat}$$

the only accurate way to determine TS is by evaporating the water from an accurately weighed sample. However, TS can be estimated from the corrected lactometer reading. The results are not likely to be very accurate because specific gravity is due to water, material less dense than water (fat) and material denser than water (SNF). Therefore, milk with high fat and SNF contents could have the same specific gravity as milk with low fat and low SNF contents.

$$\text{TS} = (L_c)/4 + (1.22 \times \text{fat\%}) + 0.72$$

$$\text{SNF} = \text{TS} - \text{fat\%}$$

or

$$\text{SNF} = L_c/4 + (0.22 \times \text{fat\%}) + 0.72 - \text{fat\%}$$

It should be noted that the relationship between L_c and TS varies with milk composition. The above formulae are called the Richmond formulae.

Table 17.4 gives the density of various fluid dairy products as a function of fat and solids-not-fat (SNF) composition.

Table 17.4 Density of various fluid dairy products as a function of fat and SNF

Product	Composition		Density (kg/L) at:			
	Fat(%)	SNF(%)	4.4°C	10.0°C	20.0°C	38.9°C
Producer milk	4.00	8.95	1.035	1.033	1.030	1.023
Homogenized milk	3.60	8.60	1.033	1.032	1.029	1.022
Skimmed milk/kg	0.02	8.90	1.036	1.035	1.033	1.026
Fortified skim	0.02	10.15	1.041	1.040	1.038	1.031
Half and half	12.25	7.75	1.027	1.025	1.020	1.010
Half and half, fort.	11.30	8.90	1.031	1.030	1.024	1.014
Light cream	20.00	7.20	1.021	1.018	1.012	1.000
Heavy cream	36.60	5.55	1.008	1.005	0.994	0.978

VISCOSITY

Viscosity of milk and milk products is important in determining the following:

- Rate of creaming
- Rates of mass and heat transfer
- Flow conditions in dairy processes

Milk and skimmed milk, excepting cooled raw milk, exhibit Newtonian behaviour, in which the viscosity is independent of the rate of shear. The viscosity of these products depends on the following.

Temperature Cooler temperatures increase viscosity. This is due to the increased voluminosity of casein micelles. Temperatures above 65°C increase viscosity due to the denaturation of whey proteins.

pH An increase or decrease in pH of milk also causes an increase in casein micelle voluminosity. Cooled raw milk and cream exhibit non-Newtonian behaviour in which the viscosity is dependent on the shear rate. Agitation may cause partial coalescence of the fat globules (partial churning), which increases viscosity. Fat globules that have undergone cold agglutination may be dispersed due to agitation, causing a decrease in viscosity.

Freezing Point

Freezing point is a colligative property, which is determined by the molarity of solutes rather than by the percentage by weight or volume. In the dairy industry, freezing point is mainly used to determine added water. It can also be used to determine lactose content in milk, to estimate whey powder contents in skimmed milk powder, and to determine water activity of cheese. The freezing point of milk is usually in the range of $-0.512°$ to $-0.550°$ C with an average of about $-0.522°$ C. Correct interpretation of freezing point data with respect to added water depends on a good understanding of the factors affecting freezing point depression. With respect to interpretation of freezing points for added water determination, the most significant variables are the nutritional status of the herd and the access to water. Underfeeding causes increased freezing points. Large temporary increases in freezing point occur after consumption of large amounts of water because milk is iso-osmotic with blood. The primary sources of non-intentional added water in milk are residual rinse water and condensation in the milking system.

Acid–Base Equilibria

Both titratable acidity and pH are used to measure milk acidity. The pH of milk at 25°C normally varies within a relatively narrow range of 6.5–6.7. The normal range for titratable acidity of herd milk is 13–20 mmol/L. There are many components in milk, which provide a buffering action. The major buffering groups of milk are caseins and phosphate.

Optical Properties

Optical properties provide the basis for many rapid, indirect methods of analysis such as proximate analysis by infrared absorbency or light scattering. Optical properties also determine the appearance of milk and milk products. Light scattering by fat globules and casein micelles causes milk to appear turbid and opaque. Light scattering occurs when the wavelength of light is near the same magnitude as the particle. Thus, smaller particles scatter light of shorter wavelengths. Skimmed milk appears slightly blue because casein micelles scatter the shorter wavelengths of visible light (blue) more than the red. The carotenoid precursor of vitamin A, β-carotene, contained in milk fat, is responsible for the "creamy" colour of milk. Riboflavin imparts a greenish colour to whey. Refractive index (RI) is normally determined at 20°C with the D-line of the sodium spectrum. The refractive index of milk is 1.3440–1.3485 and can be used to estimate total solids.

EFFECT OF HEAT ON MILK

On heating the whey, proteins lactalbumin and lactoglobulin become insoluble and precipitate. Lactalbumin starts coagulating at 66°C (150°F). The amount of coagulation increases with the increasing temperature and duration of heating. This appears initially as small particles and collects at the bottom and sides of the pan in which the milk is heated. This is also called as scorching of milk. However, the abundant milk protein casein does not coagulate at the usual temperature and duration of heating used in the food preparation. But it coagulates when heated at or above 100°C for various periods, e.g. at 100°C casein coagulates when heated for 12 hours, at 135°C for 1 hour, at 155°C for 3 minutes. The resistance of casein to heat is found to be due to its combination with definite amounts of various minerals like Ca^{2+}, Mg^{2+}, phosphate and citrate present in the milk. Duration required for casein to coagulate is shorter when the concentration of casein is higher than the normal fluid milk, e.g. during sterilization of canned evaporated milk. Thus, steps are to be taken to prevent the casein coagulation. This is commonly done by pre-warming milk prior to the sterilization. When milk is heated albumin forms a flocculant that settles on the sides of the container. Coagulation of milk protein is enhanced by increasing the acidity.

During boiling of milk in open containers, a layer of fat otherwise called as scum forms, which is the result of breaking down of the films of proteins surrounding the fat globules. The fat globules then coalesce to a fat layer. The fat layer scum gets toughened with the increase in temperature of heating. The scum formed rises up when pressure develops under the scum layer resulting in ceasing of boiling of milk. The formation of scum can be prevented by heating the milk in covered pans, diluting the milk, using milk boilers, and stirring the milk while heating.

When milk is heated, its acidity initially decreases due to the release of dissolved carbon dioxide and then increases because of the liberation of hydrogen ions when calcium and phosphate form insoluble compounds. A balance between these two maintains the pH of the milk. During heating, loss of mineral contents is also observed particularly, iodine loss is noticed. The dispersion of calcium phosphate in milk is decreased by heating as part of calcium is precipitated. Some amount of calcium collects at the bottom of the pan with the coagulum of albumin, and some are entangled in the scum.

MILK PROCESSING

To produce milk containing low bacterial count, good flavour, and to retain other essential qualities it is subjected to three operations namely, clarification, pasteurization, and homogenization.

Clarification

The primary step in the processing of milk is clarification. In this, the suspended impurities, cells from the udder, and other bacteria are removed by passing through a centrifugal clarifier. The clarified milk then is subjected to pasteurization.

Pasteurization

It is the process of heating food for the purpose of killing harmful organisms such as bacteria, viruses, protozoa, molds and yeasts.

The process was named after its inventor, the French scientist, Louis Pasteur. Unlike sterilization, pasteurization is not intended to kill all microorganisms in the food but aims to achieve a "log reduction" in the number of viable organisms, reducing their number, and therefore, are unlikely to cause disease. Pasteurization is typically associated with milk. The most common form of heat treatment, pasteurization, kills harmful bacteria found in untreated milk without affecting the nutritional value or taste of the milk. This also helps in extending the quality of milk. The shelf life of pasteurized milk is always dependent on the quality of the raw milk and can be improved by milk processing. Pasteurized milk can be kept fresh for 5 days or more in the refrigerator. The holding or batch pasteurization step, which is cheap at a large scale, is often performed prior to standard pasteurization. Batch pasteurized milk is often called "raw milk" or, "unpasteurized milk." It cannot be called "pasteurized," even though a significant number of pathogens are destroyed during the process. In this, the milk is heated up to 65°C and hold at that point for at least 30 min. followed by rapid cooling. A higher temperature is also used at times but the time of holding is shortened. The holding method reduces efficiently the bacterial count without affecting the cream line.

Pasteurization methods are usually standardized and controlled by National Food Safety Agencies. These agencies stress that milk has to be HTST pasteurized in order to qualify for the "pasteurized" label. There are different standards adopted for various dairy products, depending on the fat content and the intended usage. For example, the pasteurization standards for cream differ from the standards for fluid milk, and the standards for pasteurizing cheese are designed to preserve the phosphatase enzyme, which aids in curing the cheese.

There are two widely used methods to pasteurize milk: high temperature/short time (HTST), and ultra-high temperature (UHT).

UHT milk   While pasteurization conditions effectively eliminate potential pathogenic microorganisms, it is not sufficient to inactivate the thermoresistant spores in milk. The term sterilization refers to the complete elimination of all microorganisms. The food industry uses the more realistic term "commercial sterilization"; a product is not necessarily free of all microorganisms, but those that survive the sterilization process are unlikely to grow during storage and cause product spoilage. In canning, we need to ensure that the "cold spot" has reached the desired temperature for the desired time. With most canned products, there is a low rate of heat penetration to the thermal centre. This leads to over-processing of some portions, and damage to nutritional and sensory characteristics, especially near the walls of the container. This implies long processing times at lower temperatures.

Milk can be made commercially sterile by subjecting it to temperatures in excess of 100°C, and packaging it in airtight containers. The milk may be packaged either before or after sterilization. The basis of UHT, or ultra-high temperature, is the sterilization of food before packaging, then filling into pre-sterilized containers in a sterile atmosphere. Milk that is processed in this way using temperatures exceeding 135°C, permits a decrease in the necessary holding time (to 2–5 s) enabling a continuous flow operation.

Ultra heat treated (UHT) milk is sterilized for a very short period and its flavour is less affected than sterilized milk. It is packaged in cartons. UHT milk can be kept for several months unopened without refrigeration. Once opened it should be treated as fresh and can be stored in the fridge for up to 5 days.

HTST milk   HTST is by far the most common method. Milk simply labelled "pasteurized" is usually treated with the HTST method, whereas milk labelled "ultra-pasteurized" must be treated with the UHT method. HTST involves holding the milk at a temperature of 161.5°F (or 72°C) for at least 15 s. UHT involves holding the milk at a temperature of 280°F (or 138°C) for at least 2 s. The HTST pasteurization standard was designed to achieve a 5-log (approximately one million-fold) reduction in the number of viable microorganisms in milk. This is considered adequate for destroying almost all yeasts, mold and common spoilage bacteria and also to ensure adequate destruction of common pathogenic heat-resistant organisms (including particularly *Mycobacterium tuberculosis*, which causes tuberculosis and *Coxiella burnetii*, which causes Q fever). HTST pasteurization processes must be designed so that the milk is heated evenly, and no part of the milk is subject to a shorter time or a lower temperature. HTST pasteurized milk typically has a refrigerated shelf life of 2–3 weeks, whereas ultra pasteurized milk can last much longer when refrigerated, sometimes two to three months. When UHT pasteurization is combined with sterile handling and container technology, it can even be stored unrefrigerated for longer periods.

In recent years, there has been some consumer interest in raw milk products, due to perceived health benefits. Advocates of raw milk, maintain correctly that, vitamins and nutrients survive much better in milk that has not been pasteurized. They also maintain that organic raw milk (most retail raw milk is also organic) is less likely to contain harmful pathogens due to better husbandry in organic dairy herds. This may be true, but it has not been proven. However, doctors (and even most raw milk advocates) acknowledge that certain people, for example, pregnant or breast-feeding mothers, those undergoing immunosuppression treatment for cancer, organ transplant or autoimmune diseases, and those who are immunocompromised due to diseases like AIDS should avoid all but UHT pasteurized dairy products.

Homogenized Milk

Homogenized milk is whole milk that has undergone treatment to break up the fat globules in the milk, so that the cream does not rise to the top. Homogenized milk is produced by mechanically forcing the milk through a small passage at high velocity. This breaks down the fat globules in milk into much smaller ones and creates a stable fat emulsion. Homogenization decreases the size of the fat globules, increases their number and surface area. A film of lipoprotein immediately surrounds each of the globules, acting as an emulsifier, and diminishes the tendency of the fat globules to clump together and coalesce into cream.

Homogenized milk coagulates readily than the non-homogenized milk. This is because of the increased protein content present as a thin film around each fat globule. It has a fat content of 4%. Bottles of homogenized whole milk have red foil caps. Regarding storage, it should be kept in fridge and used within 3 days of opening.

Advantages

- Uniform distribution of fat with no cream layer
- Full-bodied flavour
- Whiter and possesses more appetizing colour
- Faster coagulation in the manufacture of rennet

Whole Milk

Also known as full-fat milk, this is milk which has not had its fat content altered. It has a fat content of about 4% and contains a range of nutrients and vitamins. Whole milk sold in bottles has a silver cap, and a blue cap and label indicates that a carton or plastic bottle contains whole milk. This is used as a popular choice for cooking, because it has a rich flavour and is also good for milkshakes and drinks. Pasteurized whole milk keeps fresh for 5 days or more in the fridge.

BASIC MILK CATEGORIES AND THEIR DEFINITIONS

Skimmed Milk Powder

Non-fat dry milk is the product resulting from the removal of fat and water from milk and contains lactose, protein and minerals in the same relative proportions as in the fresh milk from which it was made. It contains not more than 5.0% moisture (by weight). The fat content is not more than 1.5% (by weight) unless otherwise indicated. Skimmed milk powder also has reduced amount of fat-soluble vitamins like vitamin A and D. For this reason it is fortified with these vitamins and supplied. This finds extensive use in confectioneries and baking industry. It is recommended for low-calorie and high-protein diets.

Whole Dry Milk Powder

Whole dry milk is the product resulting from the removal of water from milk and contains not less than 26% and not more than 40% of milk fat and not more than 5.0% moisture (as determined by weight of moisture on a milk solids-not-fat basis).

Buttermilk Powder

Dry buttermilk is the product resulting from the removal of water from liquid buttermilk derived from the churning of butter. It contains not less than 4.5% milk fat (by weight) and not more than 5% moisture (by weight). Buttermilk powder must have a protein content of not less than 30% (by weight).

	Skimmed milk powder	Whole milk powder	Buttermilk powder
Protein	34.0–37.0	24.5–27.0	32.0–34.5
Lactose	49.5–52.0	36.0–38.5	49.5–50.5
Fat	0.6–1.25	26.0–28.5	5.5–6.0
Ash	8.2–8.6	5.5–6.5	7.5–8.0
Moisture	3.0–4.0 (non-instant) 3.5–4.5 (instant)	2.0–4.5	3.0–4.0

If raw milk were allowed to stand, the fat globules would begin to rise to the surface in a phenomenon called creaming.

Manufacture of Dry Whole Milk Powder

Advancement in the technology has facilitated these processes of converting dairy products into a variety of dry powders. Transformation of liquid milk into dry powder involves complete removal of water. During this, changes in physical appearance, properties, and structure occur significantly.

Spray-drying   This process consists of pre-heating, concentration, spray-drying and pneumatic cooling. Ordinary milk powder obtained by this process is very fine, dusty, and hygroscopic and therefore is prone to caking. Hygroscopicity and caking are influenced also by local climatic conditions. The hygroscopicity, caking and all the problems associated with the stickiness of milk powder are mainly due to lactose being present in an amorphous glassy state. In the spray-drying of milk products, lactose is in an amorphous state and is not stable in atmospheric air or normal humidity. The only form of lactose that is stable to humidity is α-lactose monohydrate. Since the lactose content of whey powder comprises more than 70% of the total solids in comparison with 30% in whole milk, the problem of the lactose content in whey powder is more severe. However, since the solubility of lactose is 17 g/100 cm^3 H$_2$O at 20°C, during the drying process, a major part of the lactose is transformed to the stable α-lactose monohydrate form.

Pre-crystallization and crystalline treatment   The basic process layout is modified by conducting a pre-crystallization treatment before spray-drying. During the pre-crystallization process, it is easy to keep ideal conditions for crystallization. Viscosity of the concentrate is reasonably low, temperatures may be exactly adjusted and controlled, displacement of used solution from the surface of crystals may be accelerated by agitation, and the required amount of suitable crystals of lactose may be ensured by proper seeding. The product made by this process is non-caking and, being agglomerated, is dustless and free-flowing. The agglomerates tend to be small and thus the bulk density is relatively high.

The main operations used for the manufacture of milk powder are as follows:

- Pre-heating
- Concentration
- Flash cooling
- Pre-crystallization
- Spray-drying
- Cooling in a vibrated fluid bed

A conventional spray dryer consists of the following main components:

- Drying chamber
- Hot air system and air distribution feed system
- Atomizing device
- Powder separation system
- Pneumatic conveying and cooling system
- Fluid bed after-drying/cooling
- Instrumentation and automation

The feed is pumped from the product feed tank to the atomizing device, which is located in the air disperser at the top of the drying chamber. The drying air is drawn from the atmosphere via a filter by a supply fan and is passed through the air heater to the air disperser. The atomized droplets meet the hot air and the evaporation takes place cooling the air at the same time. After the drying of the spray in the chamber, the majority of the dried product falls to the bottom of the chamber and enters a pneumatic conveying and cooling system. The fines, which are the particles with small diameter, will remain entrained in the air, and it is therefore necessary to pass the air through cyclones for separation of fines. The fines leave the cyclone at the bottom via a locking device and enter the pneumatic system, too. The air passes from the cyclone to the atmosphere. The two fractions of powder are collected in the pneumatic system for conveying and cooling and are passed through a cyclone for separation, after which they are bagged off. The instrumentation comprises indication of the temperature of the inlet and outlet air, as well as automatic control of the inlet temperature by altering the steam pressure, amount of oil or gas to the air heater, and automatic control of the outlet temperature by altering the amount of feed pumped to the atomizing device.

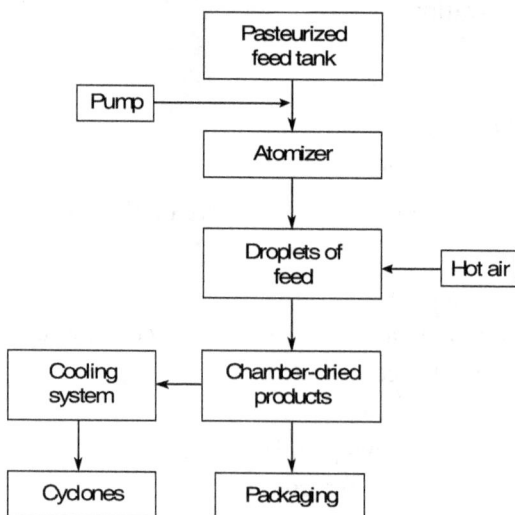

Figure 17.3   Flow chart showing various operations in drying of milk

BUTTER

The rich, oily, and yellowish part of milk, which, when the milk stands unagitated, rises, and collects on the surface is the part of milk from which butter is obtained. Butter is one of the most highly concentrated forms of fluid milk.

Twenty litres of whole milk are needed to produce 1 kg of butter. This process leaves approximately 18 L of skimmed milk and buttermilk. Commercial butter is 80–82% milk fat, 16–17 percent water, and 1–2 percent milk solids other than fat (sometimes referred to as curd). It may contain salt, added directly to the butter in concentrations of 1 to 2%. Unsalted butter is often referred to as "sweet" butter. This should not be confused with "sweet cream" butter, which may or may not be salted. Reduced-fat, or "light," butter usually contains about 40 percent milk fat. Butter also contains protein, calcium and phosphorus (about 1.2%) and fat-soluble vitamins A, D and E.

Although there are over 120 different compounds that contribute to butter's unique flavour, the five primary factors responsible for butter's flavour include: fatty acids, lactones, methyl ketones, diacetyl and dimethyl sulphide.

Chemically butterfat consists essentially of a mixture of triglycerides; particularly those derived from fatty acids (Figure 17.4), such as palmitic, oleic, myristic and stearic acids. The fatty acid composition of butterfat varies according to the animal's diet that produces it. A measure of the amount of these acids, the Reichert–Meissl, or Reichert–Wollny number is important in the analysis of butterfat. Milk fat from cows that are fed diets higher in stearic acid produced softer butter than milk fat from cows fed diets higher in palmitic acid. The change in butter softness was associated with changes in fatty acid and triglyceride structure of the milk fat.

Milk fat is comprised mostly of triglycerides, with small amounts of mono- and diglycerides, phospholipids, glycolipids and lipoproteins. The trigylcerides (98% of milk fat) are of diverse composition with respect to their component fatty acids, approximately 40% of which is unsaturated, fat firmness varies with chain length, degree of unsaturation, and position of the fatty acids on the glycerol.

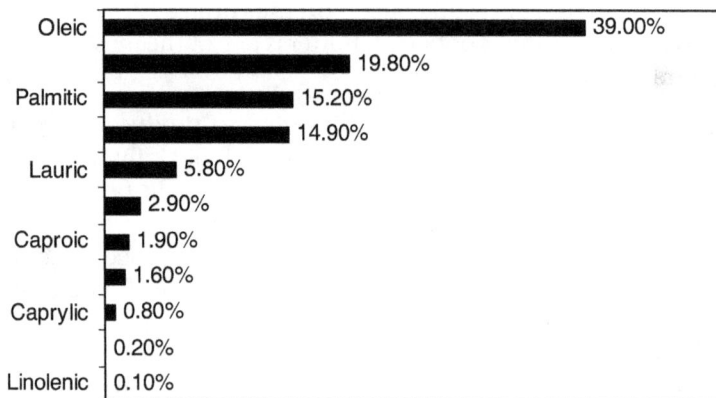

Figure 17.4   The composition of fatty acids in butter

Butteroil (clarified butter), anhydrous butter oil, ghee are products prepared form butter or cream by the removal of most of the water and solids-not-fat content. The compositional standards for butteroil and anhydrous butteroil are given below.

Butteroil (clarified butter) and ghee—anhydrous butter oil, ghee contain not less than 99.3% milk fat and not more than 0.5% water.

Anhydrous butter oil—contains not less than 99% milk fat and not more than 0.1% water.

Flavourful fatty acids play an important role in the flavour of butter and are present in varied concentrations. Although long-chain fatty acids are present at higher concentrations in butter, they do not make a significant contribution to flavour. Short-chain fatty acids (SCFA), on the other hand, do play an important role in butter's flavour. Butyric acid as a glyceride, makes up 3–4% percent of butter; the disagreeable odour of rancid butter is that of butyric acid resulting from hydrolysis of the glyceride.

Typically, SCFA are found in the serum portion of butter (aqueous solution of all non-fat components) where their flavour potential is stronger. They occur below their flavour threshold value

(FTV) which is the minimum concentration level below which aroma or taste is imperceptible. Despite low concentrations, SCFA react in a synergistic and additive manner to provide characteristic flavours found in butter. Butyric acid is the most widely known and most potent SCFA and is attributed to providing intensity to fatty acid type flavours associated with butter. Butter also contains a variety of fatty acid precursors of 4-*cis*-heptenal, a compound which provides butter with a creamy flavour.

It is a curious feature of fats that once melted, they have to be cooled to well below their melting point to resolidify them. Butter, for example, melts at about 35°C (96°F) but has to be cooled to about 23°C (73°F) to solidify it.

Compositional Standards for Cream and Whipping Cream

Preparation and composition   It is the fatty liquid prepared from milk by separating the milk constituents in such a manner as to increase the milk fat content. The cream shall contain a minimum of 10% milk fat and the whipping cream shall contain a minimum of 32% milk fat.

Permitted ingredients and additives   These consist of a pH adjusting agent, a stabilizing agent. In the case of whipping cream which has been heat-treated (thermized) to a temperature greater than 100°C, the following ingredients and food-permitted additives are present:

- skimmed milk powder in an amount not exceeding 0.25%
- glucose solids in an amount not exceeding 0.1%
- calcium sulphate in an amount not exceeding 0.005%
- xantham gum in an amount not exceeding 0.02%
- microcrystalline cellulose in an amount not exceeding 0.2%.

GHEE OR CLARIFIED BUTTER

This is the rendered milk fat. Unlike butter, ghee can be stored for extended periods without refrigeration, provided it is kept in an airtight container to prevent oxidation. Also unlike butter, ghee can be heated up to its smoke point without discolouring or developing a burnt taste, making it ideal for deep frying.

Ice Cream

Ice cream has the following composition:

- It has a value greater than 10% milk fat by legal definition, and usually between 10% and as high as 16% fat in some premium ice creams.
- The milk solids-not-fat include 9–12%. This component, also known as the serum solids, contains proteins (caseins and whey proteins) and carbohydrates (lactose) found in milk.
- The sweeteners, usually 12 to 16% is a combination of sucrose and glucose-based corn syrup sweeteners.
- The stabilizers and emulsifiers range from 0.2 to 0.5%.
- The water ratio ranges from 55% to 64% which comes from the milk or other ingredients.

These percentages are by weight, either in the mix or in the frozen ice cream. However, when frozen, about one-half of the volume of ice cream is air. So by volume in ice cream, these numbers can be reduced by one-half. Since air does not contribute to weight, the composition of ice cream is taken on a weight basis. All ice cream, with the possible exception of chocolate, is made from a basic white mix.

Formulations can be derived from a number of different starting points. Ice milk is very similar to the composition of ice cream but must contain between 3% and 5% milk fat by legal definition. The ingredients to supply the desired components are chosen on the basis of availability, cost, and desired quality. These ingredients will now be examined in more detail.

Milk Fat (or Butterfat)

Milk fat is important to ice cream for the following reasons:

- Increases the richness of flavour in ice cream.
- Produces a characteristic smooth texture by lubricating the palate.
- Helps to give body to the ice cream, due to its role in fat destabilization.
- Aids in good melting properties, also due to its role in fat destabilization.
- Aids in lubricating the freezer barrel during manufacturing (Non-fat mixes are extremely hard on the freezing equipment).

The limitations of excessive use of butterfat in a mix includes:

- cost
- hindered whipping ability
- decreased consumption due to excessive richness
- high calorific value

The best source of butterfat in ice cream for high quality flavour is fresh sweet cream from fresh sweet milk. The triglycerides in milk fat have a wide melting range of, $+40°C$ to $-40°C$, and thus there is always a combination of liquid and crystalline fat. Alteration of this solid : liquid ratio can affect the amount of fat destabilization that occurs. Duplicating this structure with other sources of fat is difficult. During freezing of ice cream, the fat emulsion which exists in the mix will partially destabilize or churn as a result of the air incorporation, ice crystallization and high shear forces of the blades.

This partial churning is necessary to set up the structure and texture in ice cream, which is very similar to the structure in whipped cream. Emulsifiers help to promote this destabilization process.

Milk Solids-not-fat

The serum solids or milk solids-not-fat (MSNF) comprises lactose, caseins, whey proteins, minerals, and ash content of the product from which they were derived. They are important ingredient for the following beneficial reasons:

1. Improve the texture of ice cream, due to the protein functionality.
2. Help to give body and chew resistance to the finished product.
3. Capable of allowing a higher over run without the characteristic snowy or flaky textures associated with high over run, also due to the protein functionality.
4. May be a cheap source of total solids, especially whey powder.

The limitations on their use include off-flavours, which may arise from some of the products, and an excess of lactose, which can lead to the defect of sandiness prevalent when the lactose crystallizes out of solution. Excessive concentrations of lactose in the serum phase may also lower the freezing point of the finished product to an unacceptable level.

The best sources of serum solids for high-quality products are (a) concentrated skimmed milk and (b) spray process low heat skimmed milk powder.

Other sources of serum solids include: sweetened, condensed whole or skimmed milk, frozen condensed skimmed milk, buttermilk powder or condensed buttermilk, condensed whole milk, or dried or condensed whey. Superheated condensed skimmed milk, in which high viscosity is promoted, is sometimes used as a stabilizing agent, which also contributes to serum solids.

It has recently become a common practice to replace the use of skimmed milk powder or condensed skim with a variety of milk powder replacers, which are blends of whey protein concentrates, caseinates, and whey powders. These are formulated with less protein than skim powder, usually 20–25% protein, and thus cost less, but are blended with an appropriate balance of whey proteins and caseins to do an adequate job. Caution must be exercised in excessive use of these powders.

The proteins, which make up approximately 4% of the mix, contribute much to the development of structure in an ice cream including:

- Emulsification properties in the mix
- Whipping properties in the ice cream
- Water holding capacity leading to enhanced viscosity and reduced iciness

Lactose Crystallization

1. A decrease in temperature favours rapid crystallization so far that it increases the supersaturation.
2. A decrease in temperature favours slow crystallization so far that it increases the viscosity, reduces the kinetic energy of the particles, and decreases the rate of transformation from β to α-lactose.

Supersaturated state can exist, however, due to extreme viscosity, it is likely that much of the lactose in ice cream is non-crystalline. Stabilizers help to hold lactose in supersaturated state due to viscosity enhancement. Fruits, nuts and candy add crystal centres which enhance lactose crystallization. Nuts pull out moisture from ice cream immediately surrounding the nut thus concentrating the mix.

Citrate and phosphate ions decrease tendency for fat coalescence (sodium citrate, disodium phosphate). They prevent churning in soft ice cream for example, producing a wetter product. These salts decrease the degree of protein aggregation. Calcium and magnesium ions have the opposite effect, and thereby promote partial coalescence. Addition of calcium sulphate, for example, results in a drier ice cream. Calcium and magnesium increase the degree of protein aggregation. Salts may also influence electrostatic interactions. Fat globules carry a small net negative charge, and these ions could increase or decrease the charge as they are attracted to or repelled from the surface.

Sweeteners

A sweet ice cream is usually desired by the consumer. As a result, sweetening agents are added to ice cream mix at a rate of usually 12–16% by weight. Sweeteners improve the texture and palatability of the ice cream, enhance flavours, and are usually the cheapest source of total solids. In addition, the sugars, including the lactose from the milk components, contribute to a depressed freezing point so that the ice cream has some unfrozen water associated with it at very low temperatures typical of their serving temperatures, –15° to –18° C. Without this unfrozen water, the ice cream would be too hard to scoop.

Sucrose is the main sweetener used because it imparts excellent flavour. It has become common in the industry to substitute all or a portion of the sucrose content with sweeteners derived from corn syrup. This sweetener is reported to contribute a firmer and chewier body to the ice cream, is an economical source of solids, and improves the shelf life of the finished product. Corn syrup in either its liquid or dry form is available in varying dextrose equivalents (DE). The DE is a measure of the reducing sugar content of the syrup calculated as dextrose and expressed as a percentage of the total dry weight. As the DE is increased by hydrolysis of the cornstarch, the sweetness of the solids is increased and the average molecular weight is decreased. This results in an increase in the freezing point depression, in foods such as ice cream, by the sweetener.

The lower DE corn syrup contains more dextrins, which tie up more water in the mix thus supplying greater stabilizing effect against coarse texture. An enzymatic hydrolysis and isomerization procedure can convert glucose to fructose, a sweeter carbohydrate, in corn syrups thus producing a blend (high fructose corn syrup, HFCS), which can be used to a greater extent as a sucrose replacement. However, these HFCS blends further reduce the freezing point producing a very soft ice cream at usual conditions of storage.

Stabilizers

The stabilizers are a group of compounds, usually polysaccharide food gums, which are responsible for adding viscosity to the mix and the unfrozen phase of the ice cream. This results in many functional benefits, listed below, and also extends the shelf life by limiting ice recrystallization during storage. Without the stabilizers, the ice cream would become coarse and icy very quickly due to the migration of free water and the growth of existing ice crystals.

The smaller the ice crystals in the ice cream, the less detectable they are to the tongue. Especially in the distribution channels of today's marketplace, ice cream has many opportunities to warm up, partially melt some of the ice, and then refreeze as the temperature is once again lowered. This process is known as heat shock and every time it happens, the ice cream becomes icier tasting. Stabilizers help to prevent this.

The functions of stabilizers in ice cream are the following.

1. *In the mix* To stabilize the emulsion to prevent creaming of fat and, in the case of carrageenan, to prevent serum separation due to incompatibility of the other polysaccharides with milk proteins, also to aid in suspension of liquid flavours.

2. *In the ice cream at draw from the scraped surface freezer* To stabilize the air bubbles and to hold the flavourings, e.g. ripple sauces, in dispersion.

3. *In the ice cream during storage* To prevent lactose crystal growth and retard or reduce ice crystal growth during storage, also to prevent shrinkage from collapse of the air bubbles and to prevent moisture migration into the package (in the case of paperboard) and sublimation from the surface.

4. *In the ice cream at the time of consumption* To provide mouth feel without being gummy, and to promote good flavour release.

Limitations on their use include:

a. Production of undesirable melting characteristics, due to too high viscosity

b. Excessive mix viscosity prior to freezing

c. Contribution to a heavy or chewy texture

The stabilizers in use today include

Locust bean gum   Soluble fibre of plant material derived from the endosperm of beans of exotic trees grown mostly in Africa (Note: locust bean gum is a synonym for carob bean gum, the beans of which were used centuries ago for weighing precious metals, a system still in use today, the word carob and Karat having similar derivation).

Guar gum   It is obtained from the endosperm of the bean of the guar bush, a member of the legume family.

Carboxymethyl cellulose (CMC)   Derived from the bulky components, or pulp cellulose of plant material, and chemically derivatized to make it water-soluble.

Xanthan gum   Produced in culture broth media by the microorganism *Xanthomonas campestris* as an exopolysaccharide. However it is used to a lesser extent.

Sodium alginate   An extract of seaweed, brown kelp, also is used to a lesser extent.

Carrageenan   An extract of Irish moss or other red algae. Each of the stabilizers has its own characteristics and often, two or more of these stabilizers are used in combination to lend synergistic properties to each other and improve their overall effectiveness. Guar, for example, is more soluble than locust bean gum at cold temperatures, thus it finds more application in HTST pasteurization systems. Carrageenan is not used by itself but rather is used as a secondary colloid to prevent the wheying of the mix, which is usually promoted, by one of the other stabilizers.

Gelatin   A protein of animal origin, was used almost exclusively in the ice cream industry as a stabilizer but has gradually been replaced with polysaccharides of plant origin due to their increased effectiveness and reduced cost.

Emulsifiers

The emulsifiers are a group of compounds in making ice cream, which aid in developing the appropriate fat structure and air distribution necessary for the smooth eating and good meltdown characteristics desired in an ice cream. Since each molecule of an emulsifier contains a hydrophilic and a lipophilic portion, they reside at the interface between fat and water. As a result, they act to reduce the interfacial tension or the force, which exists between the two phases of the emulsion.

The emulsifiers actually promote a destabilization of the fat emulsion, which leads to a smooth, dry product with good meltdown properties. The original ice cream emulsifier was egg yolk, which was used in most of the original recipes. Today, two emulsifiers predominate most ice cream formulations. They are mono- and diglycerides, derived from the partial hydrolysis of fats or oils of animal or vegetable origin.

Polysorbate 80

This is sorbitan ester consisting of a glucose alcohol (sorbitol) molecule bound to a fatty acid, oleic acid, with oxyethylene groups added for further water solubility.

Other possible sources of emulsifiers include buttermilk and glycerol esters. All of these compounds are either fats or carbohydrates, which are important components in most of the foods we eat. Together, the stabilizers and emulsifiers make up less than one half percent by weight of ice cream. They are all compounds which have been exhaustively tested for safety and have received the "generally recognized as safe" (GRAS) status.

18

LEATHER CHEMISTRY

INTRODUCTION

The hide of an animal, is the thick tough outer cover or pelt of a large animal. The skin of an animal, is the outer cover of a small animal, such as calf, sheep, goat. Leather is cured animal hide. The pelt of an animal, which has been transformed by tanning into a non-putrescible, useful material, is called leather.

The outer covering of animal, usually a mammal, is tanned or otherwise dressed and prepared in such a manner as to render it usable and resistant to putrefaction, even when wet. The changes made to the animal skins in the process of making leather results in a durable, pliant and practicable material. Leather is a unique and flexible sheet-like material that is analogous to textiles, and may be considered as the first natural fabric. Leather occupies an unique position among the covering materials used by bookbinders. Its structure gives it the desirable softness and strength, while its chemical nature gives it the property of adhering well to paper, board, linen, etc. Its outstanding characteristics include its durability when properly prepared and cared for, suppleness, porosity, beauty, temper, and feel, in addition to its strength and softness. The little over the years when one stops to consider that the animal skin itself has not changed.

The unique characteristics of leather are largely due to its structure, which is an interwoven, three-dimensional network of fibres inherent in the natural raw materials—hides and skins. This raw material is principally a fibrous protein called collagen and is composed of one continuous network of fibres.

In the raw skin, at least four distinct structures can be distinguished:

1. Thin outermost layer termed the epidermis.
2. Grain layer or dermal surface.
3. Juncture between the grain layer and the dermis or corium.
4. Major portion of the skin—the dermis or corium, which is the part converted into leather. In addition, there is the flesh layer, or hypodermis, which is the structure adjacent to the body tissues.

Dry, raw skin (Figure 18.1) is horny and hard due to tightly packed collagen. When exposed to water in general, dry skin begins to putrefy and rot; when exposed to hot water, glue and sizing can be made from it. Leather, however, requires human skill to be processed and it is the essence of skin, stripped of what is unnecessary and preserved, in its peak of usefulness. Leather can be oiled and softened; it can be exposed to water without rotting, and even made waterproof.

Before tanning, the approximate composition of a freshly flayed hide is:

Water	64%
Protein	33%
Fats	2%
Mineral salts	0.5%
Other substances (pigments, etc.)	0.5%

The 33%, which is protein, consists of

1. **Structural proteins** Elastin an yellow fibre woven in the collagen fibre 0.3%. Collagen, which tans to give leather, 29%; Keratin, protein of the hair and epidermis 2%.

2. **Non-structural proteins** Albumins or globulins-soluble, non-fibrous proteins 1%; Mucins or mucoids, mucous materials associated with fibres 0.7%.

While all mammalian skins are made up of these constituents, the percentage of keratin will vary widely, depending on the amount of hair present; the percentage of fat will also vary. The distribution between albumin and mucins is debatable.

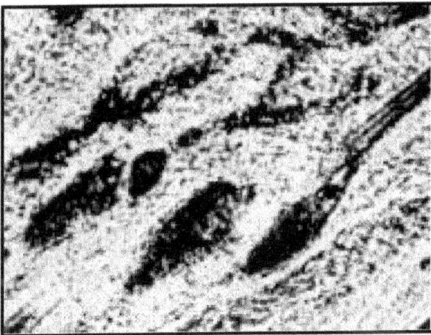

Figure 18.1   A piece of raw, dry animal hide

CONSTITUENTS OF ANIMAL SKIN

All mammalian skins vary tremendously in size, for example, from the hide of an elephant or an ox to the skin of a rabbit or mouse, and they vary considerably in shape and thickness. In addition, some animals have but little hair or wool and a thick epidermal layer, for example, the pig, while others, such as the sheep, have heavy fleece with curly wool and curly hair follicles but a relatively thin epidermis. The state of development of the animal is also important. A calf, for example, has finer structured hair than a full-grown cow, consequently, leather made from the skin of a calf is relatively smooth and very fine-grained, while that of a cow is rougher and has a very pronounced grain pattern. The more natural the animal's feeding and living conditions, the better the quality of the resultant leather. For example, overfeeding produces greasier, weaker skins, while starvation results in thin, weak, misshapen skins showing skeleton markings. The skin of the female usually, is more fine-grained than that of the male, and has a looser fibre structure, especially in the flanks, giving somewhat softer and tensile leather. The lesser the hair or wool on the animal, the tougher and stronger is the resultant leather, especially in the grain layer. Heavily wooled sheep are inferior in this respect to goats and pigs. The hairs are embedded in the skin, each in a sheath of epidermis known as the hair follicle and each with a hair-root at its end, are fed by a tiny blood vessel. Chemically, the hairs consist of the protein keratin, and penetrate deeply into the papillary layer of the dermis. Most animals have hair of two types, primary and secondary. The positions which these hairs occupy relative to each other as they enter the surface of the skin, together with their different thickness, determine the characteristic marking or grain of the dermal surface, i.e., the grain pattern of the leather, which is exposed upon removal of the hair and other epidermal structures.

Skins, though, can be used generically to describe all animal outer coverings consist of three layers of body cells. The top layer is called the epidermis, the middle layer of skin, the derma, or corium, and the fatty bottom layer is known as adipose or flesh.

The epidermis is a protective, hard layer of keratinous cells. Those on the outside are dead and, upon drying and shrinking, fall off the skin. On the underside, adjacent to the skin proper, they consist of soft, jelly-like living cells, which have little resistance and are readily attacked and degraded by bacterial action or enzymes which occurs with stale skins. They are disintegrated by alkalis or alkaline compounds especially sodium sulphide or sodium hydrogen sulphide.

The skin proper or dermis (corium) is not composed of true cells, but is rather a network of collagen fibrils (Figure 18.2), very intimately woven and joined together. In the grain layer, these fibres become thin and tightly woven and are so interlaced that there are no loose ends on the surface beneath the epidermis. Thus, when the epidermis is removed, a smooth layer is revealed which gives the characteristic grain surface of the leather. Towards the centre of the dermis, the fibres are coarser and stronger, and the predominant angle at which they are woven indicates the properties of the resultant leather. If the fibres are upright and tightly woven, the leather will be firm and hard, with little stretch, while if they are more horizontal and loosely woven, the leather will be softer and stretchier. The dermis is also the strongest part of the skin. It is this dermal layer which the leather maker seeks to retain for his leather.

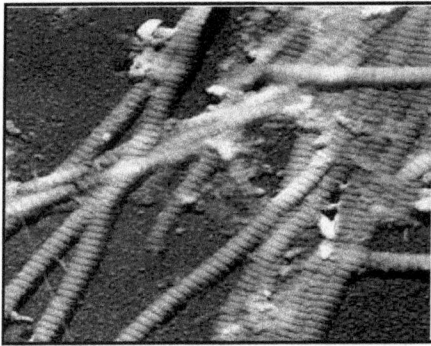

Figure 18.2   Dermal collagen fibres of human skin

In the living skin, the collagen fibres (Figure 18.2) and cells are embedded in a watery jelly of protein-like substance called the ground substance. The living collagen fibres are formed from this substance which ranges in constitution from the blood sugars to substances which are almost collagen. The latter fibres have been called "inter-fibrillary" proteins, also known as non-structural proteins, or pro-collagens. These are essential for the growth of the skin and render the fibre structure non-porous.

When the skin is dried (as in some forms of curing), a hard, glue-like substance, which cements all of the corium fibres together makes the skin hard and horny. In producing leather, which is to be soft or supple, it is essential that these inter-fibrillary proteins be removed. The corium fibres are composed of rope-like bundles of smaller fibrils, which in turn consist of bundles of sub-microscopic micelles. These in turn are made up of very long, thread-like molecules of collagen twisted together. Together, this gives a very tough, strong, flexible, three-dimensional structure, forming a network, which decides the characteristics of the leather. It is this structure, which makes the leather unique and is unquestionably the basis for the remarkably high tensile strength of leather.

The flesh of the dermis is the layer next to the body wall of the animal namely the fatty, adipose tissue. The skins of certain animals (at certain times of their lives) also contain considerable quantities of fat in globular cells, which lie approximately in the centre of the dermis. For example, the sheepskin contains fat of this type, which is in the interior of the skin and not merely on the flesh layer, amounting to 25% of the weight of the skin. Such excessive growth of fat cells disrupts and weakens the dermal fibre structure to such an extent that some sheepskins can be split into two layers along the line where the fat is located. In general, the younger the animal at the time of slaughter, the thinner and smaller the skin, the smoother and finer the grain structure, and the lesser the likelihood of damage.

PREPARING SKINS AND HIDES

Before a piece of animal skin can become a workable piece of leather, it must be treated. "Skins" refer specifically to the outer covering of a smaller animal, such as a pig, calf or sheep, while the term "hides" refer to the outer covering of a larger animal, such as a cow or horse. In order to separate the corium from the epidermis and the flesh, so that the skin or hide may become a piece of useful leather, several steps are required.

Cleaning and Soaking

During this step, rehydration, softening and restoration of the skin to its original degree of swelling occurs. Surfactants are used to improve the re-absorption of water into hides and skins and to remove surface soils. First, if the animal skin is fresh from slaughter, the piece of skin is washed and cleaned. Nowadays most animal hides are shipped from slaughterhouses to tanneries, and so they are usually cured after slaughtering to arrest putrefaction. Cleaning the skin involves soaking it in water for up to two days, to remove all impurities and, if necessary, with the help of curing agents such as salt the last remnants are also purified.

Liming and Degreasing

This process which is preliminary to tanning serves the following purposes.

    i. Removal of the outer layers from the raw hide and elimination of natural fats and oils.

    ii. To plump or swell the fibres suitable for the action of tan liquors.

Surfactants improve penetration to extract and to emulsify natural fats and oils. Removal of hair is done by first immersing the skins in strong solutions of lime for several days to loosen and soften the hair roots and epidermis, and then scraping off the loosened layer. The modern process of removing the epidermis involves scraping the skin over a beam made of metal or wood. Some modern tanneries, however, use machines with blunt blades to scrape away the hair and cells. The process of lime soaking makes the skin more receptive to the agents and dyes used during the tanning process.

Fleshing

The final cell layer to be removed is the flesh, and the process of removing it is appropriately called fleshing. While fleshing is a process requiring considerably more skill than removing the epidermis, the skin need not be removed from the scraping beam for this process and the general principle is the same as scraping. In the cutting away of the fatty tissue, a single slip of the sharp knife can render an entire skin useless. Today, most skins are fleshed by machines with quickly revolving helical blades, and the possibility of human error in fleshing is eliminated.

Shaving

The last step in the pre-leather making process is shaving or splitting during which the thickness of the skin is adjusted for tanning. Cattle hide is normally about 4.5 mm in thickness, but for some purposes, the leather needs to be half this thickness, or even less. Shaving, like fleshing, requires great precision and skill. It also requires very specific tools. The skin is placed over a specially shaped beam and the skin is cut with a currier's knife, which has a double edge. The currier's knife is held at right angles to the flesh-side of the leather, and small pieces are shaved off until the skin has the correct thickness needed for the tanning process. This is called "skiving." The skin can be split by this process into two useful sheets, whereas the small pieces shaved off by the Roman technique were useless.

After shaving, the skins are given a final cleaning, to remove every last bit of dirt and hair possible. In ancient times, for a skin to be particularly soft, at this stage infusions of hen or pigeon dung, followed by dog dung, were applied to the skin, setting off a fermentation by bacteria, which reduced the skin to the desired softness. These processes often took several days, or even weeks. It was important to halt the reaction at the perfect time in order to keep the skin from being destroyed by fermentation, and the judgment of whether or not a skin was properly tempered was made entirely by feel. Today, softening compounds are chemical rather than natural, and the skins are given a last bath of sulphuric acid and common salt, to remove the last traces of lime and to impart an overall degree of acidity to the skin, which is helpful in the tanning process to follow.

Leather Tanning

"Leather tanning" is a general term for the numerous processing steps involved in converting animal hides or skins into finished leather. Hides and skins have the ability to absorb tannic acid and other chemical substances that prevent them from decaying, make them resistant to wetting, and keep them supple and durable. The surface of hides and skins containing hair and oil glands is known as the grain side. Tanning is essentially the reaction of collagen fibres in the hide with tannins, chromium, alum or other chemical agents. The most common tanning agents used are the trivalent chromium and vegetable tannins extracted from specific tree barks. Alum, syntans (man-made chemicals), formaldehyde, glutaraldehyde and heavy oils are other tanning agents.

Once a skin has been prepared, there are three possible methods to turn it into leather.

- Tanning
- Chamoising
- Tawing

While these processes differ considerably, they are grouped together under the term "tanning." Only two methods of leather making—tanning and tawing—have been found to be in use in ancient times. Chamoising is a process that results in a soft, pale yellow, oiled type of leather. Tanning gets its name from the tannins that are present in all vegetable matter, and these tannins are used to effect a permanent chemical change in the collagen content present in raw skins. This is also the reason why tanned leather is more specifically referred to as vegetable-tanned leather (Figure 18.3). Skins that are to be tanned are immersed in vats filled with water and a large amount of chopped bark—the material from which leather making tannins are derived. Present tanning practices place skins in rotating barrels full of water and tannins, thereby reducing the tanning time from several months to a few days. There are also many variations of tanning, each tailored to the specific products, the leather to be obtained. Leather for shoe uppers, for example, have a slightly different tanning process from that for handbags.

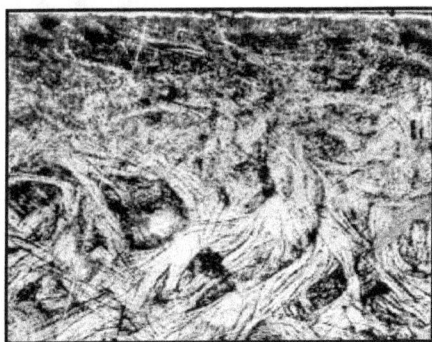

Figure 18.3   A piece of tanned leather

To finish tanned leather a process called currying is done. It involves applying grease, such as cod oil mixed with tallow, directly into the fibres of the leather with a "slicker" of stone, metal, or glass, as a means of waterproofing the leather and ensuring flexibility and strength. The drawback of the process is that waterproofing the leather eliminates its breathability.

Vegetable-tanned leathers can be given smooth, grained, bright, or dull finishes depending upon how the leather is hammered, folded, or rolled. Waterproofing changes the final appearance of leather and sometimes is used as more of an aesthetic choice than a practical one.

The other process in the leather making is tawing, or mineral tannage. The white, porous leather resulting from this process bears little resemblance to the sturdy, hefty leathers produced by chamoising or vegetable tanning. The main step in tawing is to soak the prepared skin in a mixture of alum and salt. To soften the salt-stiff skin when it is pulled from the alum solution, the tawed leather is pulled while damp over a wooden or metal stake in a process called, "staking." The resultant leather is so white and open-pored that if it were to be rinsed in warm water, the alum and salt would wash out, leaving the finished piece of material with a texture very similar to that of the untreated skin when the entire tanning process began. In order to give tawed leather the sturdiness required for common use, it must be rubbed with supportive substances, such as fat, grease, flour, or egg yolks, which can fill the large pores.

Once leather is tanned, it needs to be dried. All leathers, and particularly vegetable-tanned leathers, are very temperamental about how they are dried. If the process is completed too slowly, molds can develop and ruin the material. If done too quickly, tanned leather can become brittle and lose the rich colour it received during the tanning process. The drying process is also impeded by exposure to the natural elements. Sun and rain are two foes of leather that is to be dried. To combat these obstacles, the ancient people built drying lofts, which were covered structures that were kept open enough to allow warm air to circulate around the pieces of drying leather. Midway through the drying process, if needed, a process called "laying the grain" could be applied to the pieces of leather, where the hides are struck with a triangular, blunt-edged tool, which would smooth the leather's grain. To test a piece of leather to make sure it is dry, modern tanneries have machines, which detect moisture. The ancients, however, had a more interesting method. To see if a piece of leather was properly dried, they held a small, very cold mirror about half an inch away from the surface of the drying leather. When moisture no longer condensed on the surface of the glass, the leather was considered dry and ready for use.

Vegetable Tanning

Heavy leathers and sole leathers are produced by the vegetable tanning process, the oldest of any process in use in the leather tanning industry. The hides are first trimmed and soaked to remove salt and other solids and to restore moisture lost during curing. Following the soaking, the hides are fleshed to remove the excess tissue, to impart uniform thickness, and to remove muscles or fat adhering to the hide. Removal of hair is done to ensure that the grain is clean and the hair follicles are free of hair roots. Liming is the most common method of hair removal, but thermal, oxidative, and chemical methods also exist. The normal procedure for liming is to use a series of pits or drums containing lime liquors (calcium hydroxide) and sharpening agents. Following liming, the hides are dehaired by scraping or by a machine. Deliming is then performed to make the skins receptive to the vegetable tanning. Bating, an enzymatic action for the removal of unwanted hide components after liming is performed to impart softness, stretch, and flexibility to the leather. Bating and deliming are usually performed together by placing the hides in an aqueous solution of an ammonium salt and proteolytic enzymes at 27–32°C (80–90°F). Pickling may also be performed by treating the hide with brine solution and sulphuric acid to adjust the acidity for preservation or tanning. In the vegetable tanning process, the concentration of the tanning materials are kept at minimum levels in the beginning and is gradually increased as the tannage proceeds. It usually takes 3 weeks for the tanning material to penetrate to the centre of the hide. The skins or hides are then wrung and may be cropped or split; heavy hides may be retanned and scrubbed. True splitting is not usually a part of the vegetable tanning process; however, an operation called levelling is used to produce a uniformly thick piece of leather. Levelling removes only the thickest portions of the underside of the hide, and no "split" is produced. The hide is oiled, which is a process similar to the fat-liquoring in chrome tanning. Following oiling, the hide is dried and then mechanically conditioned.

For sole leather, the hides are commonly dipped in vats or drums containing sodium bicarbonate or sulphuric acid for bleaching and removal of surface tannins. Materials such as lignosulphate, corn sugar, oils and speciality chemicals may be added to the leather. The leather is then set out to smooth and dry and may then undergo further finishing steps. However, a high percentage of vegetable-tanned leathers do not undergo retanning, colouring, fat liquoring or finishing. Leather may be dried by any of the common methods. Air-drying is the simplest method. The leather is hung or placed on racks and dried by the natural circulation of air around it. A toggling unit consists of a number of screens placed in a dryer that has controlled temperature and humidity. In a pasting unit, leathers are pasted on large sheets of plate glass, porcelain, or metal and sent through a tunnel dryer with several zones of controlled temperature and humidity. In vacuum drying, the leather is spread out, grain down, on a smooth surface to which heat is applied. A vacuum hood is placed over the surface, and a vacuum is applied to aid in drying the leather. High frequency drying involves the use of a high frequency electromagnetic field to dry the leather.

Chrome Tanning/Mineral Tanning

The mineral tanning process is also known as chrome tanning because the tanning agent that is frequently used is a salt of chromium. This is a method of tannage stemming back to the discovery in 1858, of leather production by treating skins with basic chromium sulphate. It is a quick, cost-effective, requiring much less time than the vegetable method and used for mainly tanning lighter leather. It produces leather with water-repellant and heat-resisting properties, of great tensile strength, but difficult to dye than that tanned by the tannic acid method.

The two basic methods employed are the one-bath and two-bath methods, the former being most often used. Chrome tanning is performed using a one-bath process that is based on the reaction between the hide and a trivalent chromium salt, usually a basic chromium sulphate. In the typical one-bath process, the hides are in a pickled state at a pH of 3 or lower, the chrome tanning materials are introduced, and the pH is raised. The widely used chemical in chrome tanning is sodium or potassium dichromate.

The process (Figure 18.4) involves treating the hides with a weak solution of dichromate to which sufficient amount of hydrochloric acid has been added to liberate chromic acid, which is the tanning principle. After the skins have taken up a bright yellow colour throughout their texture, they are drained and transferred to a bath of hyposulphite of soda to which hydrochloric acid is added again. This liberates sulphurous acid which reduces the excess of chromic acid to green chromic oxide, which coats each fibre as a casing to preserve it, and resulting in leather with a characteristic pale bluish-green colour. To prepare the stock for chrome tanning, the bated skins are pickled in a solution of salt and acid. The skins are then immersed in a basic chromium sulphate solution within a large revolving drum that tumbles the skins. This type of liquor penetrates the skins so rapidly that tannage is accomplished in less than a day. The chrome process originally involved the use of two different liquors, both solutions of compounds of chromium, and required substantially more time. The two-bath process is still used for some varieties of leather. Aluminium or zirconium compounds may be used in place of chromium in the production of white leather. Alum, formaldehyde, gluteraldehyde and synthetic tannins (syntans) are also used to impart special characteristics. As in vegetable tanned leather, the degree of control exercised in the tanning process has great influence on the nature of the leather produced. If, for example, the final pH of a chrome-tanned leather is too low, the leather will be flat, hard and wet, and may show grease spots on the surface; but if it is too high, the leather will become plump, loose and dry, and may have a drawn grain or may be too soft.

The two-bath method has almost been completely superseded by the one-bath tannage, except in certain cases where the older two-bath process is thought to give a particularly uniform tannage and a deposit of colloidal sulphur in the leather. The major characteristics of chrome-tanned leather are its blue-green colour and absence of filling power, i.e., an empty tannage. Chrome-tanned leather tends to be softer and stretchier than vegetable-tanned leather, and is very stable in water. Unlike vegetable-tanned or alum-tawed skins, chrome-tanned leather can withstand boiling water and has a shrinkage temperature higher at times than 100°C; however, it does not resist perspiration of organic acids well and is difficult to emboss. In addition, it does not take gold tooling well and is difficult to fabricate in operations such as turning-in, etc. It is, on the other hand, very durable leather. Chrome-tanned leather tends to be softer and more pliable than vegetable-tanned leather, it has higher thermal stability, is very stable in water, and takes less time to produce than vegetable-tanned leather. Almost all leather made from lighter-weight cattle hides and from the skin of sheep, lambs, goats, and pigs are chrome tanned. The production of chromium sulphate is given by the following equation.

$$Na_2Cr_2O_7 + 3SO_2 + H_2O \longrightarrow Na_2SO_4 + 2Cr(OH)SO_4$$

However, in chrome tanning, the additional processes such as retanning, dyeing and fat liquoring are usually performed to produce usable leathers and a preliminary degreasing step may be

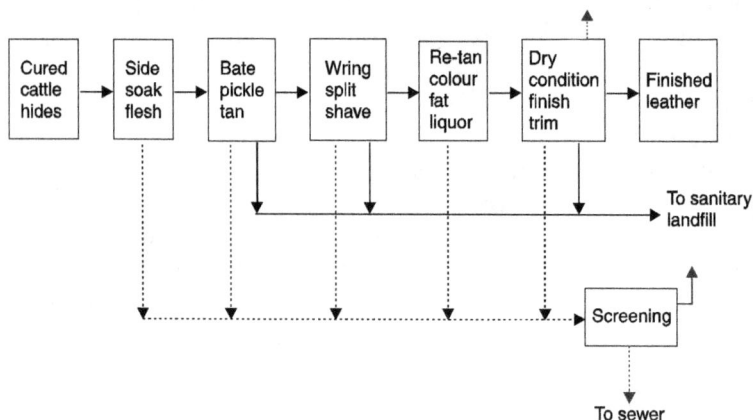

Figure 18.4   Process flow diagram of a chrome tannery

necessary when using animal skins, such as sheepskin. Following tanning, the chrome-tanned leather is piled down, wrung, and graded for the thickness and quality, split into flesh and grain layers, and shaved to the desired thickness. The grain leathers from the shaving machine are then separated for retanning, dyeing and fat liquoring. For other types of leather, i.e., shoe leather, the dye must penetrate further into the leather. Typical dyestuff are aniline-based compounds that combine with the skin to form an insoluble compound.

Dyeing and Fat Liquoring

The history of the dyeing of leather goes right back to ancient times, when natural dyes such as dyewood extracts laked with metal salts were used. The processes for using these products were complicated, and the range of colours was limited. With the advent of the "aniline" (synthetic) dyes at the end of the nineteenth century, dyeing became simpler and it became possible to dye virtually any shade. Currently the anionic dyes are used, to give improved physical properties (especially light fastness).

Fat liquoring is the process of introducing oil into the skin before the leather is dried. This is to replace the natural oils lost in beam houses and tan yard processes. Fat liquoring is usually performed in a drum using an oil emulsion at temperatures of about 60–66°C (140–150° F) for 30–40 mins. In the production of leather, fat liquoring is usually the last operation in the aqueous phase before drying. This process is carried out using either fish oils or synthetic oils that have been emulsified to allow their use in aqueous solutions. Like the retannage, it is of importance for the quality and properties of the leather. The fat liquoring process largely determines the mechanical and physical properties of the leather. If the leather is dried without fat liquoring, it becomes hard and thin, because the fibres are not lubricated. After fat liquoring, the leather is wrung, set out, dried, and finished. The function of the fat liquoring is to separate the fibres in the wet state so that they do not stick together too much during drying. After the dyes and fat liquors have been given sufficient time to penetrate the leather, they are then fixed to the leather by acidifying the leather to around pH 3 using formic acid. The use of formic acid is important as this enhances the effect of the dyes. Finally, the skins are lifted from the drum, and dried.

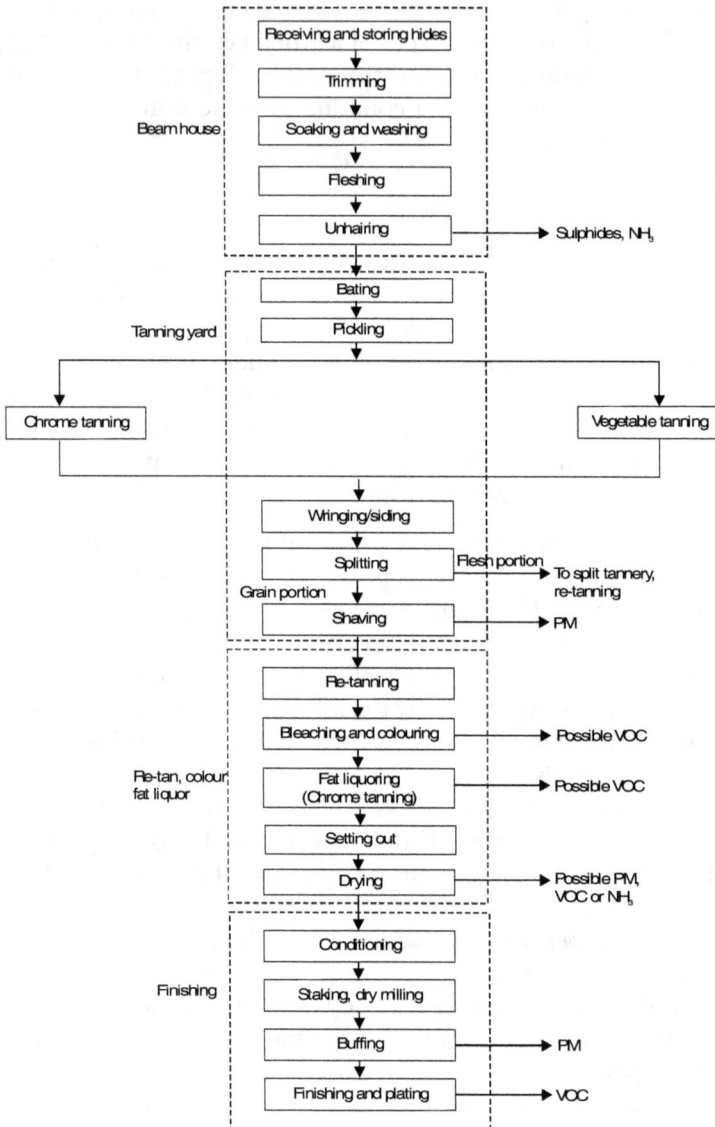

Figure 18.5   General flow diagram for leather tanning and finishing process

Leather Finishing

The finishing (Figure 18.5) process refers to all the steps that are carried out after drying. Leathers may be finished in a variety of ways: buffed with fine abrasives to produce a suede finish, waxed, shellacked, or treated with pigments, dyes, and resins to achieve a smooth, polished surface and the desired colour; or lacquered with urethane for a glossy patent leather. Water-based or solvent-based finishes may also be applied to the leather. Plating is then used to smooth the surface of the coating materials and bond them to the grain. Hides may also be embossed.

TANNERY EFFLUENTS—POLLUTION AND CONTROL

The treatment and processing of animal skins and hides is a source of considerable environmental impact. The tanning industry discharges different types of waste into the environment, primarily in the form of liquid effluents containing organic matters, chromium, sulphide, ammonium and other salts. Pollution becomes acute when tanneries are concentrated in clusters in arid areas. Solid wastes and some atmospheric emissions may also arise. The major public concern over tanneries has traditionally been about odours and water pollution from untreated discharges. Other issues have arisen more recently from the increasing use of synthetic chemicals such as pesticides, solvents, dyes, finishing agents and new processing chemicals, which introduce problems of toxicity and persistence. Simple measures intended to control pollution can themselves create secondary cross-media environmental impacts such as ground water pollution, soil contamination, sludge dumping and chemical poisoning. Tanning technology that is now available, based on a lower chemical and water consumption, has less impact on the environment than traditional processes. However, many obstacles remain to its widespread application.

The different wastes and environmental impacts associated with the various processes used in the tanning industry is presented in Figure 18.6.

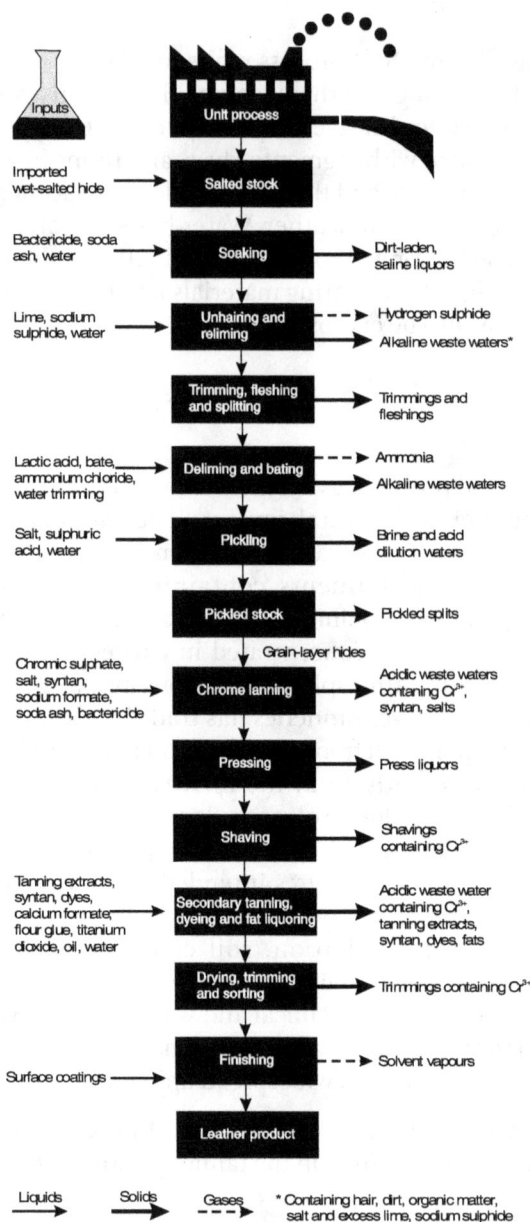

Figure 18.6   Wastes associated with processes in tanning industries

Pollution Control

Water pollution control Untreated tannery wastes in surface waters can bring about a rapid deterioration of their physical, chemical and biological properties. Simple end-of-pipe effluent treatment processes can reduce over 50% of suspended solids and biochemical oxygen demand (BOD) of effluents. More sophisticated measures are capable of higher levels of treatment. As tannery effluents contain several chemical constituents that need to be treated, a sequence of treatment processes in turn must be used. Flow segregation is useful to allow separate treatment of concentrated waste streams.

Technological choices available for treatment of tannery effluents are summarized in Table 18.1.

Table 18.1 Methods for the treatment of tannery effluents

Pre-treatment settling	Mechanical screening to remove coarse material
	Flow equalization (balancing)
Primary treatment	Sulphide removal from beam house effluents
	Chromium removal from tanning effluents
	Physical–chemical treatment for BOD reduction and neutralization
Secondary treatment	Biological treatment
	Activated sludge (oxidation ditch)
	Activated sludge (conventional)
	Lagooning (aerated, facultative or anaerobic)
Tertiary treatment	Nitrification and denitrification
Sedimentation and sludge handling	Different shapes and dimensions of tanks and basins

Air pollution control Air emissions fall into three broad groups: odours, solvent vapours from finishing operations and gas emissions from the incineration of wastes. Biological decomposition of organic

Table 18.2   Emissions of toxic air pollutants from a typical tannery

Emission point	Pollutants	Emission rate kg/hr	Control methods
Solvent receiving	Methyl ethyl ketone	22.58	
	Methyl isobutyl ketone	1.67	
	Toluene	10.04	
	Xylol	1.17	
Mixing vault	Methyl ethyl ketone	0.52	
Supply drum	Methyl ethyl ketone	0.52	
Spray chamber	Diacetone alcohol	1.89	
	Glycol ether EB	11.85	
	Glycol ether PMA	7.6	
	Methyl ethyl ketone	75.72	Incineration
	Methyl isobutyl	59.05	
	Toluene	95.78	
	Xylol	33.38	
Dryer	Diacetone alcohol	1.89	Process modification (e.g. water-based process instead of solvent-based process)
	Glycol ether EB	11.85	
	Glycol ether PMA	7.6	
	Methyl ethyl ketone	75.72	
	Methyl isobutyl	59.05	
	Toluene	95.78	
	Xylol	33.38	
Receiving recycled solvents	Acetone	0.61	
	Methyl ethyl ketone	0.98	
	Toluene	0.61	
Cleaning operation	Less than 1kg/hr of each pollutant		
Waste solvent storage	Less than 1kg/hr of each pollutant		

matter as well as sulphide and ammonia emissions from waste waters is responsible for the characteristic objectionable odours arising from tanneries. The siting of installations has been an issue because of the odours that have historically been associated with tanneries. Reduction of these odours is more a question of operational maintenance than of technology. Solvent and other vapours from the finishing operations vary with the type of chemicals used and various technical methods are employed to reduce their generation and release. Up to 30% of the solvent used may be wasted through emissions, while modern processes are available to reduce this to around 3% in many cases. The practice by many tanneries of incinerating solid wastes and off cuts raises the importance of adopting good incinerator design and following careful operating practices. Table 18.2 describes a list of toxic air pollutants.

Waste management   Treatment of sludge constitutes the largest disposal problem, apart from effluents. Sludges of organic composition, if free from chrome or sulphides, have value as soil conditioners as well as fertilizers due to the nitrogenous compounds contained in them. These benefits are best realized by ploughing immediately after application. Agricultural use of chrome-containing soils has been a matter of controversy in various jurisdictions, where guidelines have determined acceptable applications. Various markets exist for the conversion of trimmings and fleshings into by-products used for a variety of purposes, including the production of gelatin, glue, leather board, tallow grease, and proteins for animal feed. Processed effluents, subject to suitable treatment and quality control, are sometimes used for irrigation where water is in short supply. To avoid problems of leachate generation and odour, only solids and dewatered sludges should be disposed off at landfill sites. Care must be taken to ensure that tannery wastes do not react with other industrial residues, such as acidic wastes, which can react to create toxic hydrogen sulphide gas. Incineration under uncontrolled conditions leads to unacceptable emissions and is not recommended. Table 18.3 describes a list of hazardous wastes from any tannery.

Table 18.3 Hazardous wastes from a typical tannery

Waste source	Pollutant	Concentration range (wet weight in mg/kg)	Disposal method
Chrome trimmings and shavings	Cr^{3+}	2,200–21,000	
Chrome fleshings	Cr^{3+}	4,000	
Unfinished chrome leather trim	Cr^{3+}	4,600–37,000	
	Cu	2.3–468	
	Pb	2.5–476	
	Zn	9.1–156	
Buffing dust	Cr^{3+}	19–22,000	Landfill
	Cu	29–1,900	
	Pb	2–924	
	Zn	160	
Finishing residues	Cr^{3+}	0.45–12,000	Dewater sludge; all waste disposed in certified hazardous waste disposal facility
	Cu	0.35–208	
	Pb	2.5–69,200	
	Zn	14–876	
Finished leather trim	Cr^{3+}	1,600–41,000	
	Pb	100–3,300	
Sewer screenings	Cr^{3+}	0.27–14,000	Landfill with leachate collection
	Pb	2–110	
	Zn	35–128	
Waste water treatment residues (sludges)	Cr^{3+}	0.33–19,400	
	Cu	0.12–8,400	
	Pb	0.75–240	
	Zn	1.2–147	

Pollution Prevention

Improving production technologies to increase environmental performance can achieve a number of objectives, such as:

- Increasing the efficiency of chemical utilization
- Reducing water or energy consumption
- Recovering or recycling rejected materials

Water consumption can vary considerably, ranging from less than 25 L/kg of raw hide to greater than 80 L/kg. Water use efficiency can be improved through the application of techniques such as increased volume control of processing waters, "batch" versus "running water" washes, low float modification of existing equipment; low float techniques using updated equipment, re-use of waste water in less critical processes and recycling of individual process liquors. Traditional soaking and removal of hair account for over 50% of the BOD and chemical oxygen demand (COD) load in typical tanning effluents. Various methods can be employed to substitute for sulphide, to recycle lime/sulphide liquors and to incorporate dehairing techniques. Reduction in chromium pollution can be achieved through measures to increase the levels of chrome that is fixed in the tanning bath and reduce the amounts that are "bled out" in subsequent processes. Other methods to reduce release of chromium are through direct recycling of used chrome liquors (which also reduces salinity of waste effluent) and the treatment of collected chrome-bearing liquors with alkali to precipitate the chromium as hydroxide, which can then be recycled.

Where vegetable tanning is employed, preconditioning of hides can enhance the penetration and fixation of tannins and contribute to decreased tannin concentrations in effluents. Other tanning agents such as titanium have been used as substitutes for chromium to produce salts of generally lower toxicity and to generate sludges that are inert and safer to handle.

19

POLYMER CHEMISTRY

HISTORY AND SIGNIFICANCE OF POLYMERS

The simplest definition of a polymer is that it is a molecule made up of many units. Think of a polymer as a chain. Each link of the chain is the "mer" or basic unit that is made of carbon, hydrogen, oxygen, and/or silicon. In the making of the chain, many links or "mers" are hooked or polymerized together. The differences in polymer properties result from how the atoms and chains are linked together in space.

All the substances referred to as *polymers* or macromolecules are giant molecules with molar masses ranging from several thousands to several millions.

Polymerization can be demonstrated by linking strips of construction paper together to make paper garlands or hooking together hundreds of paper clips to form chains.

In chemistry, a long molecule made up of a chain of smaller, simpler molecules is termed as polymer. The word polymer comes from the Greek word *polumeres* meaning "having many parts".

Polymers are a very important class of materials. They occur naturally in the form of:

1. Proteins—silk, collagen and keratin
2. Carbohydrates—cellulose, starch and glycogen
3. DNA–RNA
4. Rubber (hydrocarbon base)

Natural polymers include things such as tar and shellac, tortoise shell and horns, as well as tree saps that produce amber and latex. These polymers were processed with heat and pressure into useful articles like hair ornaments and jewellery. Natural polymers began to be chemically modified during the 1800s to produce many materials. The most famous of these were vulcanized rubber, gun cotton and celluloid. The first synthetic polymer produced was bakelite in 1909 and was soon followed by the first semi-synthetic fibre, rayon, which was developed in 1911.

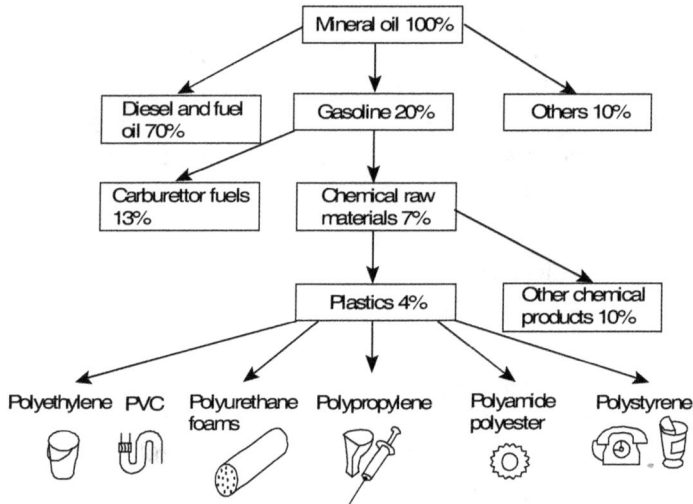

Figure 19.1   Main source for polymers: mineral oil (previously coal)

Even with these developments, it was not until World War II that significant changes took place in the polymer industry. Prior to World War II, natural substances were generally available; therefore, synthetics that were being developed were not a necessity.

During the war, the natural sources of latex, wool, silk and other materials were cut off, which marked the making of synthetics critical. About 4% of mineral oil was used in the production of plastics such as PE, PVC, etc. During this period, the use of nylon, acrylic, neoprene, SBR, polyethylene, and many more polymers took over the place of natural materials that were no longer available. Since then, the polymer industry has continued to grow and has evolved into one of the fastest growing industries in the world.

Engineering polymers, however, deal with usually synthetic polymers. The main advantages of engineering polymers are their manufacturability, recyclability, mechanical properties, and lower cost as compared to many alloys and ceramics. Also, the macromolecular structure of synthetic polymers provides good biocompatibility and allows them to perform many biomimetic tasks, which include drug delivery, use as grafts for arteries and veins and use in artificial tendons, ligaments and joints.

The unit forming the repetitive pattern is called a "mer" or "monomer".

Plastics are polymers, molecules that form long chains, repeating themselves like pearls in a necklace.

Figure 19.2   Ball and stick model of a polymer

PLASTICS AND POLYMERS

The word polymer is applied when there are more than 50 mers stacked together. Polymers that have a 1-D structure will have different properties than those that have either a 2-D or 3-D structure. Most of the plastics are polymers. The origin of the word "plastics" comes from Greek. Its original Greek root means, "to form."

Historically, polymers have mostly been used to make solid plastics where the chains virtually do not move. Now the search is for newer applications of polymer liquids where fluctuations (Brownian motion) and interactions (the sticking together or association of different types of molecules) can play a more important role.

Thus, a polymer is a long, repeating chain of atoms, formed through the linkage of many molecules called monomers. The monomers can be identical, or in complex polymers such as proteins, the monomers have one or more substituted chemical groups. Although most of them are organic (based on carbon chains), there are also many inorganic polymers.

One-dimensional polymers   One-dimensional polymers are most common. They can occur whenever reacting molecules join to make a chain. If the long chains pack regularly, side-by-side, they tend to form crystalline polymers. If the long chain molecules are irregularly tangled, the polymer is amorphous, since there is no long-range order. This type of polymer is glassy.

Two-dimensional polymers   Two-dimensional polymers are rare. The best example would be graphite. It is the structure which is responsible for the great lubricating capability of graphite. The condition to form a planar structure is to have three or more active groups all directed in the same plane and capable of forming a planar network. This structure offers low shear strength and good lubricating properties.

Three-dimensional polymers   Crystalline diamond is an example of the three-dimensional crystalline polymer in which carbon is linked to four corners of the tetrahedron and these are packed with long-range order in space to form a lattice. Because of this, diamonds have properties which are much more like ceramics than polymers in terms of mechanical behaviour (high melting point, modulus, hardness, strength and fracture behaviour).

Structure of Polymers

Many common classes of polymers are composed of hydrocarbons. These polymers are specifically made of small units bonded into long chains. Carbon atoms make up the backbone of the molecule and hydrogen atoms are bonded along the backbone. The structure of polyethylene, the simplest polymer is shown in Figure 19.2.

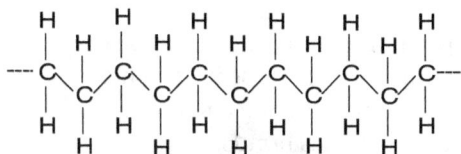

Figure 19.3   Structure of polyethylene

There are other polymers that contain only carbon and hydrogen, for example, polypropylene, polybutylene, polystyrene and polymethylpentene. Even though the basic make-up of many polymers is carbon and hydrogen, other elements can also be involved. Oxygen, chlorine, fluorine, nitrogen, silicon, phosphorus, and sulphur are other elements that are found in the molecular make-up of polymers. Polyvinyl chloride (PVC) contains chlorine. Nylon contains nitrogen and oxygen. Teflon contains fluorine. Polyesters and polycarbonates contain oxygen. Vulcanized rubber and thiokol contains sulphur. There are also some polymers having silicon or phosphorus as backbones. These are considered inorganic polymers. The most famous silicon-based polymers are silicones.

CLASSIFICATION OF POLYMERS

CLASSIFICATION BASED ON PHYSICAL PROPERTY

Based on the physical property related to the heat response, polymers are classified as discussed below.

Thermoplastic Polymers

Technically, thermoplastics are true plastics. The term "plastic" is commonly applied to all synthetic polymers. Plastics that soften when heated and become firm again when cooled are thermoplastic polymers. This is the most popular type of plastic because the heating and cooling may be repeated. These are linear, one-dimensional, have strong intramolecular covalent bonds and weak intermolecular van der Waals bonds. At elevated temperatures, it is easy to "melt" these bonds and have molecular chains readily slide past one another. These polymers are capable of flowing at elevated temperatures, and can be remoulded into different forms and in general, are dissolvable. A thermoplastic, under the application of appropriate heat, can be melted into a "liquid" state.

> Chewing gum is a thermoplastic that becomes extremely brittle when the outside temperatures drop below its glass transition temperature. This is a useful property to use in order to remove chewing gum adhered to clothes. Once warmed above T_g, however, the gum quickly softens and regains its flexibility.

Thermoplastics can be repeatedly melted and solidified without damage. The thermoplastic substance contains long, thin molecules, which form tangled chains, and is rigid at lower temperature known as its glass transition temperature (T_g). Below T_g, the substance is brittle, having the characteristic properties of a glass; above T_g, the substance becomes flexible and soft. Some thermoplastics such as polystyrene melt before reaching their glass transition temperatures and remain as rigid materials up to their melting points. Thermoplastic polymers are used frequently in items such as food storage containers and toys that are not exposed to high temperatures. Additionally, thermoplastic polymers can be moulded, pressed and extruded.

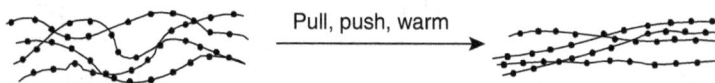
Pull, push, warm

After "mers" are polymerized together, thermoplastics are extruded into strings of plastics that are chopped into pellets. These pellets can be mixed with colourants and additives that are then remelted into other shapes. The extrusion process is customarily designed to make a variety of common objects. Plastic pellets are fed into an extruder through the hopper. The plastic travels into a heated barrel where one or two screws turn continuously, mixing, blending and melting the plastic. The screw also moves the plastic through the barrel to the end where the die is located. Depending on the shape of the die, a variety of objects are developed. Some common extruded products include drinking straws, moulding strips, hose and tubing, seamless gutters, window frames and vinyl siding. Frozen foods, dry goods, meat and vegetable wrap, boil-in-bags and microwave wrap are examples where filmed polymers are used and contribute to the health and safety of household. Plastic films have revolutionized the food packaging industry. Fibres and filaments are also produced. The specially shaped strands of molten plastics are cooled, then drawn and stretched. The drawing process will orient the molecule chains to produce the desired strength, while the shape of the fibre will affect other characteristics. All natural fibres have distinguishing cross-sections that fibre producers mimic to create a synthetic that resembles a natural fibre.

Thermosetting Polymers

These are three-dimensional amorphous polymers, highly cross-linked (strong, covalent intermolecular bonds) networks with no long-range order. Thermosetting polymers are those resins, which are "set" through a chemical reaction resulting in cross-linking of the structure into one large three-dimensional molecular network. Once the polymerization is complete, the polymer becomes hard, infusible, insoluble material, which cannot be softened, melted, or moulded non-destructively when reheated. A good example of a thermosetting plastic is a two-part epoxy system in which a resin and hardener (both in a viscous state) are mixed and within several minutes, the polymerization is complete resulting in a hard epoxy plastic. Another example is bakelite, which is used in toasters, handles for pots and pans, dishes, electrical outlets and billiard balls.

Thermosetting plastics are often employed in high-temperature environments, such as for electrical insulation in electric motors and gasoline engines.

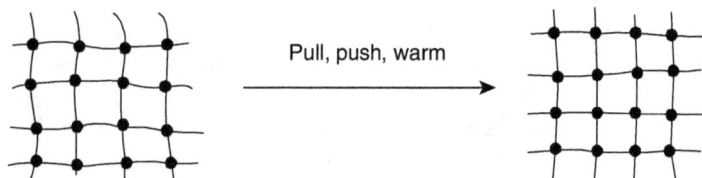

CLASSIFICATION BASED ON COMPOSITION

Based on composition, the polymers are classified as:

Homopolymers

Homopolymers are composed of one species of monomer. They consist of chains with identical bonding linkages to each monomer unit. This usually implies that the polymer is made from all identical monomer molecules. These may be represented as: –[A-A-A-A-A-A]–

Examples

Polyethylene (PE) $(-CH_2-CH_2-)_n$

Polystyrene (PS) $(-CH_2-CH-)_n$
                   |
                   C_6H_5

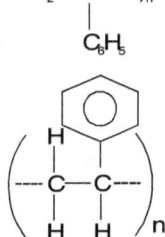

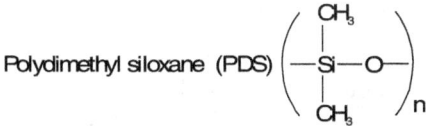

Polyethylene oxide (PEO) $\quad (-CH_2-CH_2-O-)_n$

Polydimethyl siloxane (PDS) $\left(\begin{array}{c} CH_3 \\ | \\ -Si-O- \\ | \\ CH_3 \end{array}\right)_n$

Copolymers

Polymers of all types in which the long chains are produced by joining two or more different kinds of monomers are termed copolymers. These may be represented as: —[A-B-A-B-A-B]—

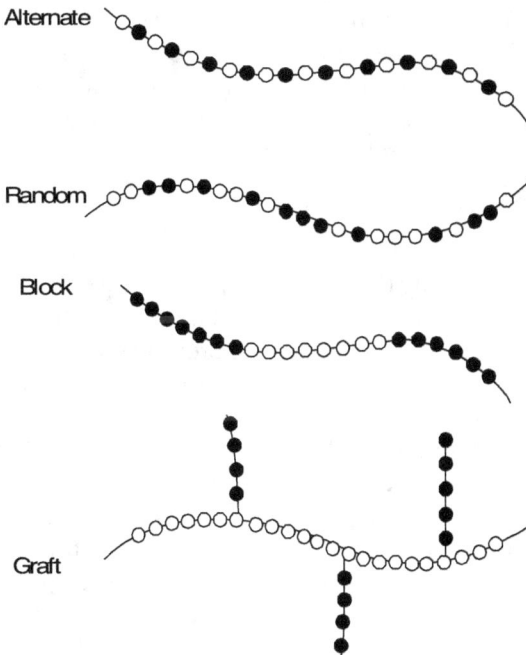

Types of copolymers

CLASSIFICATION BASED ON REACTION MODE OF POLYMERIZATION

Polymers are further classified by the reaction mode of polymerization. The initial compound that is used to form polymers is the "mer" or monomer. Monomers are chemically joined together in either of the two ways: addition polymerization or condensation polymerization. During polymerization, the entropy of mixing reduces drastically.

Addition Polymers

The monomer molecules bond to each other at unsaturated carbon atoms; without loss of any other atoms or molecules. During polymerization, the double bonds between the pair of carbon atoms "open up" and the carbon atoms of separate alkene molecules join to form a molecule of polyalkene. Alkene monomers are the biggest groups of polymers in this class. Table 19.1 describes the uses of some of the important addition polymers.

Addition polymerization involves three basic steps: (1) Initiation   (2) Propagation and   (3) Termination.

For example, during the initiation phase of the polymerization of polyethylene, the double bonds in the ethylene "mers" break and begin to bond together. A catalyst or promoter may be necessary to begin or speed up the reaction. The second step, propagation, involves the continued addition of monomers together into chains. And finally during termination, all monomers may be used, causing the reaction to cease. Polyethylene has the simplest structure. Even though the backbone of other polymers will be similarly formed by breaking the double bond between olefinic carbons, the remaining carbons in the "mer" will form a functional group whose orientation about the backbone will affect the physical nature of the resulting polymer. For example, propylene is the "mer" that will form polypropylene.

Free radical addition polymerization of ethylene takes place at high temperatures and pressures, approximately 300–2000°C. While most other free radical polymerizations do not require such extreme temperatures and pressures, they do tend to lack control.

One effect of this lack of control is a high degree of branching. In addition polymerization, as termination occurs randomly, when two chains collide, it is impossible to control the length of individual chains. A newer method of polymerization similar to free radical, but allowing more control involves the Ziegler–Natta catalyst.

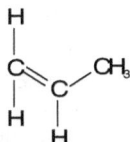

Propylene monomer

Polymerization will be initiated by the double bond breaking followed by joining of the monomers. Therefore, the methyl group on the propylene "mer" has the potential to be located at various points along the backbone. If the methyl group (CH_3) is oriented repeatedly on one side of the chain on alternate carbons, it is called *isotactic*. Majority of the polypropylene polymers have this configuration.

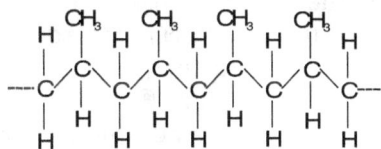

Isotactic configuration

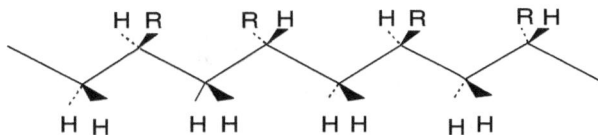

Syndiotactic configuration

During the propagation of polymer chains, branching can occur. In radical polymerization, this occurs when a chain curls back and bonds to an earlier part of the chain. When this curl breaks, it leaves small chains sprouting from the main carbon backbone. Branched carbon chains cannot line up as close to each other as unbranched chains can. This causes less contact between atoms of different chains, and fewer opportunities for induced or permanent dipoles to occur. A low density results from the chains being further apart. Lower melting points and tensile strengths are evident, because the intermolecular bonds are weaker and require less energy to break. Developed by Karl Ziegler and Giulio Natta in the 1950s, Ziegler–Natta catalysts (triethylaluminium in the presence of a metal (IV) chloride) largely solved this problem. Instead of a free radical reaction, the initial propene monomer inserts between the aluminium atom and one of the ethyl groups in the catalyst. The polymer is then able to grow out from the aluminium atom and results in almost totally unbranched chains. With the new catalysts, the tacticity of the polypropylene chain, the alignment of alkyl groups, could also be controlled. Different metal chlorides allowed the selective production of each form, i.e., syndiotactic, isotactic and atactic polymer chains could be selectively created. However, there are further complications to be solved. If the Ziegler–Natta catalyst was poisoned or damaged, then the chain stopped growing. In addition, the use of Ziegler–Natta catalyst demanded monomers to be small, and it was still impossible to control the molecular mass of the polymer chains. Again, new catalysts like, the metallocenes, were developed to tackle these problems. Due to their structure, they have less premature chain termination and branching. Other forms of addition polymerization include cationic and anionic polymerization. While, not used largely in industry, yet due to stringent reaction conditions such as lack of water and oxygen, these methods provide ways to polymerize some monomers that cannot be polymerized by free radical methods such as polypropylene. Cationic and anionic mechanisms are also more ideally suited for living polymerizations, although free radical living polymerizations have also been developed. A polymerization process is considered a "living" if there is absence of termination reaction (or) a dead chain end can be re-initiated.

Stereoregularity or tacticity describes the isomeric arrangement of functional groups on the backbone of carbon chains. Isotactic chains are defined as having substituent groups aligned in one direction. This enables them to line-up close to each other, creating crystalline areas and resulting in highly rigid polymers. In contrast, atactic chains have randomly aligned substituent groups. The chains do not fit together well and the intermolecular forces are low. This leads to a low density and tensile strength, but a high degree of flexibility. Syndiotactic substituent groups alternate regularly in opposite directions. Because of this regularity, syndiotactic chains can position themselves close to each other, though not as close as isotactic polymers. Syndiotactic polymers have better impact on strength than isotactic polymers because of the higher flexibility resulting from their weaker intermolecular forces.

A polymerization reaction can cease by quenching the reaction. Polymers formed by addition polymerization include acrylic, polyethylene and polystyrene, to name a few.

> Addition polymerization is described as the process of "mers" joining by each one adding on to the end of the last "mer". A simple visual of the process is paper clips joined together to form a long chain.

Polymers formed by addition polymerization are often thermoplastic in nature. Thermoplastics are like hot melt glue sticks that can be heated and made soft and then become hard when cooled. Thermoplastic polymers are easily processed and reprocessed or recycled. The majority of polymers used today are thermoplastics.

The first polymerization of ethylene was accomplished in 1933 by the use of very high pressure (1000 atm) and oxygen as a catalyst. Nowadays, with the development of the use of powerful catalysts, addition can occur at atmospheric pressure. Polymethyl methacrylate, also called lucite or plexiglas (originally developed as an unbreakable substitute for glass in airplane canopies), belongs to this group of addition polymers. The polymerization is initiated by a variety of substances (such as benzoyl peroxide) that can form a free radical with the unsaturated carbon atom. The resulting addition polymer is described as a branched polymer.

Condensation Polymers

This is another group of polymers formed by condensation polymerization. During condensation polymerization, a small molecule is eliminated as the monomers join. In the process of condensation, two compounds with reactive atoms at the end of their molecules react together, usually with the release of a small molecular unit such as water or hydrogen chloride. Common polymers in this group include nylons, some polyesters, urea-formaldehyde resins and urethanes (See Table 19.2). The presence of two or more functional groups in the monomer usually leads to the production of a cross-linked polymer. Glyptal resin, formed by the reaction between phthalic acid and glycerol, is a condensation copolymer, since two different types of monomers combine to form the chain. If monomers are linked through ester $-\underset{\underset{O}{\|}}{C}-O-CH_2-$ bonds; the resin is termed as polyester. The more reactive phthalic anhydride is often used in the place of phthalic acid in this reaction. These polymers can be thermoplastic or thermosetting in nature. Once a thermoset polymer is formed, it cannot be melted and reformed. All plastics flow at some time during their processing and are solid in the finished state, but once a thermoset is processed, it is dramatically different and cannot be reformed. Linear silicones or polysiloxanes, are comparatively new polymers based upon silicon–oxygen–silicon linkages. These polymers may be cross-linked to various degrees by additional $-Si-O-Si-$ bonding between adjacent chains.

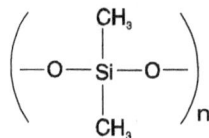

$$\left(-O-\underset{\underset{CH_3}{|}}{\overset{\overset{CH_3}{|}}{Si}}-O- \right)_n$$

Polydimethyl siloxane

The R group is generally a hydrocarbon group such as $-CH_3$ (methyl), $-CH_2CH_3$ (ethyl) or $-C_6H_5$ (phenyl). Silicones are stable at much higher temperatures than carbon-based polymers, yet they remain flexible even at exceedingly low temperatures. Among such silicones is the cross-linked polymerization product

Table 19.1 Some of the uses of important addition polymers

Example	Monomer(s)	Polymer	Use
Polyethylene (Common polymer)	$CH_2=CH_2$	$(-CH_2-CH_2-)_n$	Bags, wire insulations, squeeze bottles
Polypropylene	$CH_2=CH_2$ $\|$ CH_3	$\left(-CH_2-CH_2-\atop \qquad\quad CH_3\right)_n$	Fibres, indoor–outdoor carpets, bottles, rope
Polystyrene	$H_2C=CH$ (with phenyl ring)	$\left(-H_2C-CH-\right)_n$ (with phenyl ring)	Styrofoam; drinking cups, building insulation, packing materials
Polyvinylchloride (PVC)	$H_2C=CH$ $\|$ Cl	$\left(-H_2C-CH-\atop \qquad\quad Cl\right)_n$	Synthetic leathers, clear bottles, floor coverings, phonograph records, water pipes

(*Contd.*)

Table 19.1 (Continued)

Example	Monomer(s)	Polymer	Use
Polytetrafluoroethylene (Teflon)	$CF_2{=}CF_2$	$(-CF_2-CF_2-)_n$	Nonstick surfaces, chemically-resistant films, cookware coatings
Polymethylmethacrylate (Lucite, Plexiglas)	$H_2C{=}C$ with CO_2CH_3 and CH_3	$\left(-CH_2-\underset{CH_3}{\overset{CO_2CH_3}{C}}-\right)_n$	Unbreakable glass, latex paints
Polyacrylonitrile (Orlon, Acrilan, Creslan)	$H_2C{=}CH$, CN	$\left(-H_2C-\underset{CN}{CH}-\right)_n$	Fibres for sweaters, blankets, carpets
Polyvinylacetate (PVA)	$H_2C{=}CH_2$, $O-C({=}O)-CH_3$	$(-CH_2-CH_2-)_n$, $O-C({=}O)-CH_3$	Adhesives, latex paints, chewing gum, textile coatings

Name	Monomer	Polymer	Notes
Natural rubber	$H_2C=C(CH_3)-CH=CH_2$	$\left(-CH_2-\underset{CH_3}{C}=CH-CH_2-\right)_n$	Cross-linked with sulphur (vulcanization)
Polychloroprene (Neoprene rubber)	$H_2C=C(Cl)-CH=CH_2$	$\left(-CH_2-\underset{Cl}{C}=CH-CH_2-\right)_n$	Cross-linked with zinc oxide; resistant to oil, gasoline
Styrene–Butadiene Rubber (SBR)	$HC(=CH_2)(C_6H_5)$; $CH_2=CHCH=CH_2$	$\left(-H_2C-CH(C_6H_5)-CH_2-CH=CH-CH_2-\right)_n$	Cross-linked with peroxides; most commonly used for tyres 25% styrene, 75% butadiene

of dimethyldichlorosilane, $(CH_3)_2Si(Cl)_2$ with the release of hydrogen chloride.

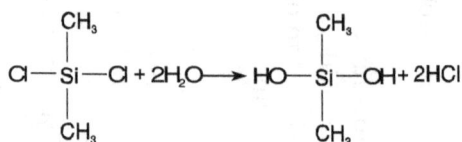

$$Cl-\underset{\underset{CH_3}{|}}{\overset{\overset{CH_3}{|}}{Si}}-Cl + 2H_2O \longrightarrow HO-\underset{\underset{CH_3}{|}}{\overset{\overset{CH_3}{|}}{Si}}-OH + 2HCl$$

The unstable dimethyldihydroxysilane condenses rapidly to form a low molecular mass, linear polymethylsiloxane polymer with the elimination of water. This material is an oily liquid. Additional heating continues the polymerization to form longer chains. The addition of boron trioxide allows three such chains to be joined to form Silly Putty.

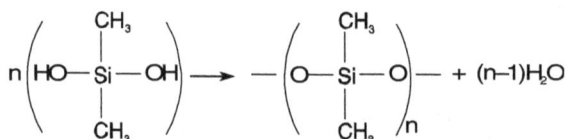

$$n\left(HO-\underset{\underset{CH_3}{|}}{\overset{\overset{CH_3}{|}}{Si}}-OH\right) \longrightarrow -\left(O-\underset{\underset{CH_3}{|}}{\overset{\overset{CH_3}{|}}{Si}}-O\right)_n- + (n-1)H_2O$$

Polydimethylsiloxane

Silly Putty with the —Si—O—Si—O— backbone terminates in Si—OH groups. These chain ends are cross-linked through hydrogen bonds with boric acid $B(OH)_3$. This polymer has physical properties between those of a fluid and an elastomer. Although it is resistant to rapid deformation, it flows easily with gradually applied stress.

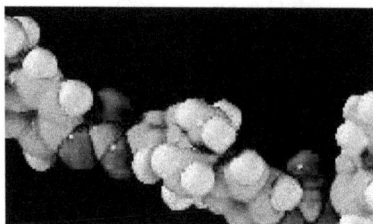

"Mers" joined together in long chains
have a linear zig-zig arrangement

Table 19.2 Some of the uses of few important condensation polymers

Example	Monomer(s)	Polymer	Use
Polyamides (Nylons)	$HOOC-(CH_2)_n-COOH$ $H_2N-(CH_2)_n-NH_2$	$\cdots C(=O)-(CH_2)_n-C(=O)-NH-(CH_2)_n-NH\cdots$	Fibres, moulded objects
Polyesters (Dacron™, Mylar™, Fortrel™)	$HOOC-\underset{}{\bigcirc}-COOH$ $HO-(CH_2)_n-OH$	$\cdots C(=O)-\underset{}{\bigcirc}-C(=O)-O-(CH_2)_n-O\cdots$	Linear polyesters, fibres, recording tape
Polyesters (Glyptal™ resin)	phthalic anhydride $HO-CH_2-\underset{OH}{CH}-CH_2OH$	cross-linked phthalate–glycerol network: $\cdots O-C(=O)-\text{(benzene ring)}-C(=O)-O-CH_2-\underset{OH}{CH}-CH_2OH$	Cross-linked polyester; paints

(*Contd.*)

Table 19.2   (Continued)

Example	Monomer(s)	Polymer	Use
Polyesters (Casting resin)	HO—C—CH=CH—C—OH (with two C=O groups) HO—(CH₂)ₙ—OH	···—C—CH=CH—C—O—(CH₂)ₙ—O—··· (with two C=O groups)	Cross-linked with styrene and peroxide, fibreglass, boat resin
Phenol formaldehyde resin (Bakelite)	phenol (OH on benzene ring) O=CH₂	cross-linked phenol-formaldehyde structure with CH₂ bridges and OH groups	Mixed with fillers, moulded electrical goods, adhesives, laminates, varnishes
Cellulose acetate	glucose ring with CH₂OH, OH, CH₃COOH, OH	glucose ring with CH₂OAC, OAC, OAC	Photographic films

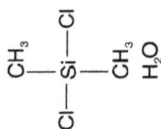

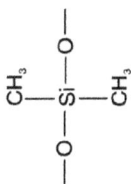

	Monomers	Polymer	Uses
Silicones	CH_3-Si with Cl, Cl, CH_3; H_2O	$-O-Si(CH_3)(CH_3)-O-$	Water-repellant coatings, temperature resistant fluids, rubbers (CH_3SiCl_3 cross-links in water)
Polyurethanes	$O=C=N$ and $N=C=O$ substituted toluene (CH_3); $HO-(CH_2)_n-OH$	$-NH-C(=O)-O-(CH_2)_n-O-$ and $-NH-C(=O)-O-(CH_2)_n-O-$ on CH_3 substituted benzene ring	Rigid and flexible foams, fibres

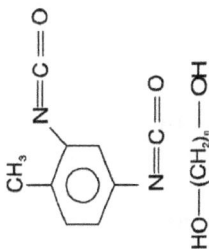

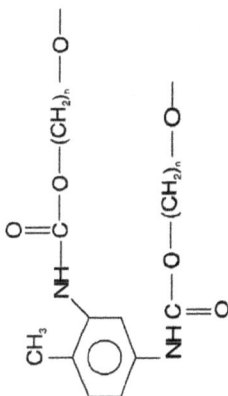

The type of polymerization will affect the thermal properties of the formed polymer; likewise, the arrangement of the "mers" within the molecule will affect the physical characteristics of the formed polymer. "Mers" joined together in long chains have a linear configuration very similar to a paper clip chain, even though in actuality tetrahedral bonds give the molecule a zig-zag arrangement. During polymerization, the "mers" not only form straight chains, but also long side chains off the main backbone. The resulting configuration is described as branched, resembling a tree branch or grape stem. A third configuration is achieved by the long chains being chemically linked together. An example would be natural rubber (isoprene) being reacted with sulphur. The sulphur bonds the chains to form a giant meshwork molecular structure that is known as vulcanized rubber. This is a cross-linked configuration.

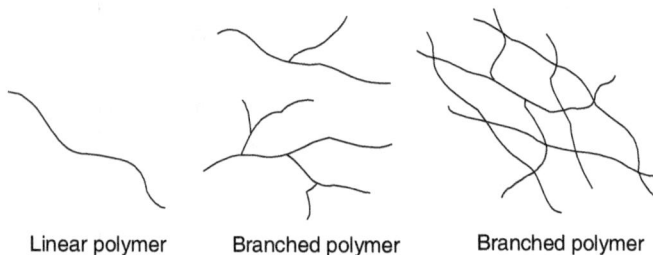

Linear polymer   Branched polymer   Branched polymer

RUBBERS AND ELASTOMERS

In general, rubber is a material, which can be stretched to at least twice its original length and rapidly contracted to its original length. Thus, rubber must be a high polymer with very long chains and its elasticity, from a molecular standpoint, is due to the coiling and uncoiling of the very long chains. To have "rubber-elastic properties", rubber material is heated above its glass transition temperature and it must be amorphous in its unstretched state, since crystallinity hinders coiling and uncoiling. Rubbers are cross-linked in order to prevent chains from slipping past one another under stress. Natural rubber is thermoplastic, and in its natural form it becomes "soft" and sticky" on hot days which is not a good property for its use in automobile tyre. Until Goodyear discovered a curing reaction with sulphur in 1839, rubbers were not cross-linked and did not have unique mechanical, rubber-elastic properties.

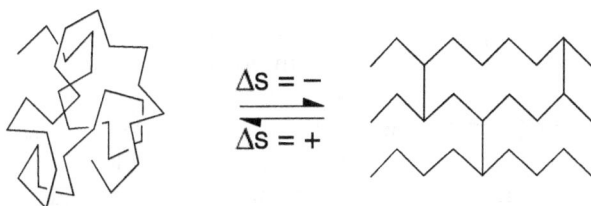

Cross-linked rubber bands are one of the best illustrations of entropy.

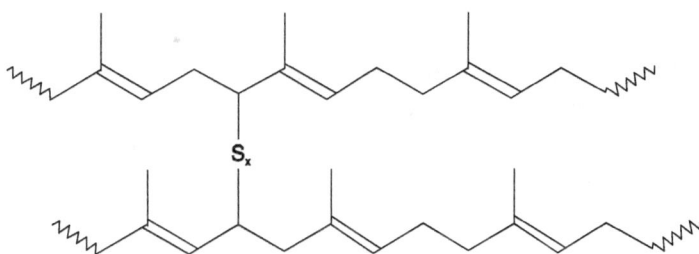

1839: Lightly cross-linked rubber (rubber-ebonite/vulcanized rubber) was discovered by Charles Goodyear.

1851: Heavily cross-linked rubber ebonite was discovered by Nelson Goodyear.

To be elastomeric, a polymer should exhibit the following characteristics:

- It must not crystallize easily
- It should have relatively free chain rotation
- It should exhibit delayed plastic deformation by cross-linking (by vulcanization)

FIBRES

Many of the polymers used as synthetic fibres are identical to those used in plastics but the two industries developed separately and employ different testing methods and terminology. A fibre is often defined as a material having an aspect ratio (length/diameter) of at least 100. Synthetic fibres are spun into continuous filaments, or chopped in shorter staple, which are then twisted into thread before weaving. The thickness of the fibre is expressed in terms of denier, which is the weight in grams of a 9000-m length of fibre. Stresses and strength of fibres are reported in terms of tenacity in units of grams/denier. In melt spinning, polymer pellets are gravity fed into an extruder and subjected to shear loading at elevated temperatures. The softened polymer is delivered to the spinneret, which has up to 1000 shaped holes for fibre formation. A molten stream of polymer is forced by pressure through shaped holes and stretched into a solid state. Then the polymer is stretched to have molecular alignment along the axial direction and crystallized in a preferred direction so that no spherulites form. Synthetic fibres include Kevlar, PE, PTFE and nylon while natural fibres include silk, cotton and wool.

MOLECULAR WEIGHT OF POLYMERS

Polymer molecular weight is important because it determines many of its physical properties. For example, at low molecular weight, the tensile strength is too low for the polymer material to be useful. At high molecular weight, the strength increases eventually saturating to the infinite molecular weight. Many properties have similar molecular weight dependencies. They start at a low value and eventually saturate at a high value that is characteristic for infinite or very large molecular weight. Similarly, for low molecular weight polymers, the temperatures for transitions from liquids to waxes to rubbers to solids and mechanical properties such as stiffness, visco-elasticity, toughness and viscosity are too low. Such polymers also have lesser commercial applications. For a polymer to be useful, it must have transition temperatures above room temperatures and it must have mechanical properties sufficient to bear design loads. Unlike small molecules, the molecular weight of a polymer is not one unique value. Rather, a given polymer will have a distribution of molecular weights. The distribution will depend on the way the polymer is produced.

All polymers are virtually mixtures of many large molecules; one must resort to averages to describe molecular weights. Among many possible ways of reporting averages, three are commonly used: the number average, weight average and z-average molecular weights.

$$\text{Number average MW} (\bar{M}_n) = \frac{\Sigma(M_i N_i)}{\Sigma(N_i)}$$

$$\text{Weight average MW} (\bar{M}_w) = \frac{\Sigma(M_i^2 N_i)}{\Sigma(M_i N_i)}$$

$$\text{Z average MW} (\bar{M}_z) = \frac{\Sigma(M_i^3 N_i)}{\Sigma(M_i^2 N_i)}$$

Weight Average Molecular Weight

The weight average is probably the most useful of the three, because it fairly accounts for the contributions of different sized chains to the overall behaviour of the polymer, and correlates best with most of the physical properties of interest. Polymer molecules, even if they belong to same type, come in different sizes (chain lengths, for linear polymers). Thus it is necessary to arrive at an average to obtain the molecular weight. The weight average molecular weight is calculated by

$$\text{Weight average molecular weight} (\bar{M}_w) = \frac{\Sigma W_i M_i}{\Sigma W_i} = \frac{\Sigma N_i M_i^2}{\Sigma N_i M_i}$$

Intuitively, if the weight average molecular weight is w, and we pick a random monomer, then the polymer it belongs to will have a weight of w on average. The weight average molecular weight can be determined by light scattering, small angle neutron scattering (SANS), X-ray scattering, and sedimentation velocity. Surprisingly, the chains may differ from one another in their molecular weight. While biopolymers have a discrete molecular weight, all synthetic polymers have molecular weight distributions.

Number Average Molecular Weight

An alternative measure of molecular weight for a polymer is the number average molecular weight. This is a property that is not influenced by the size of any particle in the mixture. The best examples of such properties are the colligative properties of solutions such as boiling point elevation, freezing point depression and osmotic pressure. For such properties, the most relevant average molecular weight is the total weight of polymer divided by the total number of polymer molecules. The number average molecular weight is the common average of the molecular weights of the individual polymers. It is determined by measuring the molecular weight of n polymer molecules, summing the weights and dividing by n.

$$(\bar{M}_n) = \frac{\sum N_i M_i}{\sum N_i}$$

where,

$\bar{M}_n$ = number average molecular weight and

N_i = Number of molecules of molecular weight M_i.

The number average molecular weight (M_n) of a polymer can be determined by osmometry, end-group titration, and from colligative properties.

A given sample of polymer contains

2 chains of polymer of mass 1,000,000 Dalton,

5 chains of polymer of mass 700,000 Dalton,

10 chains of polymer of mass 400,000 Dalton,

4 chains of polymer of mass 100,000 Dalton and

2 chains of polymer of mass 50,000 Dalton

Calculate M_w and M_n of polymer of mass 400,000 Dalton

N_i	M_i	N_iM_i
2	1000000	2000000
5	700000	3500000
10	400000	4000000
4	100000	400000
2	50000	100000
$\Sigma N_i = 23$		$\Sigma M_iN_i = 10000000$

By definition the number average molecular weight

$$M_n = \Sigma M_i N_i / \Sigma N_i$$
$$= 10000000/23$$
$$= 435,000 \text{ Dalton}$$

N_i	M_i	N_iM_i	$N_iM_i^2$
2	1000000	2000000	2×10^{12}
5	700000	3500000	2.45×10^{12}
10	400000	4000000	1.6×10^{12}
4	100000	400000	4×10^{10}
2	50000	100000	5×10^{9}
$\Sigma N_i = 23$		$\Sigma M_iN_i = 10^7$	$\Sigma N_iM_i^2 = 6.095\times10^{12}$

By definition the weight average molecular weight

$$M_w = \Sigma N_i M_i^2 / \Sigma N_iM_i$$
$$= 6.095\times10^{12}/10^7$$
$$= 609500 \text{ Dalton}$$

Glass Transition Temperature (T_g)

The temperature (actually a broad range of temperatures) at which a glassy polymer softens into a viscous liquid or rubbery phase is T_g. On the molecular level, it is the temperature at which chains in amorphous (i.e., disordered) regions of the polymer gain enough thermal energy to begin sliding past one another at a noticeable rate. For an amorphous polymer, the T_g reports the minimum processing temperature. The T_g is strongly dependant on polymer structure (including stereochemistry), and ranges from far below 0°C for very flexible chains, and above 400°C for very stiff chains.

Melting Temperature (T_m)

This is the temperature (actually a narrow range of temperatures) at which the ordered regions of a crystalline polymer melt, similar to a small molecule. Crystallization is essential for many high-performance polymers because it greatly increases the strength of the material.

Like the T_g, the T_m is detectable by DSC, TMA, and other techniques.

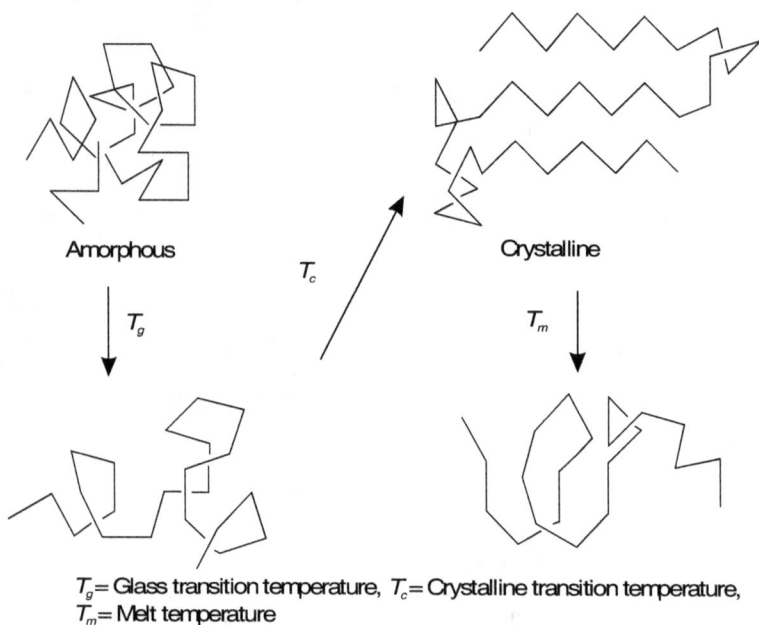

Amorphous

Crystalline

T_c

T_g

T_m

T_g= Glass transition temperature, T_c= Crystalline transition temperature, T_m= Melt temperature

Decomposition Temperature (T_d)

This is the temperature above which chemical degradation occurs. This temperature is conveniently measured by thermo-gravimetric analysis (TGA), a technique in which one simply weighs the sample continuously while heating it. Once decomposition begins, small molecular fragments are released which distill away, and the sample loses weight.

Modulus

It is the proportionality constant between stress and strain, and therefore can be thought of as a measure of stiffness. The modulus of a material like polymers is its ability to resist deformation under load. When a load is placed on a polymer material it will result in an initial deformation, but with the load remaining over time, permanent deformation will occur but without break in the material. A high-modulus polymer, polysulphone is used as a porous bone implant material. The tensile modulus decreases with increasing temperature.

Polydispersity index (PDI) represents the broadness of a molecular weight distribution. It is the ratio of the number average molecular weight (M_n to the weight average molecular weight (M_w).

$$PDI = M_n + M_w$$

The value of PDI is unity for a monodisperse polymer. Monodispersity is a subset of the term polydispersity. When PDI equals to one, (which is common for natural proteins) then M_n equals M_w and the polymer is said to be monodisperse. As M_n changes with M_w, the PDI changes and it is always greater than one. A typical commercial polymer has PDI value of two or greater. Here is a typical molecular weight distribution curve, measured by size exclusion chromatography (SEC).

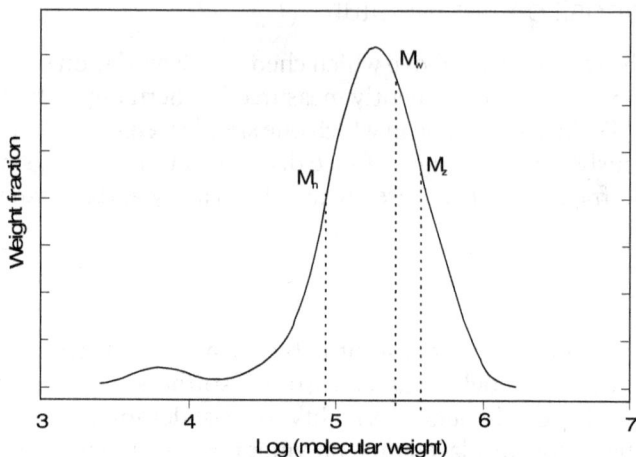

Many polymer properties of interest (T_g, modulus, tensile strength, etc.) follow a peculiar pattern with increasing molecular weight. Small molecules have small values, and then there is a sharp rise in properties as the chains grow to intermediate size (oligomers), and then the properties level off as the chains become long enough to be true polymers.

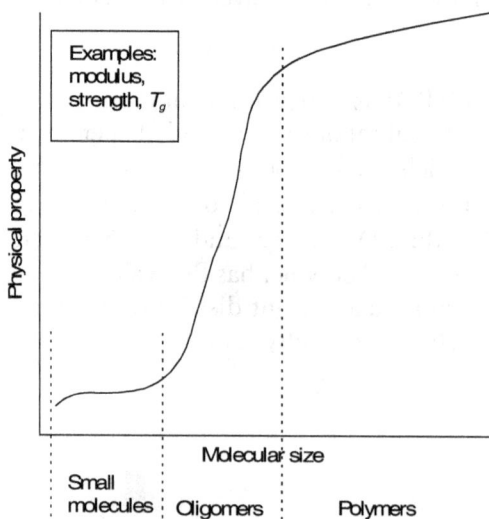

However, a few properties important for polymer processing, like melt viscosity and solution viscosity, increase monotonically with molecular weight. This means that the goal of polymer synthesis is not to make the largest possible molecules, but rather, to make molecules large enough to get onto the plateau region. Increasing the molecular weight beyond this does not improve the physical properties much, but makes processing more difficult. A few properties are dictated by the repeat units alone, and therefore these are not changed much by molecular weight. For example, colour, dielectric constant and refractive index.

Molecular Arrangement of Polymers

The molecular arrangement of polymers is of significance in determining its properties and their applications. How spaghetti noodles look on a plate? This is similar to how polymers can be arranged if they are amorphous. An amorphous arrangement of molecules has no long-range order or form in which the polymer chains arrange themselves. Amorphous polymers are generally transparent. This is an important characteristic for many applications such as food wrappers, headlights and contact lenses. Controlling and quenching the polymerization process can result in amorphous organization.

How spaghetti noodles look on a plate is similar
to how polymers can be arranged if they are amorphous

Not all polymers are transparent. The polymer chains in objects that are translucent and opaque are in a crystalline arrangement. By definition, a crystalline arrangement has atoms, ions, or, in this case, molecules in a distinct pattern. It is common to think of crystalline structures in salt and gemstones, but not in plastics. Just as quenching can produce amorphous arrangements, processing can be controlled to produce the degree of crystallinity desired. The higher the degree of crystallinity, the lesser the light can pass through the polymer. Therefore, the degree of translucence or opaqueness of the polymer is directly affected by its crystallinity. Crystallinity also affects the melting point of a polymer. The more crystalline the pattern of the molecules, the more energy is needed to cause the molecules to separate, melt and flow. Amorphous polymers, on the other hand, will have lower melting points. Care must be taken to retain the degree of crystallinity in a polymer. Reprocessing, recycling, overheating, fabricating, machining, UV light, or heat exposure in service, use or storage can potentially affect the crystallinity of a plastic. As a polymer becomes more crystalline, its melting point and strength increase. However, its strength can increase to the point that the polymer becomes brittle and loses characteristics it was originally made to have.

By manipulating factors on the molecular level that affect the final polymer produced, engineers constantly are challenged to produce better-suited materials for a wide variety of old and new applications to improve the quality of life. Manufacturers and processors introduce various fillers, reinforcements and additives into the base polymers, expanding product possibilities.

CHARACTERISTICS OF POLYMERS

Polymers have the following characteristics:

- Good chemical resistivity at room temperature (acids, alkali)
- Low density and Young's modulus
- Brittleness at low temperatures
- Low strength and chemical resistivity at high temperatures
- Processable at relatively low temperatures

- Generally good insulators (except conductive polymers)
- Often transparent and good optical properties
- Gas permeability
- Recyclable if unblended, otherwise used for fuel production.

Every polymer has very distinct characteristics but most polymers have the following general attributes.

1. Polymers are resistant to chemicals. Consider all the cleaning fluids used at home that are packaged in plastic. Reading the warning labels that describe what happens when the chemical comes in contact with skin or eyes or is ingested will emphasize the chemical resistance of these materials.

2. Polymers can be both thermal and electrical insulators. All the appliances, cords, electrical outlets, and general wiring that are used in houses and other constructions are made up of or covered with polymeric materials. Thermal resistance is evident from the kitchen with pot and pan-handles made of polymers, the foam core of refrigerators and freezers, and insulated cups, coolers and microwave cookware. The thermal underwear that many skiers wear is made of polypropylene, and the fibre-fill in a winter jacket can be made from polypropylene or polyester fibre.

3. Generally, polymers are very light in mass with varying degrees of strength. Consider the range of applications from a toy to the frame structure of space stations, or from delicate nylon fibre used to make pantyhose to Kevlar, which is used in bulletproof vests.

It would take seven tru ckloads of paper bags to deliver the equivalent of one truckload of plastic bags. The fact is that plastic is light weight and helps lower transportation costs and conserve natural resources.

4. Polymers can be processed in various ways to produce thin fibres or very intricate parts. Plastics can be moulded into bottles or the body of a car, or can be mixed with solvents to become an adhesive or paint. Elastomers and some

plastics are very flexible. Other polymers can be foamed, like polystyrene (Styrofoam) and urethane, to give a few examples.

Identification of Polymers

Polymers are materials with a seemingly limitless range of possible characteristics. Polymers have many inherent properties that can be enhanced by a wide range of additives to broaden their use and application. The ability to design or engineer the polymer for each specific application makes plastics unique among basic material types.

Light penetration qualities are dependent on the degree of crystallization of the polymer and the presence of additives

Each polymer resin can be identified by fundamental identification tests. Melting point, burning properties, solubility, relative density and halogen tests can be used to identify resins in a laboratory. Of the most common polymers, polyethylene (PE) and polypropylene (PP) have a translucent, waxy texture, and are the only non-foam plastics that float in water. The burning properties will discriminate between the two. PE burns rapidly, drips flames, smells like candle wax and when extinguished, will produce a white smoke. PP on the other hand, burns more slowly, smells like burning fuel, and does not drip flames while burning. PE is impervious to chemical solvents while PP will dissolve in hot toluene.

Other common plastics can be identified by their burning properties. Polyvinyl chloride (PVC) can be ignited but will self-extinguish as soon as the fire source is removed. PVC has a very acidic odour when burning because hydrogen chloride is produced as a by-product. A halogen test can be performed to identify the presence of chlorine in PVC. A copper wire is heated in a laboratory burner flame to incandescence and then touched to the PVC. When the wire is returned to the burner, a brilliant green flame is produced. Rigid PVC will become rubbery in the presence of benzene or will dissolve in methyl ethyl ketone. Polystyrene (PS) on the other hand, burns rapidly, has a strong gas odour, and produces tremendous amounts of soot. PS will swell readily in acetone.

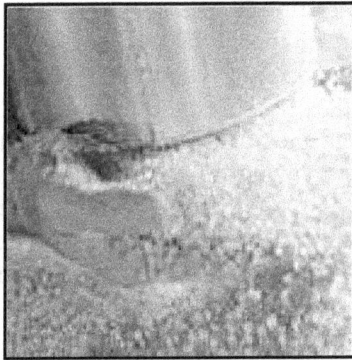

Polyvinyl chloride (PVC) pellets

These identification tests are not conclusive determinations for polymer content, but are a solid guidelines. Polymers with colourants, additives and stabilizers will naturally have different behaviours. Infrared spectrophotometry, gas chromatography and X-rays analysis are some of the analytical identification tests that can be performed to positively identify polymeric materials. Elastomers are unique polymers, which include rubber, synthetic rubber and thermoplastic elastomers and are characterized by their elasticity and flexibility. Elastomeric materials stretch and have the ability to recover with limited permanent set or distortion. These materials are generally distinguished by similar basic identification tests like those for plastics mentioned earlier.

However, commercial products are usually compounded with more than one base elastomer, making identification by analytical instrumentation a necessity.

Polymers as Adhesives

Polymers are used as adhesives that bond materials together. Cement is one type of adhesive that, by definition, is a liquid plastic in a solvent base. As the solvent evaporates, the cement hardens, which is commonly called drying.

Many types of adhesives and cements are generally referred to as "glue". Probably the most familiar adhesive is a suspension of polyvinyl acetate in a water base, better known as white glue. By common definition, white glue is cement. The solvent (water) evaporates, resulting in the polyvinyl acetate hardening. Such adhesives form mechanical bonds between materials. A mechanical bond is characterized by the interlocking of surfaces by secondary bonds. A stronger bond can be made with chemical bond adhesives. A chemical bond results when the adhesive causes the molecules of the two materials to blend together. Welding or heating forms chemical bonds. Using cement with a solvent that can dissolve the plastic, causes the intermingling of molecules between the two surfaces forming the bond. Plastic can also be polymerized between two materials, causing the intermeshing of surface molecules. Some examples of solvent-based cements that form chemical bonds include epoxy and acrylic adhesives. Other polymers are rising in popularity as effective mechanical adhesives. Hot melt adhesives that are applied with a glue gun, are used in some areas because of easy application. The plastic is melted and extruded into place, bonding two materials together or filling and sealing cracks and seams.

The molten plastics can be controlled so that a thin or heavy stream is formed depending on the specifications. As the plastic cools, the bond is formed instantly. Other types of adhesives, which are applied in similar fashion to hot melts, are caulks, sealants and putty compounds. These come in squeeze tubes or in a tipped cartridge. The cartridge is inserted into an extruder gun and the substance is forced through a tip. These compounds are used to fill cracks or produce a seal between two materials such as a joint, window edge, or bathroom fixture. The main purpose of these adhesives is to keep out moisture and air.

Equally common in use are adhesive-backed tapes or labels. These adhesives include masking tape, invisible tape, duct tape, Post-it notes, Band-aids, pressure-sensitive labels, and stickers. Heat-sensitive patches or iron-on patches are also included in this group. In each case, the adhesive is on the back of a substrate such as paper, film or fabric and can be adhered to another surface by pressure, moisture or heat.

Polymers in suspension are also used in the family of coatings. Paint, wood finishes and vinyl-coated tool handles are examples. Coatings are permanently applied to a substrate. Polymeric dispersions, used to fill a mould are considered as castings. Various additives broaden the spectrum of forms that polymeric materials can take.

Fillers

Fillers and reinforcements added to polymeric material constitute composites. Fillers are usually small particles or flakes of organic or inorganic materials that are added to polymers to extend the material, lower the costs and improve the physical properties. For example, application of metallic flakes to PVC has brought the development of electrically conductive plastics to the marketplace.

An example of vinyl piping, reinforced
with special additives

Reinforcements

Reinforcements are additives that improve physical properties, especially the tensile and impact strength of the material. Glass fibre is a common reinforcement used to fill plastics. Mats or yarns of polymer or metal are additional examples of reinforcements used in the industry. Fillers and reinforcements are not only dispersed through the polymer, but can be layered, forming composite structures more commonly called laminates. Laminate structures are used in aircraft and car bodies, foam-backed carpet and counter tops.

Additives are also used to stabilize the final polymeric material produced or to aid the processing of the material. Additives are available to curtail the effects of oxidation, ozone or ultraviolet radiation. Anti-static agents are used to prevent the build-up of electrostatic charge on the surface of the polymer. Synthetic fibre clothing is a common example of static charge building up on the material, causing the fabric to cling to other materials. Fabrics, such as children's sleepwear and carpet, are often stabilized with flame-retardants.

Polymers, especially those with plasticizers, are susceptible to microbial growth. Antibacterial or fungicidal additives, when incorporated into a polymer, will limit the growth of bacteria, fungus or mildew on a shower curtain, in a waterbed, or in tubing for medical applications. Processing additives range from accelerators and catalysts to retardants, which are critical in elastomer compounding. Waxes, lubricants and plasticizers are additives that reduce friction in processing and result in flexible plastic or elastomeric materials. Blowing or foaming agents are additives that give polymers a specific form. Cellulose sponges and polystyrene are examples of the two different foams that are produced. Open-celled foams, like the sponge, hold water and have cells or polymers that are not completely enclosed. Polystyrene is indicative of closed-celled foams that do not hold water and have completely encapsulated cells.

With the range of inherent characteristics of polymeric materials and the possible modifications from fillers, reinforcements and additives, the chemical and engineering potential of plastics and elastomers is limitless.

COMMON PLASTIC POLYMERS
USED IN PACKAGING

Polyethylene Terephthalate (PET or PETE)

PET is clear, tough and has good gas and moisture barrier properties. The vast majority of this plastic is used in soft drink bottles and blow moulded containers, although sheet applications are increasing. In addition, a small volume of PET is now used outside the packaging industry for the production of injection-moulded components such as bicycle mud guards. Cleaned, recycled PET flakes and pellets are in great demand for spinning fibre for carpet yarns and producing fibrefill and geotextiles. Other outlets include strapping, molding compounds and both food and non-food containers.

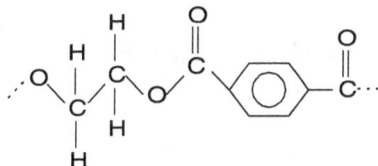

Properties   Clarity, strength/toughness, barrier to gas, resistance to grease/oil, stiffness and resistance to heat.

Uses   (1) Plastic soft drink bottles (2) mouthwash bottles (3) peanut butter and salad dressing containers.

Recycled products   Tote bags, dishwashing liquid containers, clamshells, laser toner cartridges, picnic tables, hiking boots, lumber, mailbox posts, fencing, furniture and sweatshirts, etc.

High Density Polyethylene (HDPE)

HDPE is a relatively straight chain structure, but, as its name implies, exhibits a higher density. It is naturally milky white in appearance and finds wide application in blow-moulded bottles for milk, water and fruit juices. HDPE, pigmented with a variety of colourants, is used for packaging toiletries, detergents, and similar products.

$$\left(\begin{array}{c} H \quad\ H \\ | \quad\ | \\ -C-C- \\ | \quad\ | \\ H \quad\ H \end{array}\right)_n$$

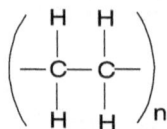

Properties Stiffness, strength/toughness, low cost, ease of forming, resistance to chemicals, permeability to gas and ease of processing.

Uses Milk, water and juice containers, grocery bags, toys and liquid detergent bottles.

Recycled products Recycling bins, benches, bird feeders, retractable pens, clipboards, fly swatters, dog houses, vitamin bottles, floor tile, liquid laundry detergent containers, etc.

Polyvinyl Chloride(PVC)

In addition to its good physical properties, PVC has excellent transparency, chemical resistance, long-term stability, flammability resistance, good flow and electrical insulation properties. The diverse vinyl products can be broadly divided into rigid and flexible materials. Rigid PVC, accounting for 60% of total vinyl production, are concentrated in construction markets, which include pipes and fittings, siding, carpet backing and windows. Bottles and packaging sheet are also major applications of rigid PVC. Flexible vinyl is used in wire and cable insulation, film and sheets, floor coverings, synthetic-leather products, coatings, blood bags, medical tubing and many more applications.

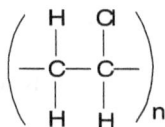

$$\left(\begin{array}{c} H \quad\ Cl \\ | \quad\ | \\ -C-C- \\ | \quad\ | \\ H \quad\ H \end{array}\right)_n$$

Properties Versatility, ease of blending, strength/toughness, resistance to grease/oil, resistance to chemicals, clarity, low cost.

Uses Clear food packaging, shampoo bottles.

Recycled products Air bubble cushioning, flying discs, decking, films, paneling, recycling containers, roadway gutters, snowplough deflectors, playground equipment, etc.

Low Density Polyethylene (LDPE)

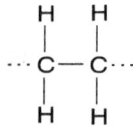

```
      H   H
      |   |
··· C — C ···
      |   |
      H   H
```

It is a plastic that is used predominantly in film applications due to its toughness, flexibility and relative transparency. Because of its lower melting point at a given density, it is used in applications where heat sealing is easily accomplished. LDPE is the preferred resin on older unconverted film extrusion equipment due to its ease of extrusion. Typically, LDPE is used to manufacture flexible films such as those used for plastic retail bags and garment dry cleaning and grocery bags. LDPE is also used to manufacture some flexible lids, and it is widely used in wire and cable applications for its good insulatory electrical properties and processing characteristics.

Properties   Ease of processing, barrier to moisture, strength/ toughness, flexibility, ease of sealing, low cost.

Uses   Bread bags, frozen food bags, grocery bags.

Recycled products   Shipping envelops, garbage can liners, floor tile, furniture, films, compost bins, paneling, trashcans, landscape timber, mud flaps, etc.

Polypropylene (PP)

Polypropylene has excellent chemical resistance, is strong and has the lowest density of the plastics used in packaging. It has a high melting point, yet is readily heat-sealable. In film forming, it may or may not be oriented (stretched). It is also relatively inexpensive. PP is found in everything from flexible and rigid packaging to fibres and large moulded parts for automotive and consumer products.

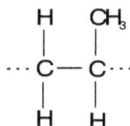

```
      H   CH₃
      |   |
··· C — C ···
      |   |
      H   H
```

Properties   Strength/toughness, resistance to chemicals, resistance to heat, barrier to moisture, low cost, versatility, ease of processing, resistance to grease/oil.

Uses   Ketchup bottles, yoghurt containers, margarine tubs, medicine bottles.

Recycled products   Signal lights, battery cables, brooms and brushes, ice scrapers, oil funnels, landscape borders, bicycle racks, etc.

Polystyrene (PS)

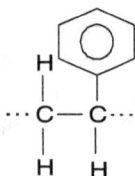

Polystyrene is a very versatile plastic that can be rigid or foamed. Generally polystyrene is clear, hard and brittle. It is a very inexpensive resin. It is a rather poor barrier to oxygen and water vapour and has relatively low melting point. Typical applications include protective packaging, containers, lids, bottles, trays and tumblers.

Properties   Versatility, insulation, ease of processing, low cost, clarity.

Uses   Videocassette cases, compact disc jackets, coffee cups, knives, spoons and forks, cafeteria trays, grocery store meat trays, fast-food sandwich containers, etc.

Recycled products   Thermometers, light switch plates, insulation, egg cartons, vents, desk trays, rulers, license plate frames, and concrete, etc.

The Society of the Plastics Industry, Inc. (SPI) introduced its resin identification coding system in 1988 at the urging of recyclers around the country. A growing number of communities were implementing recycling programmes in an effort to decrease the volume of waste subject to rising tipping fees at landfills. In some cases, these programmes were driven by state-level recycling mandates. The SPI code was developed to meet recyclers' needs while providing manufacturers a consistent, uniform system that could apply nationwide. Because municipal recycling programmes traditionally have targeted packaging, primarily containers, the SPI coding system offered a means of identifying the resin content of bottles and containers commonly found in the residential waste stream. Recycling firms have varying standards for the plastics they accept. Some firms may require that the plastics be sorted by type and separated from other recyclables; some may specify that mixed plastics are acceptable if they are separated from other recyclables; while others may accept all material mixed together. Not all types of plastics are generally recycled, and recycling facilities may not be available in some areas.

SPECIAL PROPERTIES OF POLYMERS

CONDUCTIVITY POLYMERS

Polymers (or plastics as they are also called) are known to have good insulating properties. They are one of the most used materials in the modern world. Their uses and application range from containers to clothing.

They are used to coat metal wires to prevent electric shocks. However, it is now recognized that there are some polymers, which have conducting properties. These are not necessarily complicated polymers. Even polyacteylene has been discovered to have these properties. This is one of the simplest chain polymers possessing conjugated double bonds. This research is considered so important that the 2000 Nobel Prize in Chemistry was awarded to three of the leading researchers in conducting polymers.

Nobel Story

For the discovery and development of conductive polymers many eminent scientists were awarded the Nobel Prize.

Alan J. Heeger (1936)	Alan G. MacDiarmid (1927)	Hideki Shirakawa (1936)
1/3 of the prize	1/3 of the prize	1/3 of the prize
USA	USA and New Zealand	Japan
University of California Santa Barbara, CA, USA	University of Pennsylvania Philadelphia, PA, USA	University of Tsukuba Tokyo, Japan

We are used to the great impact scientific discoveries have on our ways of thinking. This year's Nobel Prize in Chemistry is no exception. What we have been taught about plastic is that it is a good insulator—otherwise we should not use it as insulation in electric wires. Now the time has come when we have to change our views. Plastic can indeed, under certain circumstances, be made to behave very much like a metal—a discovery for which Alan J. Heeger, Alan G. MacDiarmid and Hideki Shirakawa received the Nobel Prize for Chemistry in the year 2000. What Heeger, MacDiarmid and Shirakawa found was that a thin film of polyacetylene could be oxidized with iodine vapour, increasing its electrical conductivity a billion times. This sensational finding was the result of their impressive work, but also of coincidences and accidental circumstances. Let us, shortly, tell the story of one of the great chemical discoveries of our time.

Revelation of Polymer Conductivity

The leading actor in this story is the hydrocarbon polyacetylene, a flat molecule with an angle of 120° between the bonds and hence existing in two different forms. The isomers *cis*-polyacetylene and *trans*-polyacetylene. At the beginning of the 1970s, the Japanese chemist Shirakawa found that it was possible to synthesize polyacetylene in a new way, in which he could control the proportions of *cis*- and *trans*-isomers in the black polyacetylene film that appeared on the inside of the reaction vessel. Once by mistake, a thousand-fold more catalyst was added. To Shirakawa's surprise, this time a beautiful silvery film appeared. Shirakawa was stimulated by this discovery. The silvery film was *trans*-polyacetylene, and the corresponding reaction at another temperature gave a copper-coloured film instead. The latter film appeared to consist of almost pure *cis*-polyacetylene. This process of varying temperature and concentration of catalyst was to become decisive for the development ahead.

In another part of the world, chemist MacDiarmid and physicist Heeger were experimenting with a metallic-looking film of the inorganic polymer sulphur nitride, $(SN)_x$. MacDiarmid referred to this at a seminar in Tokyo. Here the story could have come to a sudden end, had not Shirakawa and MacDiarmid happened to meet, accidentally, during a coffee break.

When MacDiarmid heard about Shirakawa's discovery of an organic polymer that also gleamed like silver, he invited Shirakawa to the University of Pennsylvania in Philadelphia. They set about modifying polyacetylene by oxidation with iodine vapour. Shirakawa knew that the optical properties changed in the oxidation process and MacDiarmid suggested that they ask Heeger to have a look at the films. One of Heeger's students measured the conductivity of the iodine-doped *trans*-polyacetylene and—eureka! The conductivity had increased ten million times! In the summer of 1977, Heeger, MacDiarmid, Shirakawa and co-workers, published their discovery in the article "Synthesis of electrically conducting organic polymers: Halogen derivatives of polyacetylene $(CH)_n$" in *The Journal of Chemical Society, Chemical Communications*. The discovery was considered a major breakthrough. Since then, the field has grown immensely.

Conductivity

Conductivity can be defined simply by Ohms Law.

$$V = IR$$

where,

R is the resistance,
I is the current and
V is the voltage.

Thus, from this relationship, conductivity is found. The conductivity depends on the number of charge carriers (number of electrons) in the material and their mobility. For example, in a metal, it is assumed that all the outer electrons are free to carry charge and the impedance to flow of charge is mainly due to the electrons "bumping" into each other. Thus for metals, as temperature is increased the resistance in the material increases as the electrons bump into each other more as they are moving faster.

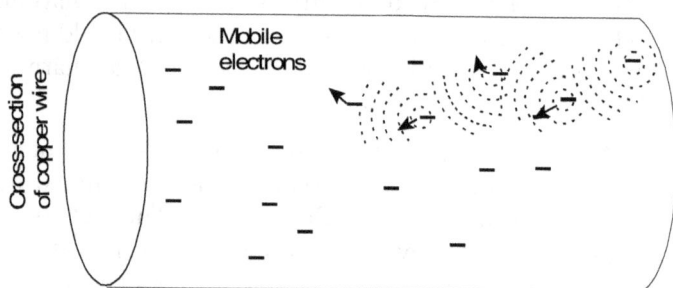

(−)   An external influence repels a nearby electron
The electron's neighbours find it repulsive. If it moves toward them, they move away, creating a chain of interactions which propagates through the material at the speed of light.

Insulators however have tightly bound electrons so that nearly no electron flow occurs so they offer high resistance to charge flow. So for conductance free electrons are needed.

Why do Polymers Conduct?

It is well known that graphite is a good conductor. Previously it was thought that polymers with substituted carbons, could not conduct. However, the knowledge of conjugated systems has enabled the discovery of conducting polymers. In becoming electrically conductive, a polymer has to imitate a metal, that is, its electrons need to be free to move and not bound to the atoms. Hence, in a conjugated system where the electrons are only loosely bound, electron flow may be possible. However, it is not enough to have conjugated double bonds. To become electrically conductive, the plastic has to be disturbed, either by removing electrons from (oxidation), or inserting them into (reduction) the material. The process is known as *doping*.

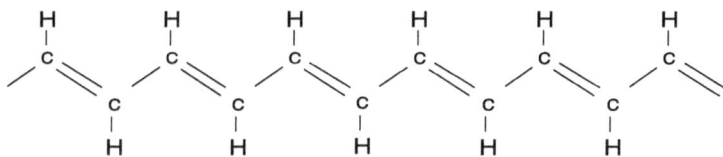

Polyacetylene

Once doping has occurred, the electrons in the π-bonds are able to "jump" around the polymer chain. As the electrons are moving along the molecule, an electric current is created. However, the conductivity of the material is limited, as the electrons have to "jump" across molecule. For better conductivity the molecules must be well ordered and closely packed to limit the distance "jumped" by the electrons. This occurs better in *trans* undoped polyactelyene. By doping, the conductivity of the polymer increases from $10^{-3}\,Sm^{-1}$ to $3000\,Sm^{-1}$. An oxidation doping (removal of electrons) can be done using iodine vapours. The iodine attracts an electron from the polymer from one of the π-bonds. Thus, the remaining electron can move along the chain.

The game in the illustration shown below is a simple model of a doped polymer. The pieces cannot move unless there is at least one empty "hole." In the polymer, each piece is an electron that jumps to a hole vacated by another one. This creates a movement along the molecule, an electric current. This model is a greatly over-simplified representation.

Doped polymer model

Uses of Conducting Polymers

Conducting polymers have many uses. The most documented are as follows:

- Corrosion inhibitors
- Compact capacitors
- Anti-static coating
- Electromagnetic shielding for computers
- "Smart Windows"

A second generation of conducting polymers have been developed which have industrial uses like:

- Transistors
- Light Emitting Diodes (LEDs)

- Lasers used in flat televisions
- Solar cells

Process of Doping

What exactly happened in the polyacetylene films? On comparing some common compounds with regard to conductivity, it was observed that the conductivities of the polymers vary considerably. Doped polyacetylene is, for example, comparable to good conductors such as copper and silver, whereas in its original form it is a semiconductor.

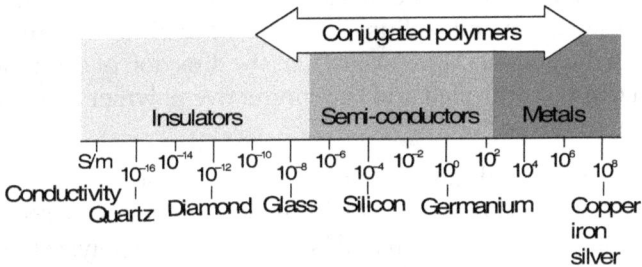

A metal wire conducts electric current because the electrons in the metal are free to move. How then do we explain the conductivity of the doped polymers? When describing polymer molecules we distinguish between σ-(sigma) bonds and π-(pi) bonds. The σ-bonds are fixed and immobile. They form the covalent bonds between the carbon atoms. The π-electrons in a conjugated double bond system are also relatively localized, though not as strongly bound as the σ-electrons. Before a current can flow along the molecule, one or more electrons have to be removed or inserted. If an electrical field is then applied, the electrons constituting the π-bonds can move rapidly along the molecule chain. The conductivity of the plastic material, which consists of many polymer chains, will be limited by the fact that the electrons have to "jump" from one molecule to the next. Hence, the chains have to be well packed in ordered rows. As mentioned earlier, there are two types of doping, oxidation, or reduction.

In the case of polyacetylene the reactions are written as follows:

Oxidation with halogen (p-doping):

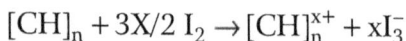

$$[CH]_n + 3X/2\ I_2 \rightarrow [CH]_n^{x+} + xI_3^-$$

Reduction with alkali metal (n-doping):

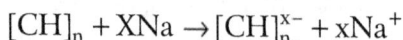

$$[CH]_n + XNa \rightarrow [CH]_n^{x-} + xNa^+$$

The doped polymer is a salt. However, it is not the movement of iodide or sodium ions that creates the current, but the movement of electrons from the conjugated double bonds, which is responsible for the current. Furthermore, if a strong enough electrical field is applied, the iodide and sodium ions can move either towards or away from the polymer. This means that the direction of the doping reaction can be controlled and the conductive polymer can easily be switched "on" or "off".

Polarons are doped carbon chains. In the first of the above reactions; during oxidation, the iodine molecule attracts an electron from the polyacetylene chain and becomes I_3^-. The polyacetylene molecule, now positively charged, is termed a radical cation, or polaron.

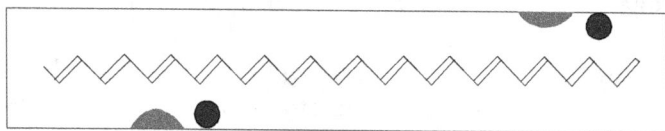

Polaron

The lonely electron of the double bond, from which an electron was removed, can move easily. Consequently, the double bond successively moves along the molecule. The positive charge, on the other hand, is fixed by electrostatic attraction to the iodide ion, which does not move so readily. If the polyacetylene chain is heavily oxidized, polarons condense pairwise into so-called solitons. These solitons are then responsible, in complicated ways, for the transport of charges along the polymer chains, as well as from chain to chain on a macroscopic scale.

Potential Applications of Conductivity Polymers

Plastic batteries are the most radical innovation in commercial batteries since the dry cell was introduced in 1890. Plastic batteries offer higher capacity, higher voltage and longer shelf life. The development of plastic batteries began with an accident. In the early 1970s, a graduate student in Japan was trying to repeat the synthesis of polyacetylene, a dark powder made by linking together the molecules of ordinary acetylene welding gas. After the chemical reaction took place, instead of a black powder, the student found a film coating the inside of his glass reaction vessel that looked much like aluminum foil. He later realized that he had inadvertently added much more than the recommended amount of catalyst to cause the acetylene molecules to link together. His further studies then ascertained that polyacetylene exhibits surprisingly high electrical conductivity. The key breakthrough leading to practical application as batteries occurred, while investigating alternative ways for doping polyacetylene. When two strips of polyacetylene are placed in a solution containing the doping ions and an electric current is passed from strip to strip as expected. The positive ions migrated to one strip and the negative ions to the other. When the current source was removed, the charge remained stored in the polyacetylene polymer. This stored charge could then be discharged if an electrical load was connected between the two strips, just as in a conventional battery.

Chemically, the plastic battery is different from conventional metal-based rechargeable batteries, in which material from one plate migrates to another plate and back in a reversible chemical reaction. In a conducting plastic battery, only the stored ions of the solution move. The plates are not consumed and reconstituted, thus offering a longer recharge cycle lifetime. One potential application for polymer batteries is its use in battery-powered automobiles. Two key measures of a battery's suitability for automotive application are the power density, which determines acceleration and hill-climbing ability, and the energy density, which determines the number of miles that can be driven between charges. Polyacetylene's power density is 12 times that of ordinary lead acid batteries. Its energy density is also higher—about 50 Watts-hours per kilogram versus 35 for lead acid batteries. Although plastic batteries are competing against other advanced development batteries with similar capability for this application, they have the unique potential to be made of low-cost and environmentally benign materials. A polymer battery can be part of the battery-powered car of the future.

Polyacetylene, however, is not an ideal battery material. It degrades in air, is chemically stable only in liquid solutions. Several other potentially suitable plastics were discovered thereafter. One such material was polyaniline. It is inexpensive and unlike polyacetylene, it is stable in both air and water. Polyaniline is the material used in the plastic batteries that first became commercially available. In just 8 years, plastic batteries went from laboratory discovery to commercial availability.

Metal wires that conduct electricity can be made to light up when a strong enough current is passing—as we are reminded of every time we switch on a light bulb. Polymers can also be made to light up, but by another principle, namely electroluminescence, which is used in photodiodes. These photodiodes are, in principal, more energy saving and generate less heat than light bulbs. In electroluminescence, light is emitted from a thin layer of the polymer when excited by an electrical field. In photodiodes, inorganic semiconductors such as gallium phosphide are traditionally used, but now one can also use semi-conductive polymers. Electroluminescence from semi-conductive polymers has been known for about 10 years. Today there is extensive commercial interest in photodiodes and in light-emitting diodes (LEDs). A LED can consist of a conductive polymer as an electrode on one side, then a semi-conductive polymer in the middle and, at the other end, a thin metal foil as electrode. When a voltage is applied between the electrodes, the semi-conductive polymer will start emitting light.

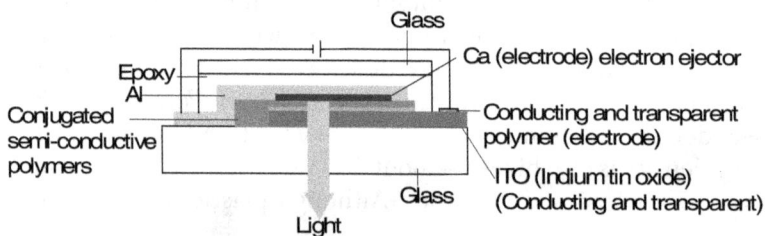

Uses of Conductivity Polymers

There are many applications of this brilliant plastic. In a few years, for example, flat television screens based on LED film, luminous traffic signs and information signs are becoming realty. Since it is relatively simple to produce large, thin layers of plastic, one can also imagine light-emitting wallpaper in our homes, and other spectacular things.

Some applications of conductivity polymers that have come onto the market, or are undergoing trials, are:

1. Polythiophene derivates, are of great commercial use in antistatic treatment of photographic film. They can also be used in devices in supermarkets for marking products. The checkouts will then automatically register what the customer has in the trolley.

2. Doped polyaniline is used as an antistatic material, e.g. in plastic carpets for offices and operating theatres, where it is important to avoid static electricity. It is also used on computer screens, protecting the user from electromagnetic radiation, and as a corrosion inhibitor.

3. Materials such as polyphenylene vinylene are used in mobile phone displays.

4. Polydialkylfluorenes are used in the development of new colour screens for video and TV.

POLYMERS IN SPECTACLE LENSES AND CONTACT LENSES

Synthetic materials have been largely replacing the traditional materials in day-to-day life. It is not surprising that in the field of optics, polymers have made inroads into the domain of optical lenses made of glass. Spectacles may still be called "glasses" or "sun-glasses," but for decades they have been manufactured from polymers such as CR39 and the transparent polymer polymethyl methacrylate (PMMA). Plastics have now advanced into new applications that glasses could not, such as contact lenses that are soft and "breath." However, it is not always possible to obtain the desired combination of optical and mechanical properties from a single polymer. At this stage, the trick of copolymerization modifies a polymer easily. In other words, the judicious selection of co-monomers, a combination of properties obtained that otherwise is not available from any single monomer. The PMMA is referred to as the workhorse of optical plastic as it is easy to mould, fabricate and has high optical transmission (92%). However, the victory of plastics over glasses is not total, there are certain physical properties where glasses retain their superiority. This includes:

- Transparency
- Specific gravity
- Scratch resistance
- Moisture resistance
- Thermal stability

Polymers offer profound advantages over traditional inorganic glass in terms of lighter weight, impact and shatter resistance, and greater focal control through higher refractive indices. Refractive index n is the ratio of the speed of light in a vacuum, c_o, to the speed of light through a given medium, c.

$$n = c_o/c$$

Carbazole-based polymers have demonstrated good optical properties, combined with ease of processing. Polymethacrylates and polyacrylates are used commercially for many optical applications, and are easily polymerized and processed.

The Making of Contact Lens

The contact lens is a device worn in the eye to correct vision, although some people wear coloured contact lens to enhance or change their eye colour. The thin plastic lens floats on a film of tears directly over the cornea. For some forms of eye disease, contact lenses correct vision better than conventional spectacles. Contact lenses are preferred over glasses for cosmetic reasons, and active sports enthusiasts prefer contact lens for the freedom it provides. There are three types of lenses: soft, hard, and gas-permeable. Soft contact lenses are usually more comfortable to wear, but they also tear more easily than hard contact lenses. Hard lenses also tend to "pop" out more frequently. Gas-permeable lenses are a compromise between the hard and soft, allowing greater comfort than hard lenses but less chance of tearing than soft lenses.

The first contact lens was made by the German physiologist Adolf Fick in 1887. Fick's lens was made of glass and was a so-called scleral lens because it covered the sclera, the white part of the eye. By 1912, another optician, Carl Zeiss, had developed a glass corneal lens, which fit over the cornea. Two scientists, Obrig and Muller, introduced a plastic scleral lens in 1938. It was made of the material commonly known as plexiglas. Because it was lighter than glass, the Plexiglas lens was easier to wear.

The first plastic corneal lens was made by Kevin Touhy in 1948. To fit these early lenses, an impression was made of the patient's eyeball, and the lens was formed in the resulting mould. This procedure was doubtlessly uncomfortable, and the lenses themselves were often problematic to wear. Scleral lenses deprived the eye of oxygen, and many of these earlier lenses slipped out of place or popped out of the eye, and were often, oddly enough, difficult to remove. Touhy's first corneal lens had a diameter of 10.5 mm and in 1954, Touhy reduced the diameter further to 9.5 mm, resulting in better wearability. Around this time, the Bausch and Lomb Company developed the keratometer, which measures the cornea, and eliminated the need for eyeball impressions.

The first successful soft contact lenses were developed by chemists in Czechoslovakia. In 1952, the Department of Plastics at the Technical University in Prague set a task of designing a new material that was optimally compatible with living tissue. By 1954, the team of Czech scientists had invented what is called a "hydrophilic" gel, a polymer plastic that was suitable for eye implants. The scientists immediately recognized the new plastic's potential as a corrective lens, and they began experimenting on animals. These efforts were met with scorn by their colleagues in the optics field, but one of the scientists, Otto Wichterle, was undaunted and began perfecting soft contact lenses. Wichterle and his wife produced 5,500 pairs of contact lenses for testing in 1961, and their success eventually received the attention of the wider scientific community. The American firm Bausch and Lomb licensed the technology and launched their Softlens in 1971. That first year alone, the firm sold about 100,000 pairs, and soft contact lenses have had great appeal with the public ever since.

The raw material for contact lenses is a plastic polymer. Hard contact lenses are made of some variant of polymethyl methacrylate (PMMA). In the 1960s, the first contact lenses that became commonly available were made of poly methyl methacrylate (PMMA). PMMA is used in plexiglas and lucite and for aquariums and hockey rink barriers. It is even found in latex paints. PMMA lenses are hard, rigid, and not very comfortable. These lenses do not allow oxygen to pass directly to the cornea, which can be detrimental to the eye. Hard lenses are not popular anymore.

In 1971, the first soft contact lenses were introduced commercially. These were made of polyacrylamide, which contained nitrogen. This polymer dissolves in water, and it is similar to polymers used to make acrylic fibres for fabric. Cross-linked polyacrylamide actually absorbs water, so it is a good material for contact lenses. Anywhere from 38% to 79% of a soft contact lens is water, and the water keeps the lens soft and flexible. Soft contact lenses that are made of a polymer such as polyhydroxyethyl methacrylate (PHEMA) has hydrophilic qualities, that is, it can soak up water and still retain its shape and optic functions. In 1979, the first rigid gas-permeable lenses (also known as RGPs) were introduced. These lenses combine PMMA with silicone and fluoropolymers, permits oxygen to pass directly through the lens to the eye. This makes the lens more comfortable for the wearer. In addition, the rigidity of RGPs can make vision crisper, and RGPs are better suited to correcting astigmatism and bifocal needs. The science of lens material is always being updated by lens manufacturers, and the specific material of any contact lens may differ depending on the maker.

As the material for contact lenses is the subject of much research, scientists are investigating chemical recipes that may give plastic of more desirable characteristics. One polymer currently being researched is a silicon–oxygen compound called siloxane. Siloxane forms a thin, flexible film and admits oxygen through to the eye 25 times better than current standard soft lenses. There are disadvantages to this compound, however, siloxane does not wet easily and it attracts lipids (fats) to its surface, causing it to cloud.

Researchers have found a way to add fluorine molecules to the siloxane compound, causing the material to resist lipids. Then they chemically attach a wetting agent, which changes its molecular shape when boiled in a saline solution, so that the material can soak up water like traditional soft lens. This material may ultimately lead to extended-wear contacts that can be worn for weeks at a time.

Researchers are also investigating new polymers that can be used for scleral lenses. For most people, corneal lenses are the norm, but the large scleral lenses are useful for patients with severely damaged corneas. Depending on the eye problem, some patients cannot regain their sight without a corneal transplant, but scleral lenses may help patients avoid eye surgery. Scleral lenses rest on the white part of the eye and form a vault over the cornea itself. This space over the cornea is filled with artificial tears, which serve to smooth out the cornea's damaged surface. In the past, scleral lenses have been uncomfortable because they do not allow enough oxygen to the eye, but investigations into new materials are focusing on lenses that are more oxygen-permeable. Material for oxygen-permeable lenses has also been experimented on the space shuttle endeavour. The designers of the experiment believe that micro gravity conditions would promote a lens material that repels debris better and processes oxygen more effectively than polymers made in traditional laboratories. If commercially feasible, a new generation of contact lenses may be manufactured in space.

Contact lenses may be produced by cutting a blank on a lathe, or by a moulding process. The forming of the lens involves shaping the plastic into specified curvatures. The major curves of the lens are named the central anterior curve (CAC) and the central posterior curve (CPC). The CAC refers to the overall curve of the side of the lens that faces out. This outer contour produces the correct refractive change to fit the patient's visual needs. The CPC is the concave inner side of the lens. This conforms to the measurements of the patient's eye. Usually these two curves are formed first, and the lens is then called semi-finished. The lens is deemed finished when peripheral and intermediate curves are formed and the edge is shaped.

The lenses are inspected after each stage of the manufacturing process. The lenses are examined under magnification for anomalies.

They are also measured by means of a shadowgraph. A magnified shadow of the lens is cast on a screen imprinted with a graph for measuring diameter and curvature. Any errors in the lens shape show up in the shadow. This process may be automatically performed by computer. Lens is sterilized by boiling in a mixture of water and salt for several hours to soften the lens. Then they are packaged. Standard packaging for lenses is a glass vial, filled with a saline solution, and stoppered with rubber or metal. The hydrophilic material of soft contact lenses soaks up the saline solution, which is similar to human tears, and becomes soft and pliable. The lenses in this state are ready to wear.

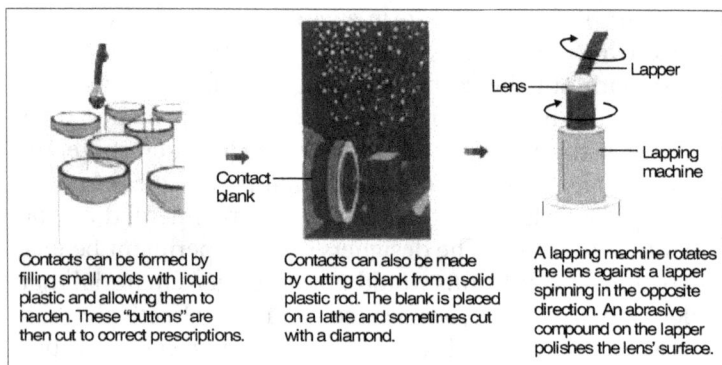

Contact blank

Lens

Lapper

Lapping machine

Contacts can be formed by filling small molds with liquid plastic and allowing them to harden. These "buttons" are then cut to correct prescriptions.

Contacts can also be made by cutting a blank from a solid plastic rod. The blank is placed on a lathe and sometimes cut with a diamond.

A lapping machine rotates the lens against a lapper spinning in the opposite direction. An abrasive compound on the lapper polishes the lens' surface.

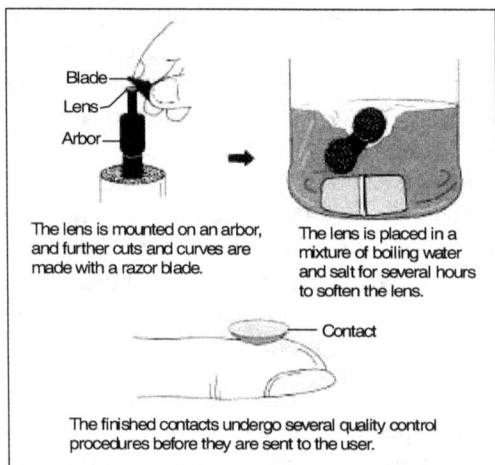

Blade

Lens

Arbor

The lens is mounted on an arbor, and further cuts and curves are made with a razor blade.

The lens is placed in a mixture of boiling water and salt for several hours to soften the lens.

Contact

The finished contacts undergo several quality control procedures before they are sent to the user.

BIODEGRADABLE POLYMERS:
A REBIRTH OF PLASTIC

Plastics are being used all over the world, from drinking cups and disposable tableware to parts for automobiles and motorcycles. The use of plastics is on a rise. Plastics are extremely important as packaging materials and are used on a large scale throughout the world. Plastics make up about 20% by volume waste per year. They have been an environmental trepidation because of the lack of degradation. Since plastics are vital in day-to-day life, production of biodegradable plastics to make plastics more compatible with the environment is necessary. Biodegradable plastics began to spark interest during the oil crisis in the 1970s. In the 1980s, products such as biodegradable films, sheets and mould-forming materials were introduced. Plant-based green materials—have become increasingly more popular. This is due, in part, to the fact that they are a renewable resource and are much more economical then they were in the past.

Biodegradable plastics are a new generation of polymers. *Biodegradable plastics* are capable of undergoing decomposition into carbon dioxide, methane, water and inorganic compounds or biomass, in which the predominant mechanism is the enzymatic action of microorganisms, that can be measured by standardized tests, in a specified period of time, reflecting available disposal condition. This leads to significant changes in the chemical structure of the material. In essence, biodegradable plastics should break down cleanly, in a defined time period, to simple molecules found in the environment such as carbon dioxide and water. Biodegradation rates are highly dependent on the thickness and geometry of the fabricated articles. Many polymers that are claimed to be "biodegradable" are in fact "bio-erodable", "hydro-biodegradable", or "photo-biodegradable". These different polymer classes all come under the broader category of "environmentally degradable polymers".

The biodegradability of plastics is dependent on the chemical structure of the material and on the constitution of the final product, not just on the raw materials used for its production.

Composition of Biodegradable Plastics

Biodegradable plastics can be natural or synthetic resins. Natural biodegradable plastics are based primarily on renewable resources such as starch and can be either naturally produced or synthesized from renewable resources. Non-renewable synthetic biodegradable plastics are petroleum-based. As any marketable plastic product must meet the performance requirements of its intended function, many natural biodegradable plastics are blended with synthetic polymers to produce plastics, which meet these functional requirements. Biodegradable plastics can be made from many different sources and materials. A number of fibres are obtained from pineapple leaves and banana stems, resins are made from microorganisms and commercial resins as well as composites are made from soybean protein and plant based fibres. Research is being done on plastics that are obtained from either starches or bacteria. The development of new materials is constantly in progress.

Based on the degradation mechanism, biodegradable plastics are classified as

1. Biodegradable
2. Compostable
3. Hydro-biodegradable
4. Photo-biodegradable
5. Bioerodable

Starch-based Plastics

Starch-based plastics are mainly harvested from wheat, potatoes, rice and corn. About 20% of starch is used for many non-food items, such as making paper, cardboard, textile sizing and adhesives. Starch-based plastics have also been processed into eating utensils, plates, cups and other products. The cornstarch-based material has the, "look, feel and flexibility" of conventional plastics and can be used for a range of items, from cellophane to plant pots and medical devices. When water is added, it completely disappears into the soil over a period of time. This is excellent for food packaging and farming. This cornstarch blend of plastic is cheap enough to compete with conventional plastic. There is a cornstarch-based organic waste bag now on the market. It is called the Biobag. It is 100% biodegradable and 100% compostable. After 10–45 days, it is completely biodegraded depending on prevailing conditions and methods. The Biobag is compliant with the FDA and EU requirements and can be printed using flexo printing. Another feature of this Biobag is that it "breathes", and reduces the weight of its contents by up to 25% in five days. Biobags are used for trash and compost disposal.

The starch-based plastics resemble many conventional plastics and are as biodegradable as pure cellulose. There are processes that change lactic acid monomer into a polymer chain called polylactic acid (PLA) or polyglycolic acid (PGA). Both PLA and PGA are crystalline polymers, but PLA is more hydrophobic than PGA. PLAs are very brittle and stiff and they require plasticizers for most applications. High gloss and clarity are other features of PLA plastics. PLA is distinctive because it is available in renewable resources such as the starches. These renewable resources are on the leading edge of technology and PLAs are being used for pharmaceuticals. PLA can also be processed like most thermoplastics into fibres, or it can be thermoformed. PLAs can be used in a wide range of applications such as packaging (wrapping film, film for dry food packaging, board lamination, etc.), stationery (pens, cartridges, pencil sharpeners, etc.), and personal care items.

A potential re-use feature of the Biobag and other starch-based composites is that they can be used as feed for farm animals. Some animal feeds are required to contain some starch along with 13–24% protein. Starch–protein plastics are used as food containers. They could then be pasteurized into animal feed, rather than sending the starch-based trash into a landfill.

Bacteria-based Plastics

Bacteria are additional resources used to create a different type of biodegradable plastics. Polyhydroxyalkanoate (PHA) a biodegradable polymer, is produced inside the bacterial cells. The bacteria are harvested after they are grown in the culture, and then converted into biodegradable plastics. The mechanical properties of their resins can be altered depending on the needs of the product. The PHA fibres "degrade aerobically and anaerobically".

Soy-based Plastics

Soy-based plastics use another alternative material for biodegradable plastics. Less than 0.5% of the available soy protein is used for industrial products. Soy proteins are used for making adhesives and coatings for paper and cardboard. In the opinion of researchers soy protein may be a first rate material for engineering plastics when a proper moisture-barrier is applied. Soybeans are composed of protein with limited amounts of fat and oil. Protein levels in soybeans range from 40–55%. The high amount of protein means that they must be properly plasticized when being formed into plastic materials and films. The films produced are normally used for food coatings, but more recently, freestanding plastics used for bottles have been formed from the plasticized soybeans.

Ford has taken advantage of the soy protein plastics and has been using it to manufacture parts for automobiles.

Dr. Amar K. Mohanty is a professor at Michigan State University whose research is primarily on the diverse types of biodegradable plastic polymers. Dr. Mohanty forecasts that the demand for biodegradable plastics will grow by 16% per annum in future.

Starch-based biodegradable plastics have been shown to degrade 10–20 times faster than traditional plastics. When traditional plastics are burnt, they create toxic fumes, which are hazardous to health and the environment. If any biodegradable films are burnt, there is little toxic chemicals or fumes released into the air. Biodegradable plastics have been proved to improve soil quality. As the microorganisms and bacteria in the soil decompose the material, it actually makes the soil more fertile. With all the advantages of biodegradable plastics, there are a number of disadvantages. Recycling helps the environment and it works well for many plastic containers such as bottles. Eventually there is a limit to how many times a piece of plastic can be recycled, so in the end there will be waste produced. The cost of recycling plastics, in terms of energy, can be significantly higher. Toxics are released from burning waste plastics in order to harness the energy for production. Many plastics that appear to be biodegradable in reality break down into minuscule bits that can affect both the soil and animals. Unfortunately, as researchers try to improve the environment with these new plastics, in essence they may be creating risks, as well.

Biodegradable Polyesters

Polyesters play a predominant role as biodegradable plastics due to their potentially hydrolysable ester bonds. The polyester family is made of two major groups—aliphatic (linear) polyesters and aromatic (aromatic rings) polyesters. The polyesters may be either naturally produced or synthetic. Naturally produced polyesters are renewable whereas synthetic polyesters are of two types— renewable and non-renewable.

Biodegradable polyesters which have been developed commercially and those which are in commercial development are as follows:

Polyhydroxyalkanoates (PHA)

Polyhydroxybutyrate (PHB)

Polyhydroxyhexanoate (PHH)

Polyhydroxyvalerate (PHV)

Polylactic acid (PLA)

Polycaprolactone (PCL)

Polybutylene succinate (PBS)

Polybutylene succinate adipate (PBSA)

Aliphatic–Aromatic copolyesters (AAC)

Polyethylene terephthalate (PET)

Polybutylene adipate/terephthalate (PBAT)

Polymethylene adipate/terephthalate (PTMAT)

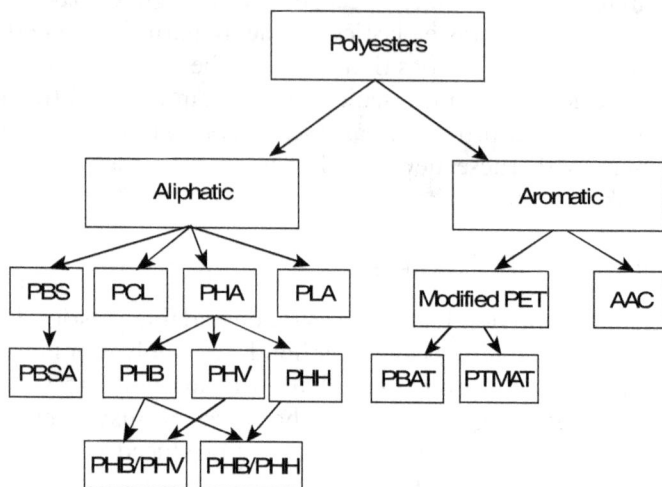

Biodegradable polyester family

While aromatic polyesters such as PET exhibit excellent material properties, they prove to be almost totally resistant to microbial attack. Aliphatic polyesters on the other hand are readily biodegradable, but lack good mechanical properties that are critical for most applications. All polyesters degrade eventually, with hydrolysis being the dominant mechanism. Synthetic aliphatic polyesters are synthesized from diols and dicarboxylic acids via condensation polymerization, and are known to be completely biodegradable in soil and water. These aliphatic polyesters are, however, much more expensive and lack mechanical strength compared with conventional plastics such as polyethylene. Many of these polyesters are blended with starch-based polymers for cost competitive biodegradable plastics applications. Aliphatic polyesters have better moisture resistance than starches, which have many hydroxyl groups. The rate of soil degradation of various biodegradable plastics has been measured. Poly-(3-hydroxy-butyrate)/valerate (PHB/PHV), PCL, PBS, PBSA and PLA were evaluated in soil burial for 12 months and samples were collected every 3 months for the measurement of weight loss. The rate of degradation of PBSA, PHB/PHV and PCL was found to be similar, with the rate of PBS and PLA respectively slower.

PHA (Naturally Produced) Polyesters

Polyhydroxyalkanoates (PHAs) are aliphatic polyesters naturally produced via a microbial process on sugar-based medium, where they act as carbon and energy storage material in bacteria. They were the first biodegradable polyesters to be utilized in plastics. The two main members of the PHA family are:

1. Polyhydroxybutyrate (PHB) and
2. Polyhydroxyvalerate (PHV).

Aliphatic polyesters such as PHAs, and more specifically homopolymers and copolymers of hydroxybutyric acid and hydroxyvaleric acid, have been proven to be readily biodegradable. Such polymers are actually synthesized by microbes, with the polymer accumulating in the microbes' cells during growth. The most common commercial PHA consists of a copolymer PHB/PHV together with a plasticizer/softener (e.g. triacetine or estaflex) and inorganic additives such as titanium dioxide and calcium carbonate. A major factor in the competition between PHAs and petroleum-based plastics is the difference in production costs.

Opportunities exist, however, for obtaining cheaper raw materials that could reduce PHA production costs. Such raw materials include corn-steeped liquor, molasses and even activated sludge. These materials are relatively inexpensive nutrient sources for the bacteria that synthesize PHAs. The PHB homopolymer is a stiff and rather brittle polymer of high crystallinity, whose mechanical properties are unlike those of polystyrene, though it is less brittle. PHB co-polymers are preferred for general purposes as the degradation rate of PHB homopolymer is high at its normal melt processing temperature. PHB and its co-polymers with PHV—polyhydroxybutyrate-co-polyhydroxyhexanoates (PHBHs) resins—are one of the newest naturally produced biodegradable polyesters. The PHBH resin derived from carbon sources such as sucrose, fatty acids or molasses via a fermentation process are semi-crystalline thermoplastics. They represent the first example of a true biodegradable thermoplastic produced via a biotechnology process. No toxic by-products are known to result from PHB or PHV. These are aliphatic–aliphatic copolyesters and are distinct from aliphatic–aromatic copolyesters. Besides being completely biodegradable, they also exhibit barrier properties similar to those exhibited by ethylene vinyl alcohol (EVOH). EVOH is another water-soluble synthetic plastic, used as an oxygen barrier layer in multilayer film packaging. The high cost of EVOH is a significant barrier to its widespread use in other biodegradable plastics applications.

PLA Polyesters

Polylactic acid (PLA) is a linear aliphatic polyester produced by poly-condensation of naturally produced lactic acid or by the catalytic ring opening of the lactide group. Lactic acid is produced via starch fermentation as a co-product of corn wet milling. The ester linkages in PLA are sensitive to both chemical hydrolysis and enzymatic chain cleavage. PLA is often blended with starch to increase biodegradability and reduce costs. However, the brittleness of the starch–PLA blend is a major drawback in many of its applications. As a remedy to this limitation, a number of low molecular weight plasticizers such as glycerol, sorbitol and triethyl citrate are used. PLA does not have full food contact approval due to its fermentation manufacturing method. The applications for PLA are thermoformed products such as drink cups, take-away food trays, containers and planter boxes. The material has good rigidity characteristics, allowing it to replace polystyrene and PET in some applications. PLA is fully biodegradable when composted in a large-scale operation with temperatures of 60°C and above. The first stage of degradation of PLA (two weeks) is via hydrolysis to water-soluble compounds and lactic acid. Rapid metabolization of these products into CO_2, water and biomass by a variety of microorganisms occurs after hydrolysis. PLA does not biodegrade readily at temperatures less than 60°C due to its glass transition temperature being close to 60°C.

PCL Polyesters

Polycaprolactone (PCL) is a biodegradable synthetic aliphatic polyester made by the ring-opening polymerization of caprolactone. PCL has a low melting point, ranging 58–60°C, low viscosity and is easy to process. Until recently, PCL was not widely used in significant quantities for biodegradable polymer applications due to cost reasons. Recently however, cost barriers have been overcome by blending the PCL with cornstarch. PCL is suited for use as food-contact foam trays, loose fill and film bags. Although not produced from renewable raw materials, PCL is fully biodegradable when composted. The low melting point of PCL makes the material suited for composting as a means of disposal, due to the temperatures obtained during composting routinely exceeding 60°C.

PBS Polyesters

Polybutylene succinate (PBS) is a biodegradable synthetic aliphatic polyester with properties similar to that of PET. PBS is generally blended with other compounds such as starch (TPS) and adipate copolymers (to form PBS-A), to make its use economical. PBS has excellent mechanical properties and can be applied to a range of end applications which include mulch film, packaging film, bags and "flushable" hygiene products. PBS is hydro-biodegradable and begins to biodegrade via a hydrolysis mechanism. Hydrolysis occurs at the ester linkages and this results in a lowering of the polymer's molecular weight, allowing for further degradation by microorganisms. The degradation rate is estimated to be 1 month for 50% degradation of 40μ thick film in garden soil.

AAC Copolymers

Aliphatic–aromatic (AAC) copolyesters combine the biodegradable properties of aliphatic polyesters with the strength and performance properties of aromatic polyesters. This class of biodegradable plastics is seen by many to be the answer to the making of fully biodegradable plastics with property profiles similar to those of commodity polymers such as polyethylene. To reduce cost, AACs are often blended with TPS. There are many types of AACs with different trade names. There are a number of specific grades under each trade name. Each grade of polymer has been designed with controlled branching and chain lengthening to match its particular application. AACs come closer than any other biodegradable plastics to equalling the properties of low density polyethylene. AACs can also meet all the functional requirements for cling film such as transparency, flexibility and anti-fogging performance, and therefore this material has great promise for use in commercial food wrap for fruit and vegetables, with the added advantage of being compostable. While being fossil fuel-based, AACs are biodegradable and compostable. ACCs fully biodegrade to carbon dioxide, water and biomass. Typically, in an active microbial environment the polymer becomes invisible to the naked eye within 12 weeks. The extent and rate of biodegradation, apart from the inherent biodegradability of the polymer itself, depends on several environmental factors such as moisture, temperature, surface area and the manufacturing method of the finished product.

Modified PET

Modified PET (polyethylene tetraphalate) is PET which contains co-monomers, such as ether, amide or aliphatic monomers, that provide *weak* linkages that are susceptible to biodegradation through hydrolysis. Depending on the application, up to three aliphatic monomers are incorporated into the PET structure. Typical modified PET materials include PBAT (polybutylene adipate/terephthalate) and PTMAT (polytetramethylene adipate/terephthalate). Modified PET is hydro-biodegradable, with biodegradation steps following an initial hydrolysis stage. It contains weak linkages which create sites for microbial attack. The mechanism involves a combination of hydrolysis of the ester linkages and enzymatic attack on ether and amide bonds. With modified PET, it is possible to adjust and control degradation rates by varying the co-monomers used. The options available for modified PET provide the opportunity to produce polymers which specifically match a range of applications based on physical properties while maintaining the ability to adjust the degradation rate by the use of co-polyesters.

NICHE MARKETS FOR BIODEGRADABLE PLASTICS

The makers of completely biodegradable polymers are continuing to develop and test the market with new consumer products where they make functional, economic and environment sensitive commodities. The following are current examples of applications of biodegradable polymers.

Medical Sutures, Pins and Dental Implants

No doctor wants to operate twice—opening the patient, a second time to remove implants or stitches—if it can possibly be avoided when the health risks are too great. More than 125 million non-toxic biodegradable polymer sutures are being used each year by surgeons in life-saving heart operations and other procedures. Easily sterilized, the sutures remain strong and intact until the surrounding tissues have healed. When their job is done, the sutures dissolve, and are readily metabolized in the body leaving no trace. Both pliable braided sutures and more wiry monofilament sutures are available, depending on the type of surgery, the knots to be used, the tensile strength needed and the potential for infection. Biodegradable plastic staples are also being used to close wounds and incisions. Plastic sutures, largely based on lactic or glycolic acid and similar materials, account for about 95% of total use, with various pins, implants and dental devices making up much of the remainder.

Biodegradable plastic pins, tacks and screws are being used to hold shattered bones together while they heal, to reattach ligaments, and for delicate reconstructive surgery on ankles, knees and hands. Fragile bones can be broken again when removing metal pins and other orthopaedic implants; a bone that slowly rebuilds around an absorbable polymer is stronger and better able to take a full load, once the cast is removed. In addition, biodegradable dental implants, made of porous polymer particles, are being used to quickly fill the hole after a tooth has been extracted. Dental surgeons are also using films of biodegradable polymer—the guided-tissue-regeneration membranes—to prevent scabbing and allow the slower-growing connective and ligament cells to re-grow.

Medical researchers are excited about other potential applications for absorbable, biodegradable polymers. Promising uses currently being investigated include, devices designed to slowly release a measured dose of medicine, various ligating clips and vascular grafts, and tissue-engineering scaffolds that serve as matrices for living cells.

A wide range of drugs can be delivered by biodegradable polymers, including hydrophobic and hydrophilic drugs. Release of the drug can be modelled by computer, and technologies for manufacturing polymer drug products to exacting clinical standards have been developed. These advances have attracted both large and small firms to this area of drug delivery. The Gliadel co-polymer (polifeprosan 20) is a random copolymer consisting of poly[bis(π - carboxyphenoxy) propane] and sebacic acid. The structure is given below:

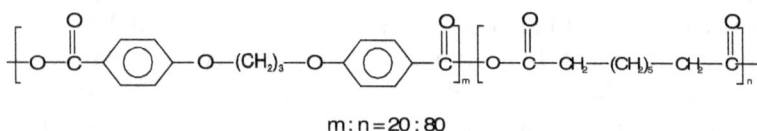

m : n = 20 : 80

Advances in biocompatible polymer technology have greatly enhanced drug delivery and have generated several possible clinical treatments. Novel biocompatible polymers have been developed as potential drug delivery vehicles, including polyanhydrides and polyphosphoesters. The polymer used in the Gliadel wafer to deliver carmustine is a polyanhydride polymer (poly[1, 3-bis (carboxyphenoxy) propane-co-sebacic acid]) (PPCP-SA). These polymers have significant advantages over earlier polymers for drug delivery. The new polymers protect the included drug from hydrolytic cleavage. The drug is delivered by surface erosion of the polymer particle and not by diffusion from the interior of the particle.

Disappearing Hospital Laundry Bags

Biodegradable plastics are not confined to the hospital operating room or dental surgery. Modern health care centres generate enormous mounds of dirty linen, soiled hospital gowns, wet towels, and similar materials every day. Water-soluble and completely biodegradable plastic laundry bags and hamper liners are making the jobs of the nurses and orderlies who collect and handle these washables both easier and much safer. The sturdy, light weight bags are simple to use; after a load of contaminated laundry is collected, the bag is sealed shut with an attached adhesive strip, which is designed to dissolve in cold water. The sealed bag is then placed in the washer where it will break down completely during the hot washing and disinfection process. Any residual polymer will biodegrade during the waste-water treatment process at the local sewage treatment plant.

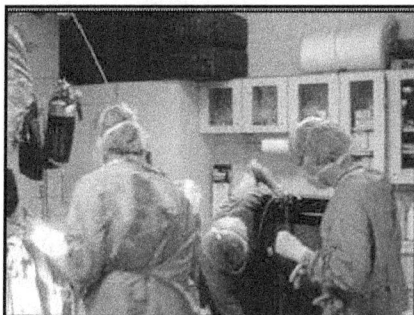

Water-soluble laundry bags
are used in hospitals

Textiles for Industrial and Institutional Uses

Both woven and non-woven fabrics made from biodegradable polymers have been developed for a number of industrial and institutional uses. They are easy to process, stand up well to ultraviolet light, are hardwearing and resilient, and offer good resistance to soiling and staining. Manufacturers have used them to develop wipes, geotextiles for erosion control and landscaping and various filters.

The biodegradable polymers represent a bridge between natural fabrics, such as wool and cotton, and the conventional synthetics. Lightweight clothing—from casual to high-end fashions—made from biodegradable polymers wears well and looks good. The polymer can be blended with wool, silk, cotton and other fibres to offer a soft silky feel, with excellent hang and drape characteristics. Easy to care for, these new fabrics resist wrinkling and handle moisture well, which makes them suitable for all kinds of sports and active wear. Biodegradable polymer fabrics and sheets also feature good wicking properties, which means they are very good at soaking up and transferring sweat, water and other fluids.

Hygiene Products

One of the strongest potential markets for biodegradable plastics is in the field of personal hygiene products. A disposable diaper made with a biodegradable plastic liner would offer all the convenience, comfort and dryness of a conventional disposable. In addition, it is economically and readily composted in facilities that are not equipped to remove conventional plastics.

Agricultural Products

Another strong potential market is the production of agricultural mulches, seeding strips and tapes made of biodegradable polymers. These new "plasticulture" initiatives can increase yields, improve crop quality and even stretch the growing season. Seeding strips and tapes contain seeds placed at regular spaces, as well as added nutrients; and they may be planted semi-automatically. The tape biodegrades in the soil as the seeds germinate and take root. Plastic mulch films are then used to give the new seedlings a head start in the spring; the mulch helps reduce evaporation and conserve moisture, increases soil temperature, and keeps competing weeds at bay. The final produce is also cleaner and less subject to decay. However, conventional plastic mulches cannot be removed until the crop is harvested—a costly and time-consuming operation. Instead, a truly biodegradable product would eliminate this step by becoming "one with the soil."

Biodegradable polymers might also replace certain difficult-to-recycle materials used on the farm. For example, chemical residues make empty fertilizer bags hard to recycle. While washing can remove the water-soluble fertilizers, the resulting rinse water can be too rich with nitrogen and phosphorous compounds for discharge to a sewage treatment system. A compostable bag would eliminate the problem and keep the nutrients on the land and out of the sewer. Other applications include silage wrappings, plant pots and seed trays.

Photo-biodegradable Plastics

Photodegradable plastics are thermoplastic synthetic polymers into which have been incorporated light-sensitive chemical additives or copolymers for the purpose of weakening the bonds of the polymer in the presence of ultraviolet radiation. Photodegradable plastics are designed to become weak and brittle when exposed to sunlight for prolonged periods. Photosensitizers used include diketones, ferrocene derivatives (aminoalkylferrocene) and other carbonyl-containing species. These plastics degrade in a two-stage process, with UV light initially breaking some bonds leaving more brittle lower molecular weight compounds that can further degrade by physical stresses such as wave action or scarring.

The future of biodegradable plastics shows great potential. Many countries around the world have already begun to integrate these materials into their markets. The Australian Government has begun its research to develop starch-based plastics. Japan has created a biodegradable plastic that is made of vegetable oil having the same strength as traditional plastics. The Mayor of Lombardy, Italy, recently announced that merchants must make biodegradable bags available to all of their customers. In America, McDonald's is now working on making biodegradable containers to use for their fast food. This increasing interest will allow the technology needed to produce more affordable biodegradable plastics. Biodegradable plastics are one of the most innovative materials being developed in the packaging industry. How widespread biodegradable plastics will be used all depends on how strongly society embraces and believes in the environmental preservation. There certainly are abundant amount of materials and resources to create more uses for biodegradable plastics. The advancement of biodegradable technology has skyrocketed and there are growing signs that the public shows a high amount of curiosity in the product.

WITH PLASTICS INTO THE FUTURE

In the 20th century, we had telephones of Bakelite, stockings of nylon, bags of polythene and thousands of other more or less essential plastic objects. Perhaps our new century will offer commercial potential of conductive and semiconductive polymers as they can be produced quickly and cheaply. Electronic components based on polymers, and polymer-based integrated circuits, will soon find their place in consumer products. The step from polymer-based electronics to real molecular-scale electronics is a large but fascinating one. Molecule-based integrated circuits could be reduced to a scale, many orders of magnitudes smaller than silicon-based electronics. While many challenges lie ahead, we stand at the threshold to a plastic–electronics revolution with exciting implications in chemistry and physics as well as information technology.

REVIEW QUESTIONS

1. Describe the observations made from the microscopic examination of milk at different magnifications.
2. Define milk. What are the different components of milk?
3. What are milk lipids?
4. Name the fatty acid that is found in the milk of ruminant animals.
5. Name the unsaturated fatty acid that is abundant in milk.
6. Discuss the features of fat globule membrane(FGM).
7. What is fat destabilization? State its significance.
8. What are coalescence and flocculation of milk?
9. What are the common terms used in describing the milk fractions?
10. What is clarification of milk?
11. Why is batch pasteurized milk known as raw milk?
12. State the functions of milk fats.
13. What is Rowland fractionation?
14. Give a short account of milk caesins.
15. Discuss the micellar structure of milk caesins.
16. What is whole milk?
17. What is the role of colloidal calcium phosphate(CCP)?
18. Enumerate the factors affecting the stability of casein micelle.
19. Explain the effect of pH and temperature on casein micelle.
20. What is age-gelation?
21. Write a short note on whey proteins.
22. Name the factors influencing the density of milk.
23. What is Millard reaction?

24. List the various vitamins and minerals present in milk.

25. What is SNF stand for?

26. Which factor controls the water content of milk?

27. What is lipoprotein lipase(LPL)?

28. What is dextrose equivalents? What is its significance in the ice cream industry?

29. Give the Richmond formula.

30. Why does homogenized milk coagulate readily?

31. Give the common range for the density of milk.

32. Name the compounds responsible for the flavour and aroma of milk.

33. State the significance of viscosity of milk.

34. What are the advantages of pasteurized milk?

35. What is the significance of the positional specificity of fatty acids in butter?

36. List and discuss the factors affecting viscosity of milk.

37. State the physical properties of milk fats.

38. Discuss the structure of milk fat globule.

39. How does the creaming of cold raw milk occur?

40. How is the milk fat protected from LPL?

41. Explain the significance of the optical properties of milk.

42. How does homogenization prevent the creaming of milk?

43. What is HFCS?

44. Explain the role of the enzyme, plasmin.

45. How is fat destabilization achieved?

46. State the functions of milk fat.

47. How does the action of heat alter the acidity of milk?

48. What happens to the mineral contents of milk on heating?

49. Name the processes that induce casein micelle aggregation.

50. Which are the substances that buffer in milk?

51. What are the parameters determined from the freezing point of milk?

52. What is scorching of milk?
53. What are the factors that affect the freezing point of milk?
54. Explain the action of heat on whey proteins.
55. What is lactose intolerance?
56. What is observed when milk is boiled in open containers? Give reasons for the observation.
57. Why is pre-warming done before sterilization?
58. Why is casein resistant to heat?
59. What is the effect of heat on milk acidity?
60. Name the operations involved in milk processing.
61. What is clarified milk?
62. Discuss the salient features of pasteurization of milk.
63. What are UHT and HTST?
64. Define and give the significance of homogenization of milk.
65. Describe suitable methods for the determination of:
    i.   Density of milk
    ii. Total solids (TS)
    iii. Solids-not-fat (SNF)
66. What is whole dry milk powder?
67. Write on:
    i. Skimmed milk powder
    ii. Butter milk powder
68. What are the basic milk categories and give their definitions.
69. Describe spray drying process with reference to the production of whole milk dry powder.
70. Define and give the composition of butter.
71. State the primary factors responsible for the flavour of butter.
72. What does Reichert–Meissl factor denote?
73. Give the compositions of:
    i.   Clarified butter   ii. Anhydrous butter oil
    iii. Ghee   iv. Cream
    v.   Whipping cream

74. What are the ingredients and additives present in whipping cream?

75. Mention the composition of ice cream.

76. What is the importance of milk fat in ice cream?

77. What limits the excessive use of butterfat?

78. Describe milk solid-not-fat (MSNF).

79. List the benefits of MSNF in ice cream.

80. Explain the role of citrate, phosphate, calcium and magnesium ions in ice cream.

81. Explain the importance of sweeteners and stabilizers in the manufacture of ice cream.

82. Name a few stabilizers used in the production of ice cream.

83. What are emulsifiers? Give illustrations.

84. What are polysorbates?

85. Define the terms:
    i.   Skin
    ii. Hide
    iii. Leather

86. What is the purpose of tanning?

87. List the uses of leather in daily life.

88. State a few special qualities of leather.

89. Give a brief description of the structure of leather.

90. Give the composition of skin.

91. List the stages involved in preparing skins and hides.

92. Discuss the following:
    i.   Cleaning and soaking
    ii. Liming and degreasing

93. How is fleshing done?

94. How is the prepared skin transformed into leather?

95. Explain vegetable and chrome tanning.

96. What is the importance of fat liquoring?

97. What are chamoising and tawing?

98. How does the leather industry cause the pollution of air and water?

99. Discuss the various measures taken in controlling tannery pollutions.

100. List the steps to prevent pollution caused by the leather industry.

101. What does the word polymer mean?

102. Define and classify polymers.

103. What does the word plastic refer to?

104. Explain the following:
    i. One-dimensional polymers
    ii. Two-dimensional polymers
    iii. Three-dimensional polymers

105. Classify polymers based on their response to heat.

106. What are thermo and thermosetting polymers? Give examples

107. What are homo and co-polymers?

108. How is an addition polymer obtained industrially?

109. What are isotactic and syndiotactic configurations in polymers?

110. Identify the following as addition/condensation or co-polymer

i.	SBR	ii.	Orlon
iii.	Teflon	iv.	PVC
v.	Nylon	vi.	Terylene
vii.	Bakelite	viii.	Buna-S
ix.	Buna-N	x.	Glyptal
xi.	Polyethylene	xii.	Lucite
xiii.	Polypropylene	xiv.	Neoprene

111. How do Zeigler–Natta catalysts help in controlling the tacticity of a polymerization reaction?

112. What is living in a polymerization process?

113. What does the term stereo regularity/tacticity describe?

114. Discuss condensation polymerization reaction using glyptal resin as the illustration.

115. What are silicones? Give a few examples.

116. What is the role of vulcanization of rubber?

117. What are the criteria for a polymer to be elastomeric?

118. Give examples of synthetic and natural fibres.

119. How does the molecular weight affect the physical properties of a polymer?

120. The molecular weight of polymers is an averaged value. Why?

121. Define number and weight average molecular mass.

122. What is PDI and discuss its significance.

123. Explain the terms T_g, T_m and T_d.

124. What are amorphous and crystalline polymers?

125. What are the characteristics of an ideal polymer?

126. Give the applications of the polymers with the following properties:
    i.   Chemical inertness
    ii. Low thermal and electrical conductance
    iii. Light weight
    iv. Fibrous

127. Discuss the role of polymers as:
    i.   Adhesives
    ii. Fillers
    iii. Reinforcing agents

128. What is resin identification coding?

129. Why and how do polymers conduct? Give a few examples of conductivity polymers.

130. What is doping? How many types of doping are known?

131. State the applications of first generation polymers.

132. List the industrial uses of second-generation polymers.

133. What are biodegradable polymers? How are they classified into?

134. Compare a chemical battery and plastic battery.
135. Write a short note on photodiodes.
136. Give a short account of LEDs.
137. Enlighten the uses of:
    i.   Polythiophene
    ii. Doped aniline
    iii. Polyphenylene vinylene
    iv. Polydialkylfluorenes
138. Explain the applications of polymers in optical and ocular lenses.
139. What are polarons?
140. What constitutes the biodegradable polyester family?
141. What is a biobag?
142. How is PHA produced?
143. What are the uses of soy-based plastics?
144. Write on
    i.   PHB
    ii. PHV
    iii. PBS
145. Discuss the importance of AAC copolymers and modified pet.
146. Discuss the potential applications of biodegradable polymers in the field of medicine and pharmacy.
147. What is the role of biodegradable polymers in agricultural industry?
148. What are photo-biodegradable polymers?
149. Write a critical account on the future role of biodegradable polymers.

PART IV

AGRICULTURAL
CHEMISTRY

20

SOIL CHEMISTRY

INTRODUCTION

Soil is the natural three-dimensional body having varied uses, the most important of which is to produce food and fibre for mankind. People are dependent on soil and to a certain extent, good soils are dependent upon people and the use they make of them. Great civilizations have invariably had good soils as one of their chief natural resources. Soil destruction and mismanagement were associated with the downfall of some of the civilizations that good soils had helped to build.

A thin layer of soil covers most of the earth's land surface, varying from a few centimetres to 2–3 m in thickness. This might appear insignificant relative to the bulk of the earth. Yet it is in this thin layer that the plant and animal kingdom meet the mineral world and establish a dynamic relationship. Thus, life is so vital to soil and so is soil to life. Liebig and other early chemists considered soil as a storehouse of plant nutrients. Early geologists considered soil to be weathered rock. All these concepts are correct but incomplete. Thus, any definition assigned to soil depends mainly on the viewpoint of the person formulating the definition.

The heterogeneity and complexity of the soil system does not provide any one easy method for its definition. It is commonly defined as "any part of earth's crust in which plants root". In other words, any part of the earth's surface that supports vegetation and has a covering is called as soil. Muddy bottoms of ponds, porous rock surfaces, etc. are also soils. However, there are limitations to all these, because soil is formed by a long-term process of complex interactions leading to the production of a mineral matrix in close association with interstitial organic matter—living as well as dead. Thus, soil is not considered as a mere physical mixture but it is called as soil complex, comprising of five components namely, mineral matter, soil organic matter (humus), soil water/soil solution, soil atmosphere and the biological system.

A comprehensive and a broader definition considers soil as a dynamic natural body developed as a result of pedogenic processes during and after weathering of rocks, consisting of minerals and organic contents, with definite chemical, physical, mineralogical and biological properties having a variable depth of earth and that provides medium for growth of land plants.

Soils are the natural bodies on which plants grow and the humankind enjoys and uses these plants. Standards of living are often determined by the quality of the soils, and the kinds and quality of plants grown on them. Indeed soils have more purpose for mankind than merely serving as a habitat for growing crops. Soils provide the foundation for building houses, factories, constructing roads, etc.

There are two basic sources on the knowledge of soils. The first one is the practical knowledge gained by the farmers through centuries of trial and error. Second is the one that provides facts about soils and their management with the advent of modern sciences. Through two centuries of scientific studies, two basic concepts of soil have been evolved out, namely **edaphology** and **pedology**.

1. Edaphology is derived from the Greek word *edapha* meaning soil or ground. This considers soil as a natural entity, a biochemically weathered and synthesized product of nature. It considers the various properties of soil edaphic

factors and plant production. The edaphic factors include the structure and composition of soil along with its physical and chemical characteristics.

2. Pedology is a branch of science that deals with the formation of soil, its classification and its description.

FORMATION OF SOIL

Soil genesis or formation of soil involves two distinct phases.

1. Weathering
2. Pedogenesis

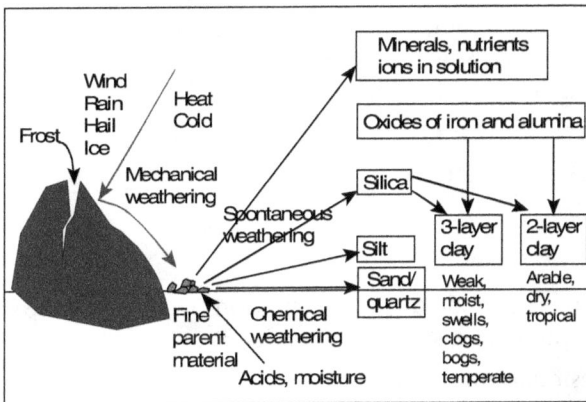

Figure 20.1 Process of weathering in the formation of soil

Weathering (Figure 20.1) is the disintegration of rocks and minerals by physical and chemical factors. However, this is not the true soil, because plants cannot grow in this. For this the weathered materials have to undergo a complex process termed as pedogenesis.

Pedogenesis is largely a biological phenomenon, in which various organic compounds, dead organic matter, living organisms such as lichens, bacteria, fungi, algae and mineral contents undergo numerous biophysical and biochemical changes. By these changes, the crusts of weathered rock debris are converted into true soil consisting of a complex mineral matrix in association with a variety of organic compounds and a rich population of microorganisms.

Therefore, during pedogenesis, various organic compounds, dead organic matter, living organisms, etc., are added to the mineral matter and this mineral matter is then gradually added to different layers of developing soil. This soil when fully developed has a number of layers, horizons of soil, known as soil profile.

CLASSIFICATION OF SOIL

Many types of classification of soil are available, each classification is based on specific parameter. Some of the important types of classification are listed.

I. The Russian soil scientist V.V. Dokuchaev and his associates were the first to develop the concept of soil as a natural body. They classified soil on the basis of strong ecological–genetic approach—a combination of climatic and vegetation data, and soil morphology as:

    i. Class A—normal/zonal soil

    ii. Class B—transitional soil

    iii. Class C—abnormal soil

II. Based on the mode of their formation particularly the origin of mineral matter, soils are classified into

    1. Residual soils

    2. Transported soils

The residual soils are those in which the entire process of soil formation, i.e., weathering and pedogenesis occurs at the place where the parent rock is present. Transported soils are those in which the weathered materials are taken away to different places by various agents and at these places, through pedological processes, soil formation is completed. Depending upon the various transporting agents, the transported soils are further classified into:

    i. Colluvial soil (transported by gravity)

    ii. Alluvial (transported by running water)

    iii. Glacial (transported by glaciers which are large masses of ice, snow)

    iv. Eolian (transported by wind)

III. The first organized attempt by the US Department of Agriculture (USDA) to classify soil was based on crop habitat and geological concepts. Based on this, soils are classified into 12 dominant soil orders. They are alfisols, andisols, aridisols, entisols, gelisols, histosols, inceptisols, mollisols, oxisols, spodosols, ultisols and vertisols.

IV. Parent material is what the soil is made from, usually from inorganic rocks, so a soil that has <20% organic matter (OM) is a mineral soil while one with >20% OM is an organic soil (peat or muck).

V. Texture refers to the proportion of sand, silt, and clay in the soil; sandy soils are called light or coarse-textured, whereas clay soils are called heavy or fine-textured. Clay tends to increase the water-holding capacity of the soil. Loamy soils have a balanced sand, silt and clay composition and are thus superior for plant growth.

VI. Structure refers to the aggregation of soil particles into platy, prismatic, blocky, spherical or crumb-like clods.

SOIL PROFILE

Soil profile is a vertical cross-section of a soil, showing the various layers or horizons, beginning at the surface with the A horizon and continuing downward through the B, C and R horizons to the parent (Figure 20.2). These horizons above the parent material are collectively referred to as solum (latin *solum* means "soil," "land"). A full soil profile develops in 2000–10,000 years, a period which is long for humans but short for the planet.

Just above the base rock is the C-horizon, containing the recently weathered and still weathering soil. It is rich in nutrients. The A-horizon is where most plant roots and all soil organisms are found. Its nutrients have been used by plants or leached downward, so it is relatively poor in nutrients, but rich in life. By comparison, the B-horizon is the zone where new material from below and nutrients from above accumulate. Sometimes an impermeable layer or pan is formed above it (podsol), denying plants to access this rejuvenating source of new nutrients. On the surface of the soil, often a thin layer rich in leaf litter and other organic materials is found.

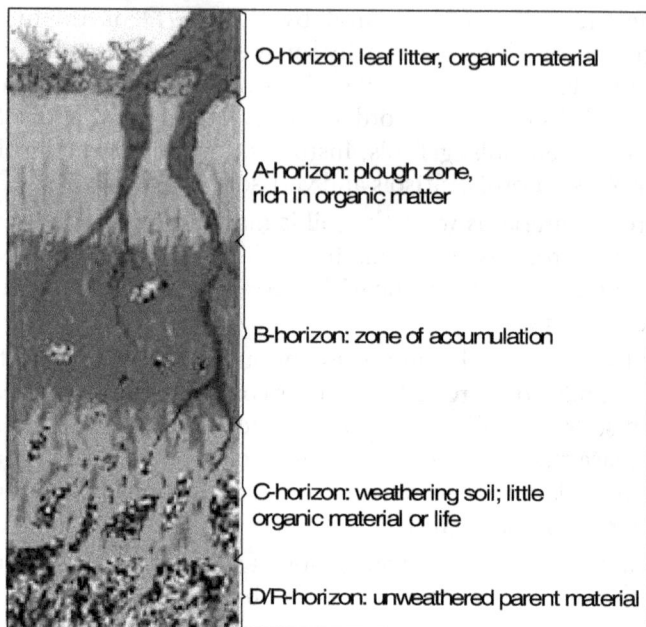

O-horizon: leaf litter, organic material

A-horizon: plough zone, rich in organic matter

B-horizon: zone of accumulation

C-horizon: weathering soil; little organic material or life

D/R-horizon: unweathered parent material

Figure 20.2   Soil profile

The five key factors in the formation of soil are

1. Type of parent material
2. Climate
3. Overlying vegetation
4. Topography or slope
5. Time

The type of parent material influences the soil pH, structure, colour, etc., profoundly. High-rainfall climates tend to have less fertile soils, due to rainwater's effect in leaching nutrients down to lower levels of the soil profile, and have more acidic soils. Low-rainfall climates tend to accumulate salts near the surface and have generally higher soil pH. Soils that form under coniferous forests tend to be more acidic than those under deciduous forests. Root action is also critical in soil formation. Soils generally have a harder time forming on steep slopes, due to run-off of soil particles during rains. The longer a soil has to form, the deeper is its profile.

The "soil textures" are based on their relative percentages of sand, silt, and clay; since these three percentages total 100, any soil sample can be plotted on a ternary graph, an equilateral triangle whose apexes represent pure sand, pure silt, and pure clay. It represents each texture by Polygon Morph, since a Polygon Morph can tell whether any specified point is within its boundary (Figure 20.3).

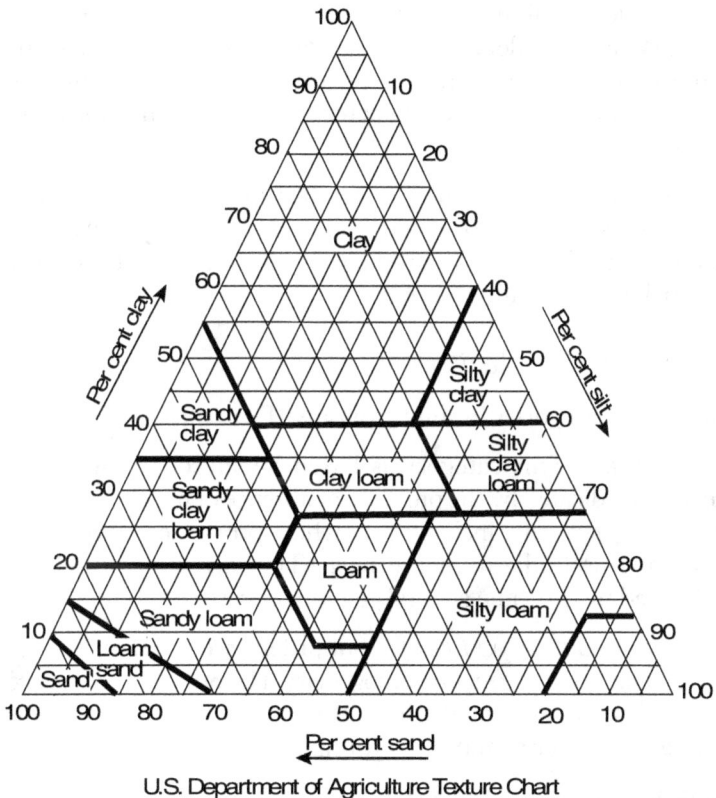

U.S. Department of Agriculture Texture Chart

Figure 20.3   Ternary diagram

Identification of the soil profile is the basis for various classifications of soil. On this basis, various classifications are provided by several countries.

The latest and a more comprehensive classification is soil taxonomy. According to this, soil, properties are the primary criteria, and the significant property used in the classification is the presence or absence of certain horizons referred to as epipedons (in Greek *epi* = over, *pedon* = soil).

Soil taxonomy is based on the properties of soils as they are found in the landscape. One objective of the system is to group soils similar in genesis, but the specific criteria used to place soils in these groups are those of soil properties. Because soil taxonomy is a hierarchical system, each soil is grouped first in the broadest category. When more details are added, lower categories are defined. Differentiating characteristics are not uniformly applied to all soils at a given categorical level, because soils have an enormous complexity.

Therefore, in soil taxonomy certain types of differentiating characteristics are applied only to certain taxa (of the level above which one is considering) to produce the desired taxa at the level with which one is dealing.

Hierarchy in Soil Taxonomy

There are six categories in soil taxonomy (Figure 20.4). They are:

Order (11 taxa)   This category is based largely on soil-forming processes as indicated by the presence or absence of major diagnostic horizons. A given order includes soils whose properties suggest that they are not dissimilar in their genesis. They are thought to have been formed by the same general genetic processes.

Suborder (60 number odd taxa)   Suborders are subdivisions of orders that emphasize genetic homogeneity. The presence or absence of properties are associated with wetness, climatic environment, major parent material and vegetation.

Great group (approximately 303)   Great groups are subdivisions of suborders according to similar kind, arrangement and diagnostic horizons. The emphasis is on the presence or absence of specific diagnostic features, base status, soil temperature and soil moisture regimes.

Subgroup (>1,200)   Subgroups are subdivisions of the great groups. The central concept of a great group makes up one group

(Typic). Other subgroups may have characteristics that are intergrades between those of the central concept and those of the orders, suborders or great groups.

Family Families are prominent in soils with a subgroup having similar physical and chemical properties affecting their response to management and especially to the penetration of plant roots. Differences in texture, mineralogy, temperature and soil depth are the basis for family differentiation.

Series Its differentiating characteristics are based primarily on the kind of arrangement of horizons, colour, texture, structure, consistency, reaction of horizons, chemical and mineralogical properties of the horizons.

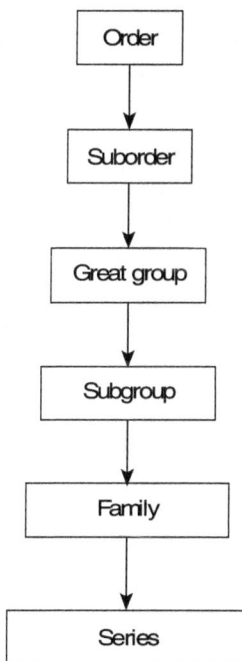

Figure 20.4   US soil taxonomy

Soil classification showing information about the 11 soil orders are listed below:

1. Alfisols—forest soils, moderately good.
2. Andisols—soils developed on volcanic materials.
3. Aridisols—desert soils, vulnerable to salinization.
4. Entisols—lack horizons, due to erosion.
5. Gelisols—soils in permafrost areas. Poor drainage and lots of freezing and thawing.
6. Histosols—swampy or marsh soils, organic matter decays slowly due to lack of oxygen.
7. Mollisols—grassland soils, excellent for farming.
8. Oxisols—tropical rainy climates, severe leaching.
9. Spodosols—cold humid climates, severe podzolization.
10. Ultisols—subtropical humid climates, some leaching.
11. Vertisols—clay soils, vulnerable to shrink, swell, bad for construction.

PROPERTIES OF SOIL

The complex mixture of soil consists of a three-phase system of solid, liquid, and gas, each of which has its own physical and chemical properties in the equilibrium or transition states. Each soil also has its own flora and fauna of bacteria, fungi, algae, etc. The liquid and gaseous phases are in small volumes and are homogeneous; but the solid phase is heterogeneous consisting of inorganic components of different sizes, such as silica, silicates, clay, metal oxides, etc. The various components of soil complex are mineral matter, soil water, soil air, soil solution, humus and soil organisms.

Soil Water

It is a dynamic solution regulating the physical, chemical and biological activities in the soil. Plants can absorb only a small quantity of rain water and dew directly. Most of the water is provided by the soil water, which is held within the pores of soil.

The soil water is available due to: (1) Infiltration of precipitated water (rain, sleet, snow and hail) and (2) Irrigation.

A definite portion of water is lost by evaporation, percolation, stream flow and transpiration. The capacity of soil to hold water depends upon the amount of water present and the size of the pores. The amount of water that a soil can store in a form available for plant use is called as the available water-holding capacity of the soil. Soil water together with its dissolved salts provides the soil solution. This is an important medium for supplying various nutrients to the growing plants.

Water is held in the soil in the following forms (Figure 20.5):

1. Gravitation water
2. Capillary water
3. Hygroscopic water
4. Water vapour
5. Combined water

Gravitation water   This is the water that temporarily occupies the aerated pore space, but drains down to a lower depth when there is a dry soil below. When soil is completely saturated with water, the excess water then displaces air from the soil pores and percolates downward under gravitational influence and finally gets accumulated in the pore spaces. This excess water is called as gravitational water. The amount of water held in the soil when all the pores are filled and drainage is restricted is called as maximum water-holding capacity.

When the gravitational water percolates down and reaches the levels of parental rock, it is called as ground water. This exerts 0.1–0.3 bar tension. The upper surface of ground water is called as water table. Ground water is of little use for plants as it is present in the soil for a shorter duration. By occupying soil pores it reduces the soil aeration. Therefore its drainage is a requisite for optimum plant growth.

Capillary water   This is the most used source of water for plants and exerts tension between 0.1–31 bars. Water that is present in the pore space of soil particles resembles water in a glass tube or capillary. The water molecules are held together by interparticle meniscus, i.e., surface tension and later on removed by gravity.

Hygroscopic water   Hygroscopic water is so tightly held such that it is considered as non-liquid and can move only in the vapour phase. Evaporation can remove more water from the soil than growing plants. However, there is a small amount of water that cannot be removed from the soil even by prolonged exposure to the air on conditions of failure of rainfall. Therefore, hygroscopic water is held tightly in the soil particles by strong cohesive and adhesive forces as a film of molecular layers of water. The tension of this water is 31 bars. The moisture content of the soil at this point is referred to as hygroscopic coefficient. Soils which are higher in colloidal contents will hold more water.

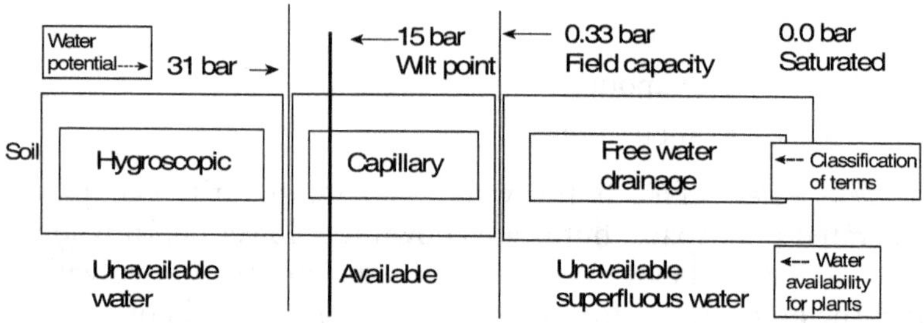

Figure 20.5   Different forms of soil water

Figure 20.5 shows the available water that can be utilized by plants, i.e., capillary water and the unavailable water that is held too tightly to sustain plant life, i.e., hygroscopic water.

Water vapour   Some amount of water is present in the soil in the form of vapour in the pore spaces between the soil particles.

Combined water   This is water that is present as hydrated oxides of aluminum, iron, silica, etc. Such chemically combined water is considered as a part of soil solids than as the liquid phase.

Terminology Used in Soil Water Status

After rainfall, the water enters the soil and fills the space between the soil particles, and the volume that can be filled is referred to as pore space. This varies with respect to texture and structure of soil.

Field capacity The moisture content of the soil after the downward movement of water has virtually ceased is termed as field capacity. This results after two or three days of rain. Field capacity is related to soil texture and influenced by the organic content, minerals and soil structure. The characteristics of adjoining layers are equally important, since more water will move down if the layer below has a greater attraction than the upper layer.

Wilting point When plants absorb water from soil, they lose most of it through evapotranspiration at the leaf surfaces. Some amount is also lost by evaporation directly from the soil surface. Both of these occur simultaneously. About half of the water contained in the soil is held tightly that it is not available to plants. The wilting point is reached when the rate at which the plant absorbs water from the soil becomes so slow that the plant wilts and cannot recover. The plant may still obtain some water content but not as fast to meet its needs. Sometimes the wilting may occur temporarily particularly on hot, dry days, but they recover at nights or when water is again supplied. The wilting point varies slightly according to the particular plant being grown and the atmospheric conditions that influence the rate of transpiration (Figure 20.6).

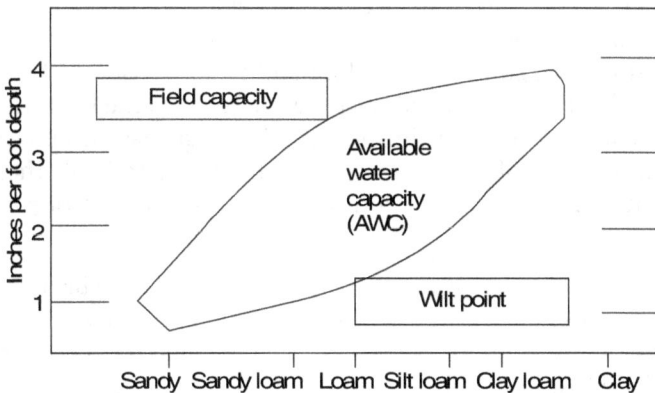

Figure 20.6 Wilting point

Soil Air

The composition of soil air depends largely on the amount of air space available along with the rate of biochemical reactions and gaseous exchange. The pore space of soil is filled with air and water. In both the poorly-drained and well-drained soils, the higher proportion of pore space is filled with water immediately after a heavy rain or irrigation. (Waterlogged soil is also referred to as anaerobic soil). If this situation prolongs, it leads to decrease in the amount of oxygen in soil, which is insufficient for plant growth. The concentration of both oxygen and carbon dioxide in soil are definitely related to the biological activity in the soil. Microbial composition of organic residue accounts for the major portion of carbon dioxide released. Presence of large quantities of manure, crop residue, etc. alters the soil air composition.

Subsoils are more deficient in oxygen because, the total pore space and the average size of pores are much less in the deeper horizons. Seasonal influences also alter the composition of soil air. This is due to the amount of soil moisture and temperature differences. Higher soil moisture lowers the level of oxygen and increases that of carbon dioxide. During summer, soils are drier, and the opportunity for gaseous exchange is more. This leads to relatively higher oxygen level. Some exceptions to this are also found. The higher summer temperature increases microbial activity thereby increasing the release of carbon dioxide. Similarly, soil with easily decomposable organic matter has higher concentration of carbon dioxide in summer. Soil flora and fauna consume O_2 and give out CO_2. Plant roots and microbes decay with the liberation of CO_2. The latter remains in soil partially in the gaseous phase. The concentration of CO_2 increases after the rains; probably there occurs an increased nitrification and decomposition of organic matter.

The composition of atmospheric air by volume is roughly taken as N_2 = 79.0%, O_2 = 20.90%, CO_2 = 0.03%. For aerobic dry soil, total composition of N_2 +Ar is roughly equal to that of the external atmosphere. The dry soil has CO_2 content between 0.5–1.0%, O_2 less than 20%.

Soil Temperature

Agriculture is the exploitation of solar energy in the presence of water and nutrients for plant growth. Soil temperature in combination with other edaphic factors influences the properties of soil itself and plants as well. This determines the geographic distribution of plants on the earth. It is affected by the air temperature, intensity of sunlight, angle at which rays of the sun strike the surface of the soil, daily duration of sunlight, amount of soil, moisture, etc. The main sources of soil heat are solar radiation and heat generated during the decomposition of dead organic matters in the soil and heat formed in the interior of earth. Climate, colour, slope, and altitude of the land and vegetation cover of the soil control temperature of the soil. The average annual temperature of soil is generally higher than that of the surrounding atmosphere. Only the surface temperature of the soil shows considerable fluctuations but the soil temperature below a certain depth remains more or less constant and is not affected by any external factors. The study of temperature has shown that in Bihar the diurnal temperature at the level 12" below the soil surface was only 1.0°C and at a depth of 24" it is nearly 0.1°C and at the depth of 3/4 feet the temperature is almost constant.

The temperature of a soil greatly affects the physical, chemical and biological process occurring in the soil. In cold soils, chemical and biological rates are slow. At lower temperatures the biological decomposition is at a near standstill, thereby limiting the rate at which the nutrients such as nitrogen, phosphorus, sulphur, and calcium are made available. For example, the oxidation of nitrogen to nitrate by the process of nitrification does not begin in spring until the soil temperature reaches 5.0°C. Soil temperature under field conditions can be altered by suitable agricultural practices such as mulching, irrigation, drainage and tillage.

Soil Minerals

Soil may be described as a three-phase system—solid, liquid and gaseous phase. When completely dry or frozen, it becomes a two-phase system, the liquid phase is either absent, or becomes a part of the solid phase. According to the international convention, soil material less than 2 mm is considered as the soil sample and the rest of the soil is rejected as unimportant.

The solid phase is composed of inorganic and organic constituents. Soils having more than 20% organic constituents is known as organic soil. When the inorganic constituents are more, it is better treated as mineral soil. Majority of the soils in India are mineral soils. The inorganic constituents forming the bulk of solid phase is made up of silicates of both primary and secondary origin, having a definite chemical composition and well-defined crystal structures. Along with these occur amorphous silicates, carbonates, soluble salts (containing calcium, magnesium, sodium and potassium; chlorides, sulphates, carbonates and bicarbonates, and in certain cases nitrates and borates also) and free oxides of iron, alumina and silica.

Primary mineral The original mineral component of a rock is known as primary mineral. Primary minerals which form the major composition and are recognized as the characteristic components of that rock are called as essential minerals. The most abundant composition includes quartz and feldspars with relatively lesser proportion of pyroxene, mica, etc. Primary minerals are concentrated in the coarse fraction.

Secondary minerals Under conditions of weathering, the primary minerals are broken down to small fragments and even to molecular species such as silica, alumina, iron oxide, etc. These undergo further natural conversions to structurally different silicates, aluminates and aluminosilicates and are called as secondary minerals. They constitute the most active ingredients of soil with respect to most of the chemical, physical and mineralogical properties. The clay fraction of the soil consists primarily of secondary minerals. The clay fractions include crystalline layer silicates, amorphous silicates, oxides of iron, silicon, aluminium, and other clay-sized primary minerals. Of these, the layer silicates are the most important and dominant ones.

21

COLLOIDAL PROPERTIES OF SOIL

A colloidal state is a two-phase system. One is a finely divided state and the other is a dispersed state. In nature, colloids are found as:

1. Emulsion—a liquid is dispersed in another liquid, e.g. milk; in this fat globules are dispersed in water.

2. Aerosol—a solid or a liquid is dispersed in a gas, e.g. smoke, a solid in gas and fog, this is a liquid dispersed in gas.

3. Gels—a solid is dispersed in liquid, e.g. gelatin and other finer fractions of soil as solids dispersed in water.

Colloidal particles are less than 1 mm in diameter. Since the clay fraction of soil is 2 mm and smaller, not all clay is strictly colloidal but even the larger clay particles possess colloid-like properties.

CLASSIFICATION OF SOIL COLLOIDS

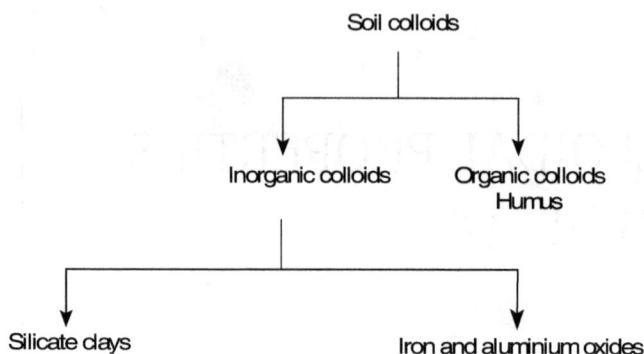

```
                        Soil colloids
                             |
            +----------------+----------------+
            |                                 |
            ▼                                 ▼
    Inorganic colloids                 Organic colloids
            |                               Humus
            |
    +-------+------------------------+
    |                                |
    ▼                                ▼
Silicate clays              Iron and aluminium oxides
```

Inorganic Colloids

The inorganic colloids are also known as mineral colloids. In soils both the organic and inorganic colloids exist together as an intimate mixture.

Silicate clays   The silicate clays are found in temperate regions and are dominant in the most developed agricultural regions of the world. In the earlier days colloidal particles were visualized as more or less spherical particles, some being crystalline and others amorphous. The latest information from electron microscope studies of soil particles have shown their occurrence as layers of plates or flakes and are crystalline, with regular internal arrangements. The silicate colloidal particles are referred to as micelles (micro cells), carrying negative charges, thereby a number of positively charged ions (cations) are attracted to each colloidal crystal, e.g. H^+, Al^{3+}, Ca^{2+}, Mg^{2+}, etc. This gives rise to an ionic double layer (Figure 21.1).

The sheet-like structure of silicate clay (micelle) shows a swarm of adsorbed cations and its innumerable negative charges. At the edge of the crystal structure is seen the negatively charged internal surfaces, to which cations and water molecules are attached.

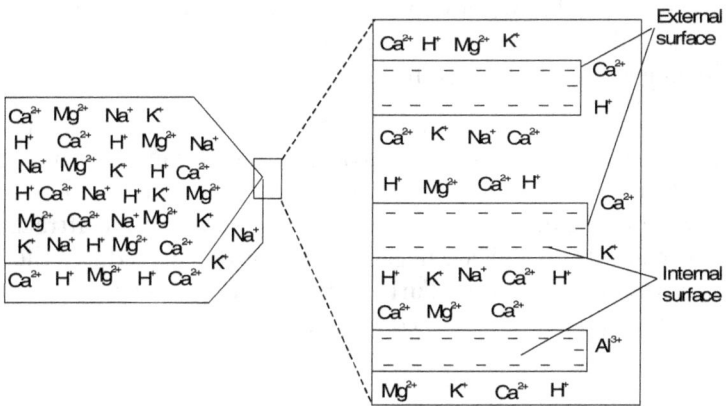

Figure 21.1 Sheet-like structure of silicate clay crystal (micelle) showing innumerable negative charges and swarm of adsorbed cations

The colloidal particles constitute the inner layer—being a huge anion, the external and internal surfaces of it are highly negative in nature. The outer layer is a swarm of loosely held cations that are attracted to the negatively charged surface. Therefore, a clay particle is accompanied by an enormous number of cations that are adsorbed or held on the particle surfaces. In association with these cation layers, a large number of water molecules remain as adsorbed on the clay particles. Some of these water molecules are carried by the adsorbed cations since most of them are hydrates. In some cases, the water molecules and cations are packed between the plates with the internal surface area constituting the clay micelle.

Oxides of iron and aluminium Clays of iron and aluminium oxides and hydroxides are found in the following:

i.   They occur in temperate regions intermixed with silicate clay.

ii. They are predominant in the highly weathered soils of tropics and semi-tropics.

The properties of red and yellow soils of tropic and semi-tropics are because of these iron and aluminium compounds. The oxides in common soil include, gibbsite ($Al_2O_3.3H_2O$), goethite ($Fe_2O_3.H_2O$). Their formulae can also be written as hydroxides.

Relatively less is known about the oxide clays. But it is known that some of their properties are in common with that of the silicates, like having a definite crystalline structure.

The clay oxides carry a small negative charge at higher pH values. Therefore, the colloidal particles as smaller micelles have lesser adsorption of cations. In the acidic condition where the pH is low, Fe and Al oxides that are positively charged, tend to attract the larger silicates which are negatively charged. This reduces the capacity to adsorb cations. Further, these clay particles are less sticky, plastic and cohesive as compared to silicates.

Organic Colloids (Humus)

The surface soils have definite amount of organic matter. It is considered to have a colloidal organization like clay. A highly charged anion (of micelle) is surrounded by adsorbed cations (see Figure 21.2).

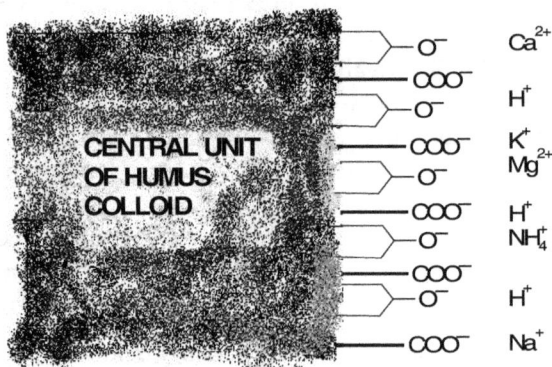

Figure 21.2   Adsorption of cations by humus colloids

The reactions of these cations are similar irrespective of their presence in inorganic or organic colloids. The definite difference observed between the two is that the humus micelle is composed of C, H and O and the inorganic ones contain Al, Si and O, etc. The organic colloids are noncrystalline, and the size of the colloidal particles are as small as clay particles. The humus colloids are dynamic and less stable than clay colloids. The structure of humus colloids is less explored. The majority of the negative charges of the colloids are due to presence of dissociated –COOH, enolic –OH and phenolic groups present with the central units. The negative charge is pH-dependent. Therefore, in acidic conditions, the hydrogen is not replaceable. The colloids then show low negative charge and less cation adsorption. At higher pH conditions, the hydrogen is easily removed and their order of removal is as —COOH, enolic or phenolic groups. This is done by the cations such as Ca^{2+}, Mg^{2+}, etc. and under this condition the cations adsorbed is far more than that absorbed by the silicate clays.

CATION EXCHANGE

This is one of the important properties of soil colloids. A colloidal complex is represented in the simplest form for each region as:

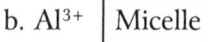

a. Ca^{2+}	e. Ca^{2+}
b. Al^{3+} $\boxed{\text{Micelle}}$	f. Mg^{2+} $\boxed{\text{Micelle}}$
c. H^+	g. K^+
d. M^{n+}	h. M^{n+}
Humid region	Arid region

where "M" refers to the small amount of metals and other base-forming cations adsorbed by the micelles and a, b, c, d, e, f, g, h refer to the number of cations that are available.

Adsorbed cations Ca^{2+}, Al^{3+} and H^+ are very prominent in the soil. It is because in the early stages of clay formation, the solution surrounding the decomposing silicates contain cations Ca^{2+}, Al^{3+}, K^+ Na^+ and Mg^{2+} which have been liberated by weathering. Not all these ions are held with equal tightness by the soil colloids.

The order of strength of adsorption is $Al^{3+} > H^+ > Ca^{2+} > Mg^{2+} > K^+ > Na^+$ when present in equal amounts. Thus, the various cations adsorbed in soil colloids are replaceable by others and this process is termed as cation exchange.

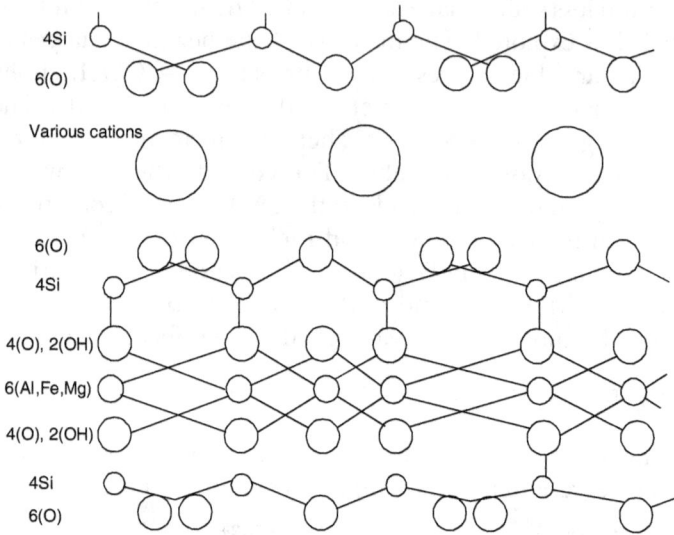

Figure 21.3   Diagrammatic representation of montmorillonite

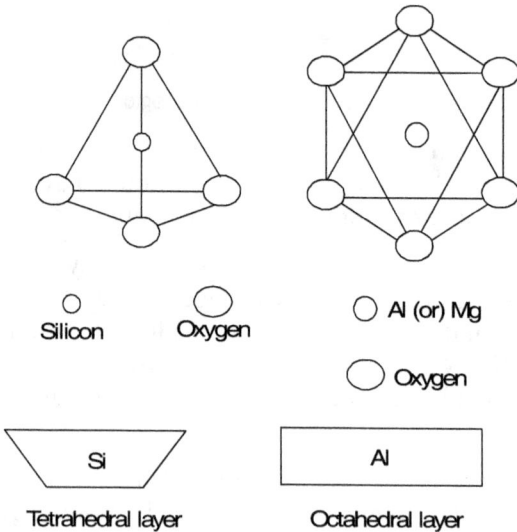

(a) Building blocks

Figure 21.4   Silicate clay minerals (*Continues*)

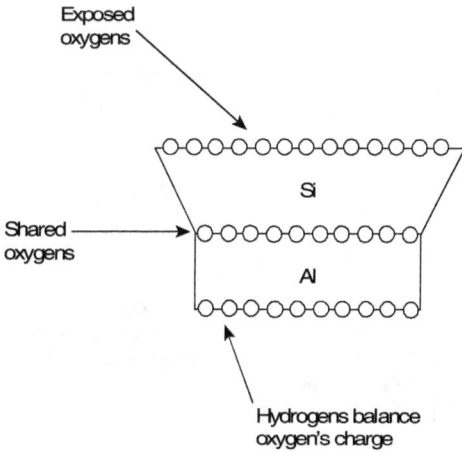

Exposed
oxygens

Si

Shared
oxygens

Al

Hydrogens balance
oxygen's charge

(b) 1:1 type minerals—general structure

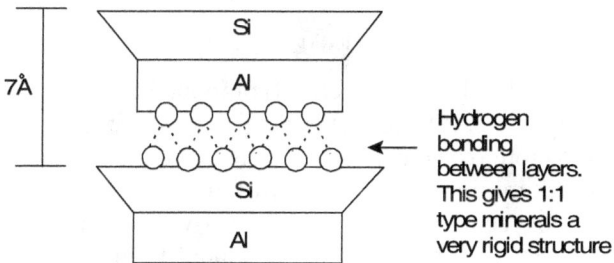

7Å

Si

Al

Si

Al

Hydrogen
bonding
between layers.
This gives 1:1
type minerals a
very rigid structure

(c) 1:1 type mineral—Kaolinite

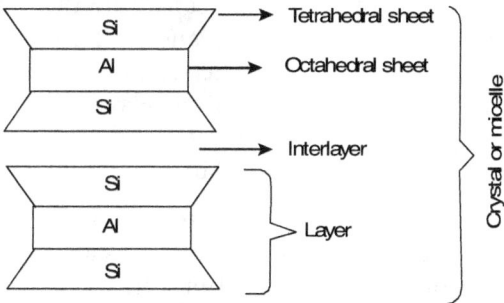

Si

Al

Si

Si

Al

Si

Tetrahedral sheet

Octahedral sheet

Interlayer

Layer

Crystal or micelle

(d) 2:1 type minerals—general structure

Figure 21.4   Silicate clay minerals (*Continues*)

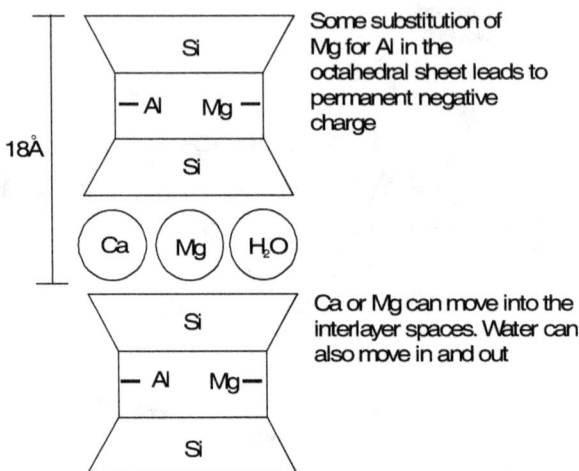

(e) 2:1 type mineral—Montmorillonite

Figure 21.4   Silicate clay minerals

The basic molecular and structural components of silicate clays (Figures 21.3 and 21.4) are: (a) A single tetrahedron of silicon surrounded by four oxygen atoms; a single eight-sided octahedron in which an aluminium or magnesium ion is surrounded by six hydroxyls or oxygens, (b) Clay crystals in which thousands of tetrahedron and octahedron units are linked to give horizontal planes of aluminium or magnesium. The planes of oxygen and hydroxyl are alternatively found. The silicon planes with their associated oxygen or hydroxyl planes constitute the tetrahedral sheet. The aluminium or magnesium planes with their oxygen or hydroxyl groups make the octahedral sheet. Different combinations of these are called as *layers*. In some silicate clays, these layers are separated by an interlayer in which water and adsorbed cations are found. These layers are also found in each crystal or micelle.

The well-drained and moderately arid, humid regions with soils rich in Ca^{2+} and Al^{3+} with less sodium ions are in agreement with the strength of adsorption of cations. In arid and semi-arid regions, Ca^{2+} and other metallic cations dominate making the pH of the soil greater than 7.0. Under these conditions, Al^{3+} forms insoluble compounds and the adsorbed H^+ ions are replaced by metallic cations. When the drainage of arid soils occurs, alkali salts accumulate increasing the amount of adsorbed sodium ions which is higher than the adsorbed Ca^{2+} ions. For a similar reason, in humid regions exchange of cations for Al^{3+} and H^+ give an aluminium–hydrogen clay. As the cations predominant in the colloidal system affect the physical and chemical properties and also has influence on plants in terms of available nutrient elements, the cation exchange receives a greater significance.

Exchange of cations between Ca^{2+} and H^+ is represented as:

$$Ca^{2+}\ \boxed{\text{Micelle}}\ + 2H^+ \rightleftharpoons {}^{H^+}_{H^+}\boxed{\text{Micelle}}\ + Ca^{2+}$$

The process is reversible and occurs rapidly. The advantage of this reversibility is illustrated further as follows.

1. When limestone or other calcium compounds are added to an acid soil, Ca^{2+} ions replace H^+ and other cations. By this, the clay becomes higher in exchangeable Ca^{2+} and lower in adsorbed H^+ and Al^{3+} and the soil pH goes up. On the other hand, addition of acid-forming chemicals such as sulphur compounds to an alkaline soil will lead to the reverse of the above to take place in order to reduce the pH of the soil.

2. The cation exchange is also observed by the liberal application of water-soluble fertilizers such as KCl and NH_4Cl. The exchange reaction may be shown as:

$$\begin{matrix}Ca_{40}\\Al_{20}\\H_{40}\\M_{20}\end{matrix}\boxed{\text{Micelle}} + 7KCl \rightleftharpoons \begin{matrix}K_7\\Ca_{38}\\Al_{20}\\H_{39}\\M_{18}\end{matrix}\boxed{\text{Micelle}} + 2CaCl_2 + HCl + 2MCl$$

Soil solids   Soil solution   Soil solids   Soil solution

The added K^+ is adsorbed on the colloids and replaces an equivalent quantity of Ca^{2+}, H^+ and other metal elements present in the soil solution. This exchange reaction is viewed as an advantage since, required nutrient, which is held tightly in the soil, cannot be leached readily and will remain available to plants.

This exchange mechanism is useful not only for nutrients already present in the soil but also for those applied in the form of commercial fertilizers and in other ways.

CATION EXCHANGE CAPACITY (CEC)

Cation exchange reaction when carried out quantitatively is called as cation exchange capacity.

It is also defined as the sum total of exchangeable cations that a soil can absorb per unit of 1 kg of dry soil. For the purpose of convenience, CEC is represented as a whole number by the expression as centimoles of +ve charges/kg of dry soil. In earlier days, CEC was expressed in meq/100 g soil. In the recent years the SI units are introduced. According to SI system, one meq/100 g soil = 1 cmol/kg soil.

Q. What is the meaning for a soil which has CEC of 10 cmol/kg ?

A. One kg of this soil can absorb 10 cmol of H^+ ions.

Q. How many cmol of Na^+ and Ca^{2+} ions can be exchanged with the above soil?

A. Irrespective, of the cations the soil is associated with 10 cmol of −ve charges. This therefore reemphasizes that the cations are adsorbed or exchanged on the basis of chemical equivalence.

[Clue: One mole of charge is provided by one mole of H^+, K^+ or any other monovalent cations or 1/2 mole of divalent cations and so on.]

Q. a. Calculate the amount of calcium required to replace 1 cmol H^+ /kg.

   b. What amount of calcium is required for one-hectare land of weight 2.2×10^6 g; also calculate the weight of limestone required for this area?

   [Recollect: 1 mole of H^+ ion (1 g) would exchange 1/2 mole of Ca^{2+} ions (20 g)]

A: a. 0.2 g calcium/kg soil. b. Weight of Ca is 440 kg; weight of limestone is 1100 kg.

With the knowledge on chemical equivalence, it becomes easy to express CEC in practical applications like agricultural fields.

For instance when an acid soil is limed, Ca^{2+} replaces part of H^+, the following reaction takes place.

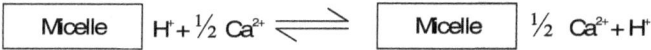

$$\boxed{\text{Micelle}}\ H^+ + \tfrac{1}{2}\,Ca^{2+} \rightleftharpoons \boxed{\text{Micelle}}\ \tfrac{1}{2}\,Ca^{2+} + H^+$$

Method of Determination of CEC

CEC is determined by the following method:

Step 1: A sample of soil containing the exchangeable cations such as Ca^{2+}, Mg^{2+}, K^+, H^+, Al^{3+} is leached with a solution containing NH_4^+ ions. The latter replaces the adsorbed cations, that are leached out and collected.

Step 2: The soil now containing adsorbed NH_4^+ ions is leached out using a solution of K^+ ions.

Step 3: The leached NH_4^+ ions collected is estimated quantitatively to give a measure of CEC of the soil, which is an estimation of negative charges on the soil colloids.

The procedure involved in the determination of CEC is given as a flow chart in Figure 21.5.

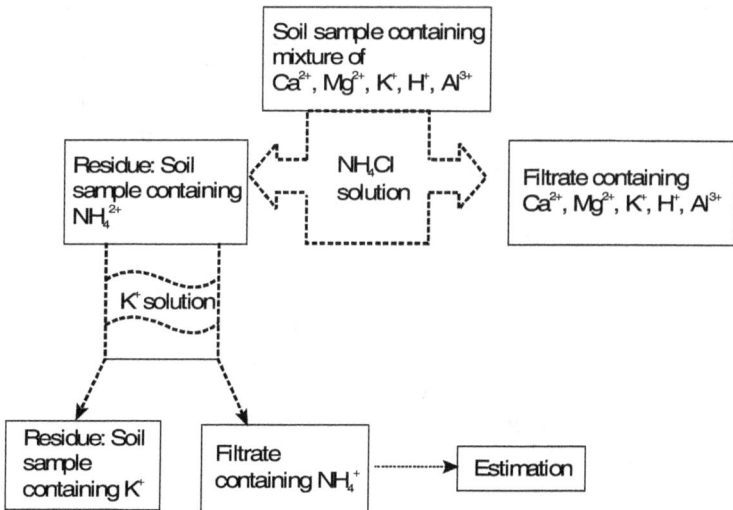

Figure 21.5   CEC determination

Table 21.1   CEC values of some colloids

Colloid	CEC cmol/Kg
Humus	200
Vermiculite	150
Montmorillonite	100
Iolite	30
Chlorite	30
Kaolinite	8
Iron oxides	4

Table 21.1 shows that a soil complex rich in kaolinite will have low CEC. Humus-rich soils have higher CEC. Factors like the texture, of the soil, (finer the texture, higher the CEC), organic content, type of clay, pH (lower the pH, lower the CEC) and climate are known to influence the CEC values.

Since only the permanent charges and a small portion of charges of organic colloid-held ions are replaced during cation exchange, with the increase in pH hydrogen, ions held by the organic colloids and silicate clays such as kaolinite, get ionized and are replaced easily. Also the adsorbed Al^{3+} ions are removed as $Al(OH)_3$. The overall result is to increase the negative charge on the colloids in turn results in high CEC. It is for this reason that the CEC determination is done at pH 7 or above.

ANION EXCHANGE CAPACITY (AEC)

Anion exchange capacity (AEC) is very similar to cation exchange. It is of importance because it provides all available nutrients readily to higher plants. The basic principle is the same as that of cation exchange except that the charges on the colloids are positive and the exchange involves negative ions. This can be pictorially represented as follows:

| Micelle | $+ NO_3^- + Cl^- \rightleftharpoons$ | Micelle | $Cl^- + NO_3^-$ |

| Soil solids | Soil solution | Soil solids | Soil solution |

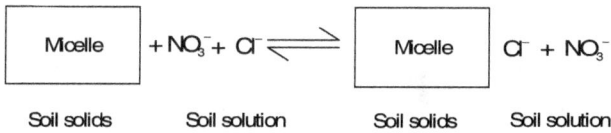

This also exhibits chemical equivalence, reversibility, etc. However, the exchange of anions such as PO_4^{3-}, SO_4^{2-} is more complex which is due to special reactions between the anions and soil constituents.

$$\rightarrow Al - \overset{+}{O}H_2 + H_2PO_4^- \longrightarrow \rightarrow Al - H_2PO_4 + H_2O$$

Soil solids   Soil solution   Soil solids   Soil solution

OTHER PROPERTIES OF COLLOIDS

Other important properties of soil colloids include:

1. Electrical properties
2. Dispersion
3. Coagulation or flocculation
4. Tyndal phenomenon
5. Brownian movement
6. Dialysis

Electrical Properties

The soil colloids as mentioned earlier, also known as *micelles* carry a definite electrical charge on them. They may be charged either positively or negatively. Thus, colloidal particles of one type of charge attract the other. In soil colloids, if the micelle has a positive charge then it is associated with a number of cations and vice versa. Colloidal particles differ from electrolytes when an electric current is passed through a colloidal suspension. Electrolytes are made up of both cations (positively charged) and anions (negatively charged), whereas the soil colloidal particles carry either positive or negative charge. The particles are attracted towards either electrode depending upon the charge carried. This phenomenon is called as electrophoresis.

Dispersion

It is a typical property of a colloidal suspension in water. The particles tend to repel each other. Therefore, each particle acts independent of the other. Increasing the pH value of the soil enhances this. The existing repulsion is attributed to each micellar particle carrying a definite charge irrespective of being positive or negative. For instance, when the charge is negative then surrounding this a number of cations gather and remain hydrated. With the increase in pH the negative charge of the micellar colloids is maximum. By this, the extent of hydration also goes up making the hydrated monovalent ions such as sodium ions, that are not tightly held by the micelles and do not effectively reduce the negative charge, help in stabilizing the colloidal dispersion. In other words, the loosely held positive ions favour the micelles to repel each other, thereby contributing to the dispersion of soil colloids.

Coagulation or Flocculation

When the dispersed colloidal particles are destabilized, aggregation begins. This is due to the increased forces of attraction between the micellar particles leading to the formation of aggregation or granules and the process is known as coagulation or flocculation. The soil colloids may be made to undergo flocculation by decreasing the negative charges associated with the micelles. This can be done by the following ways:

- Lowering the pH decreases the negative charge on the micelles. During this, replacement of some sodium ions and other monovalent ions take place.

- Divalent and trivalent cations such as calcium, magnesium and aluminium are also introduced to exchange with the monovalent cations. As these multi-charged cations are tightly held by the micelles, the overall effect is in the reduction of negative charge of the micelle. This results in the increased forces of attraction between the colloidal particles.

- Addition of simple salts also increases the positive charge around the micelle.

The methods stated above in effecting flocculation find greater benefits in the agro industry since it is the first step in the formation of stable aggregates or granules.

The ability of cations exchanged to effect flocculation is in the order $Al^{3+} > H^+ > Ca^{2+} > Mg^{2+} > K^+ > Na^+$. This is a favourable trend because the colloidal complexes of humid, sub-humid regions have colloids abundant in Al^{3+}, H^+, and Ca^{2+} and those in the semi-arid regions are rich in calcium ions. Thus, in all these soils flocculation will be successful.

The following activity performed demonstrates flocculation and deflocculation.

Activity   To a sample of clay suspended in water, a little amount of NH_4OH is added. After few minutes large particles of clay start settling down while the finer particles continue to be in suspension. Following this, little amount of lime water is added after which the finer suspended particles increase in size and form small floccules and settle down. This is therefore due to the flocculation property of the soil. To the same sample if some mineral acid is added, the floccules are found to be broken and the clay particles return to the original state. Then, it is said to deflocculate.

Tyndal Phenomenon

When a strong beam of light is passed through a colloidal suspension and viewed at an angle perpendicular to the direction of light, the colloidal particles appear strongly illuminated and larger in size. This is known as Tyndal phenomenon.

Brownian Movement

Colloidal particles when suspended in a dispersion medium, show a characteristic continuous zig-zag movement, called Brownian movement. This type of movement was first observed by an English botanist, Robert Brown, hence this phenomenon is named as Brownian movement. As there is a continuous collision of one particle with the other, this prevents them from settling down. As soil particles also exhibit this phenomenon, they are characterized as colloids.

Dialysis

Because of the larger size of the particles, the colloids do not pass through a parchment membrane whose pore size is lesser than the colloidal particles, and are therefore retained. This property helps separation of particles having smaller particle size than colloids. Soil particles like any other colloids have larger particle size and shows dialysis property.

22

SOIL REACTIONS

Soil pH is a basic soil property that affects many chemical and biological activities in the soil. The degree of acidity or alkalinity in soil is known as the "soil reaction."

Soil pH range	Soil reaction
3.0–4.0	Very strongly acidic
4.0–5.0	Strongly acidic
6.0–7.0	Slightly acidic
7.0–8.0	Slightly alkaline
8.0–9.0	Moderately alkaline
9.0–10.0	Very strongly alkaline

SOIL ACIDITY

Soil acidity is common in the regions where precipitation is high enough to leach the exchangeable cations from the surface layers of soils.

Two adsorbed cations H^+ and Al^{3+} are responsible for the acidity of soil. The mechanism by which these two ions act depends upon the level of acidity, the source and the nature of the negative charges on the clay micelle. There are two types of negative charges on the clay micelle—the permanent and the variable (pH-dependent) negative charges. Both influence soil acidity and the association of H^+ and Al^{3+} ions with soil colloids. The permanent charges because of isomorphous substitution (other than kaolinite clay) in silicate clays provide exchange sites in all pH ranges. Therefore, H^+, Al^{3+} and other base forming cations are exchangeable at these permanent charge sites. However, the magnitude of the variable negative charges depends upon the soil pH.

Under strongly acidic conditions, most of the aluminium becomes soluble and is present as Al^{3+} ions or $Al(OH)_3$. The Al^{3+} and $Al(OH)_3$ thus formed are adsorbed in preference to H^+ ions by the permanent charges of soil colloids and remain in equilibrium with the Al^{3+} ions present in the soil solution. The latter is largely responsible for making the soil acidic by undergoing hydrolysis.

The reactions involved in the soil acidity are shown by the following reaction.

$$\boxed{\text{Micelle}}\;\; Al^{3+} \; \rightleftharpoons \; Al^{3+}$$

Adsorbed aluminium ion   Aluminium ion in
                                       soil solution

[Al^{3+} ions in solution actually remain in the highly hydrated form as $Al\,(H_2O)_6{}^{3+}$]

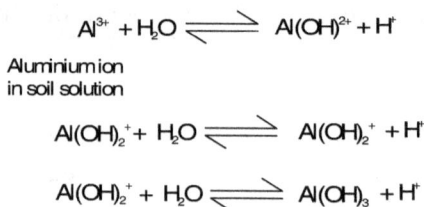

$$Al^{3+} + H_2O \rightleftharpoons Al(OH)^{2+} + H^+$$

Aluminium ion
in soil solution

$$Al(OH)_2{}^+ + H_2O \rightleftharpoons Al(OH)_2{}^+ + H^+$$

$$Al(OH)_2{}^+ + H_2O \rightleftharpoons Al(OH)_3 + H^+$$

The liberated H^+ ions enhance the soil acidity and are the major source of H^+ ions in acid soils. The micellar adsorbed H^+ ions are only the minor source of hydrogen ions in the acid soils. On strong acid groups of humus and some of the permanent charge exchange sites of clays, H^+ is held in the exchangeable form. This H^+ is in equilibrium with the H^+ of soil solution and is shown in the following reaction.

| Micelle | H⁺ | ⇌ | H⁺ |

Adsorbed hydrogen ion   Hydrogen ion in solution

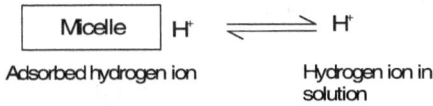

These reactions show the role of Al^{3+} and H^+ in the acidity of soil.

Causes or Sources of Acidity

Some soils are acidic because of the composition of the parent material (rocks) from which they are formed. Soils also become acidic by a number of processes. Cropping and use of nitrogen fertilizers are two main sources of soil acidity while another contributor is rainfall. The leaching of bases is the main cause for the formation of acidic soils. Leaching is in the regions of high rainfall. The mean annual temperature, types of vegetation and parent rock material also govern the acidity of soils. The net result is that hydrogen, aluminium and iron (acidic cations) replace calcium, magnesium, potassium and sodium (basic cations) on the soil cation exchange complex.

Cropping Calcium, magnesium and potassium are the essential nutrients for plant growth. Their uptake by plants and subsequent removal through harvest can have an acidifying effect on soils. The amount of these nutrients removed by cropping depends on (a) crop grown, (b) part of crop harvested and (c) stage of growth at harvest.

Fertilizers Nitrogen fertilizers have a greater acidifying effect on soils than other fertilizers. There are two processes that are involved. Firstly, commonly used nitrogen fertilizers contain ammonium ions, e.g. urea is an ammonium ion-forming fertilizer. Soil bacteria convert ammonium (NH_4^+) to nitrate (NO_3^-) through a biochemical process called nitrification. Hydrogen ion (H^+) is released in this process, causing an increase in acidity. Secondly, acidifying effect comes from nitrate that is not taken up by the growing crop. Nitrates are highly soluble and, if not taken up by plants, will move downward with soil water and may be carried below the root zone. They take with them other nutrients that have a positive charge—most likely calcium and magnesium—and their removal in this manner has the same acidifying effect on soils as removal by a crop.

As water moves down through the soil profile, it has a slow but persistent acidifying effect. Weak acids like vinegar are produced in the soil when plant residues and organic matter decompose. These weak acids react and combine with nutrients such as calcium, magnesium, potassium and sodium ions as the soil solution (water) moves down through and below the root zone (leaching). If soil pH is less than 5.2, hydrogen or aluminum ions replacing the basic cations cause the soil in the leached zone to become more acidic.

Soil acidity has a direct effect upon availability of most essential plant nutrients. The best pH range for most nutrients is 6.0–7.0. Deficiencies can be observed at both low and high pH. Manganese and iron exhibit toxicity at low pH levels and deficiency at high pH levels. Although aluminium is not an essential nutrient, it is important because it rapidly increases in solubility as the soil pH drops below 5.0. Too much aluminum in solution will restrict root and plant development. Figure 22.1 shows the general effect of pH on plant nutrient availability.

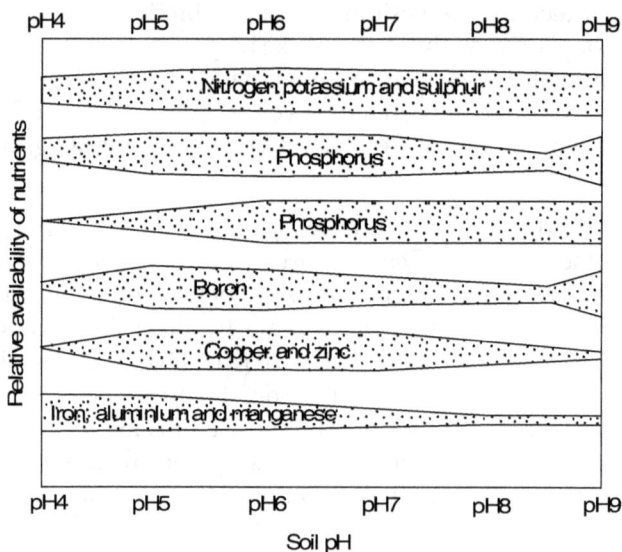

Figure 22.1   Effects of soil pH on the availability of plant nutrients

Rainfall Soils can become acid even in the absence of crop removal or fertilizer applications. Rainfall is considered as a natural cause of acidity because of the downward movement of water through the soil profile and the removal of nutrients due to surface run-off and erosion. In the regions of high rainfall, soils are acidic because the soluble basic salts such as those of calcium, magnesium, potassium and sodium are leached away by draining water and accumulation of insoluble acidic residues consisting of oxides and silicates of iron, silicon and aluminium results.

Soil microorganisms do not function effectively in acid soils. As soil pH levels decline, so does the activity of the organisms, which break down (decompose) organic matter, releasing nutrients to plants. Although these organisms function best at soil pH levels of 8.0, their effectiveness does not drop rapidly until pH levels drop below 6.0. Decomposition of organic matter also contributes to aggregation (clumping) of soil particles, which provides for good soil tilt, aeration and drainage.

Effectiveness of the bacteria, which enter legume roots and fix nitrogen (nodulation), is highest at pH levels of 6.5 to 7.0 and declines rapidly when pH levels fall under 6.0.

SOIL ALKALINITY

Soil alkalinity is of importance in arid and semi-arid regions where high concentration of soluble salts and carbonates of calcium, magnesium and sodium are found.

Formation of alkaline soils can be attributed to one or more of the following reasons.

1. In arid areas there is not enough precipitation to leach the soil.

2. Ground water carries bases into soils and leaves the bases when it evaporates.

3. Certain parent materials provide very large amount of bases and retains them when young soils are formed.

Alkaline soils are of greater concern in agriculture because the valleys of arid regions are more irrigated.

Alkaline soils are of four types: (1) high-lime soil, (2) saline soil and (3) sodic soil, (4) saline–sodic soil.

High-lime This type of soil is found in humid regions and the others are in arid and semi-arid regions. Most of the high-lime soils are freshly formed from the parental rock containing higher amount of calcium. These soils are not saline and the pH of the soil is in the range of 7.5–8.0 due to the lower solubility and the buffering action of calcium carbonate.

Saline soils They have higher concentrations of soluble salts mainly chlorides and sulphates of sodium, calcium and magnesium. The primary source of their formation is by weathering. During the process of rock weathering and soil formation, soluble salts are formed. In the humid regions where the rainfall is sufficient, most of the soluble salts are leached down into ground water or to some areas of lower depth. In arid and semi-arid regions with deficient rainfall and higher evaporation, the incrustation of the salts results. Fluctuation in the depth of water level is also a cause for the salinity. Saline soils can be reclaimed by leaching.

Sodic soil/alkali soil These are soils containing high exchangeable sodium content. They are the most alkaline of all soils. The pH of sodic soil is above 8.5. It is hardest to reclaim the sodic soil. They have dispersed colloids, very low permeability and they support very little plant growth.

Saline–sodic soil These soils have the maximum salt content with high percentage of sodium. The pH value is between 8.0 and 8.5. Their other properties are similar to that of saline soil except that leaching will convert them into sodic soils. This cannot be reclaimed without soil amendment. The most common amendment is adding gypsum ($CaSO_4 \cdot 2H_2O$). Gypsum exchanges most of the sodium ions into soluble sulphates, which in turn are leached away.

Buffering of Soils

The two factors that contribute to the acidity of soil are:

- The hydrogen ion concentration of the soil solution referred to as active acidity.

- Presence of adsorbed hydrogen and aluminium ions on the soil colloids known as exchange acidity as shown in the equation below:

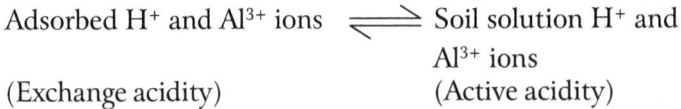

Adsorbed H^+ and Al^{3+} ions $\rightleftharpoons$ Soil solution H^+ and Al^{3+} ions

(Exchange acidity)   (Active acidity)

The exchange acidity is also known as reserve acidity indicating its role. This equilibrium reaction that exists between the active and exchange acidities is largely responsible for the distinct resistance of soil to change the pH of the soil solution. Thus, when decrease in the concentration of hydrogen ions occurs, it is replenished from the exchange acidity.

If sufficient quantity of liming material were added to neutralize the hydrogen ions in the soil solution, the above equilibrium would be immediately shifted to the right resulting in the movement of more hydrogen ions from exchange acidity into the soil solution. Therefore, no major change in the pH of the soil solution results. However, when the quantity of liming material is sufficient enough to deplete the reserve acidity, there is an increase in pH. The buffering action of soil is equally effective when certain biochemical changes take place in the soil leading to a temporary change in the pH.

The buffering capacity is high when the exchange capacity of the soil is high. This is because, more reserve acidity must be neutralized to resist the pH change. For heavier textured soil and soil containing higher organic matter, larger amount of liming material is required.

Amending the Soil

It is the process of modifying the soil and making it suitable for the intended use. Commonly, it is done to effect general changes in physical and chemical properties of soil. There are extreme cases were the soil is remade. The direct effect of reclamation is to change the soil pH and the indirect effect is changing several properties that vary with pH.

Reclamation of Acid Soil

Soils found in humid region are acidic in nature. They also occur where the basic salts are leached away by percolation of water. Amending such soil involves raising the pH to near neutrality. This is done by adding sufficient quantity of liming material, and the process is known as liming. Normally in humid regions, liming and fertilization are done as one process.

Liming Agents

A satisfactory material for raising the pH requires:

1. A mild alkalizing effect to raise the pH to near neutral. Excess of the material added should cause no harmful effect.
2. Ability to produce required proportion of cations adsorbed on the cation exchange sites. The cations to be added include mainly calcium and some magnesium. Addition of sodium ions is to be prevented.
3. Being favourable to the soil structure.
4. Being less expensive.

Limestone, which is mainly calcium carbonate, is the frequently used liming agent. Chemicals such as calcium chloride or gypsum are not recommended because they are neutral salts and will have no effect on increasing the pH. The neutralization effect of carbonates of calcium and magnesium, although mild, produces no undesirable side effects.

Reclamation of Alkaline or Sodic Soils

Alkaline soils require good drainage. Presence of excess water causes recurrence of alkaline condition, accumulation of soluble salts and prevention of root formation. The alkalinity of the soil leads to reduced availability of phosphorus, potassium and several micronutrients. Iron deficiency is also frequently encountered. Alkaline soils containing soluble salts have poor water-absorbing capacity and increase the osmotic pressure and reduce response to the added fertilizers.

A high-lime soil with adequate drainage and proper fertilization has nearly the same productivity as non-alkaline soil. The choice of the fertilizer depends on the type of crop grown. For example, growing corn on high-alkaline soil requires potassium ions as they are driven into non-exchangeable positions. The high-lime soils favour microbial activity; this in turn increases the decomposition of organic matter releasing more nitrates. Therefore, for a saline soil lesser amount of nitrogen fertilizers is recommended.

Manures, the sources of potassium ions are also capable of producing organic acids and if applied to alkaline soils, can neutralize to lower the pH.

23

SOIL FERTILITY AND PRODUCTIVITY

Soil is a natural medium for plant growth because it supplies nutrients to plants. Some soils support luxuriant growth of plants with little human effort, but other soils cannot. Two terms are frequently used in the growth of plants supported by the soil, namely; soil productivity and soil fertility.

Soil productivity is the resultant of several factors such as availability of water supply, suitable climate, etc. influencing crop production. Thus, soil productivity is an economic concept.

For soil to be productive it must:

- be easily tillable and fertile.
- contain all essential elements in sufficient amounts and in the readily available form.
- be physically good to support plants and contain just the right amount of water and air for proper growth of roots.

Soil fertility can be defined as, the inherent capacity of soil to provide all essential plant nutrients in available forms, in adequate amounts and in suitable balance when other growth factors such as sunlight, water and temperature, and the physical conditions of the soil are favourable.

Therefore, soil fertility is an aspect of the soil–plant relationship.

A soil can be highly fertile. It may have ready supply of nutrients, yet it may not be highly productive. Waterlogged soils may be highly fertile but may not produce good crops because of the unfavourable physical conditions. A fertile soil may be highly saline or alkaline which may not be good for agriculture. Sandy soil may be poor in fertility but with the use of fertilizers and water, it may be made more productive. Hence, soil fertility denotes the status of plant nutrients in the soil. There are no standards set for either soil fertility or productivity, as both depend on the crops to be grown. A soil that is productive for potatoes may not be productive for other crops. Soil fertility is the most important asset of a nation. Therefore, its maintenance is an important aspect in agriculture.

Soil fertility is of two types.

1. *Permanent fertility* It is the property of the soil itself. It cannot be improved, maintained, or corrected by soil management practices.

2. *Temporary fertility* It is acquired by suitable soil management, but the response of such fertility is highly dependent on the degree of permanent fertility existing in the soil.

Soil fertility and agricultural production can be maintained by efficient and judicious management of nutrient addition to the soil from external sources. The two most widely used sources of nutrients are the organic manures and the synthetic and naturally occurring chemical fertilizers, commonly called as fertilizers.

Nutrients are continuously removed from the soil by repeated cultivation and losses by leaching and erosion. Effective soil and crop management methods are adopted to improve and maintain soil fertility and soil physical conditions for the purpose of sustained crop production.

When a complete analysis of the plant is made, a large number of elements are detected. But only those that provide nourishment to the plants and take part in the plant metabolism are essential. Therefore an element is said to be essential:

- if the plant cannot complete its life cycle without it.
- if the element is specific in its physiological functions in plants.
- if the malady that develop in plants in its absence can be remedied only by that element.

In practice, often it becomes difficult to fulfil all the criteria for an element to be essential, and to overcome this difficulty, Nicholas suggested the term functional or metabolism nutrients which include a mineral as nutrient element that functions in plant metabolism whether their action is specific or not.

Of the 90 or so chemical elements forming the earth's crust, 16 are known to be essential for plant growth and reproduction. These mineral nutrients are conveniently classified (on the basis of the relative quantity that is normally present and not on their relative importance) into the following types.

- *Macronutrients*   These are required in larger amounts and include hydrogen, oxygen, nitrogen, carbon from air and water, phosphorus, potassium and calcium from soil (with the exception of hydrogen, carbon and oxygen all other inorganic plant requirements that are obtained directly or indirectly from soil are referred to as mineral nutrients).

- *Micronutrients*   The other nine essential elements which are required in minor amounts are called as secondary or micronutrients. They include zinc, magnesium, sulphur, boron, copper, iron, manganese, molybdenum and chlorine as chloride.

The requirement of micronutrients in a very low concentration suggests that they all function as catalysts or are closely involved in such catalytic activities taking place in plants. Evidences are available for copper, zinc and manganese are components of certain biological oxidation–reduction systems. Some of the micronutrients are essential for and related to the activity of certain enzymes. Manganese performs some functions in photosynthesis, acting as a regulator of the intake state of oxidation of certain elements.

Zinc is involved in the oxidation–reduction potential regulation within the cells and in the functioning of cysteine. It is also related to the growth-promoting substance, auxin. Copper is an essential component of the oxidizing enzyme polyphenol oxidase. Molybdenum is a constituent of nitrate reductase and nitrogenase and therefore is involved in nitrogen utilization and its fixation.

ORGANIC MANURES

They are derived from animal, human and plant residues, which contain nutrients in complex organic forms. Since the early 20th century, when straw was converted into "artificial humus" or "synthetic humus", scientific interest in this subject received more attraction. The main technique was the microbial decomposition of the organic materials.

Organic manures improve the soil fertility in the following manner:

1. They modify the physical properties by increasing the granulation of the soil, which in turn increases the permeability and moisture-holding capacity of the soil.
2. They host soil microbes and thus enhance microbial activities.
3. The decomposition products of organic manure help to bring mineral constituents of soil into solution.
4. They improve physico-chemical properties of the soil, such as cation exchange capacity and buffering action.

Organic manures are of several kinds. Some of them are discussed below:

Farmyard Manure (FYM)

Solid and liquid excreta of all farm animals when allowed to decompose under favourable conditions by the microbes already present in the excreta, both aerobic and anaerobic reactions occur simultaneously. The decomposition is carried out in pits or trenches 1 m in depth. The top of the trench is plastered with slurry of cattle dung and soil. After nearly 2 months, the materials are ready for use as manure. This supplies nitrogen, phosphorus and potassium. However, there is variation in the composition depending upon excreta of various animals, size of the trench, duration of manuring, etc. During decomposition, mineralization of carbon and nitrogen takes place. There is considerable loss of organic carbon leading to a reduction in the volume. If the heaps of manure are exposed to sunlight, loss of nitrogen and ammonia takes place. By the action of rain, loss by leaching is observed. However, addition of superphosphate to the decomposing material prevents such a loss. The heaps of farmyard manure are suitable for all types of soils and plants.

Compost

Composting, as in the case of farmyard manure, is also microbial decomposition of organic residues and mixtures of soil collected either from rural or urban areas. In compost-making, a variety of carbonaceous and nitrogenous organic waste materials are used in addition to animal excreta. In the preparation of compost, the decomposition is carried out in trenches/pits, a layer of refuse is initially spreaded, followed by a layer of animal waste and the process is continued until the level of dumping is above the ground. The top most layer is that of refuses which is then covered with a layer of soil. Nearly after 4 months, the compost is ready for use.

The manure obtained from rural resources is called rural compost and the manure of urban sources as urban compost. Compost provides gardeners with manure well adapted to their needs. During the preparation of compost, much of the carbon content is lost, some amount of nitrogen is also lost, but the percentage of nitrogen in it is higher than that of carbon. During the preparation of compost, some farmers add chemicals such as lime, superphosphate or any other fertilizers to increase the nutrient content. When the compost is added to the soil, it gradually releases the plant nutrients by decomposition. This released nutrient provides better-balanced nutrients than the fresh manure. It is for this reason that gardeners prefer to add large quantities of compost. The latter also results in humus formation and promotes fertility of the soil.

Oil Cakes

Oil cakes are the most important concentrated manures. Their manurial value lies mainly in the nitrogen content, although other nutrients are present in smaller amounts. Unlike bulky organic manures, oil cakes, because of their low C/N ratio (3–5), nitrify very rapidly in soil. Thus, nitrogen becomes readily available. Powdered oil cakes are applied a few days before sowing. They are used for top dressing also.

Bone Meal

Bones consist of calcium phosphate along with some organic matter, mainly fats and proteins. They serve as rich sources of calcium and phosphorus. The crushed bones are used either in the raw form or after steam sterilization. The glue separated from bones has commercial value and the residue in the powdered form is used either as manure or as cattle feed. The sources of bones are from carcasses of dead animals, slaughterhouses and meat-processing industries. Bone meal is used as a phosphatic fertilizer. It is slow-acting, suitable for acidic soils, long duration crops, and coarse-textured, well-drained soils.

Blood Meal

It is the coagulated, dried and powdered form of blood, a by-product from slaughterhouse. The very low C/N ratio (3–4) makes the blood meal susceptible to rapid microbial decomposition and nitrification. It can be used at any time during the crop growth and is effective on all soils.

Meat Meal

It is prepared after removing bones, by boiling waste meat and flesh of dead animals, followed by drying, and powdering. Horn and hoof meals are also prepared in a similar manner. Manufacture of blood and meat meal is not yet commercialized in India.

Fish Meal

It is prepared by drying and crushing non-edible fish, carcasses, and fish waste. This is also a quick acting manure.

Green Manure

Green manuring is the practice in which leguminous plants are grown, particularly during monsoon season and before sowing of crops. The grown plants are ploughed into the soil and allowed to decompose *in situ* under favourable soil and water conditions. This improves the organic content and nitrogenous content of the soil to a definite extent. In our country, the commonly grown legume plants include, sunhemp (sanai), dhainchaphaseolus mungo, cowpea, etc. Non-leguminous plants are also used in green manuring. Another practice in green manuring is that leaves from shrubs and trees are collected from outside and ploughed in to the soil under cultivation.

The advantages of green manuring include the following

- Increase in mobilization of minerals.
- Reduction in loss of nutrients by erosion, leaching, and percolation.
- Improvement in the physical, chemical and biological activities of the soil.
- Improvement in soil aeration and drainage.

However, the main disadvantage of green manuring is the land on which the green plants are grown could have been profitably used for growing crops of economic importance. In addition to this, for the *in situ* decomposition, adequate soil water, supplied either by rainfall or irrigation, is required.

CHEMICAL FERTILIZERS

Fertilizers commonly referred to as chemical fertilizers or commercial fertilizers are the synthetic or naturally occurring chemical compounds applied to soil to supply the nutrients for crop growth. Unlike manures, the nutrients in fertilizers are present in higher concentration and can either be readily used by the plants or used after transformation. Fertilizers are never considered as substitutes or alternatives to manures, which have their own role to play. The role of fertilizers is to introduce additional supply of nutrients into cycles of plant growth and decay thereby improving the soil fertility.

As the fertilizers are concentrated in nutrient content, they have certain advantages over manure.

- Handled in smaller quantities in contrast to manures of bulkier quantity
- Economical
- Easy to transport
- Easy to store
- Easy to handle

The quantity of fertilizers required can be adjusted to suit the requirement as determined by soil fertility evaluation. The high nutrient content in easily available form has led to the worldwide adoption of fertilizer use for increased and sustained crop production. It is well established that the proper use of fertilizers greatly increases the crop production. Proper benefits from these commercial chemicals can be realized by using improved variety of crops, better cultivating processes, improved soil and water conditions, better plant protection and crop management.

Requisites of a Good Fertilizer

The following are the basic requisites of a good fertilizer.

- They must supply necessary nutrients for plant growth.
- They must be slightly basic (pH~7–8).
- They should not produce any acidity.
- They must be capable of dissolving in soil moisture, and thus assimilated easily by the plants.
- They must not alter the soil texture.
- They must be sufficiently stable in soil so that plants derive nutrient elements over a longer period.
- They should not be leached away from the soil readily.

Classification of Fertilizers

Fertilizers form a wide group of agrochemicals. Based on the supply of single or more than one nutrient element, their chemical nature, and commercial mode of supply, they are classified into:

- *Straight fertilizers* These include nitrogenous fertilizers, e.g. urea, calcium ammonium nitrate (CAN), ammonium sulphates, etc. Phosphatic fertilizers, e.g. superphosphate (single), superphosphate (triple), dicalcium phosphate, etc. Potash fertilizers, e.g. potassium chloride (Muriate of potash), potassium sulphates, etc.
- *Complex fertilizers* Diammonium phosphate, ammonium phosphate, sulphates, urea-ammonium phosphate, nitro phosphate with potash and NPK complex fertilizers.
- *Mixed fertilizers* This is a mixture of fertilizers.

Straight Fertilizers

Nitrogenous fertilizers It is available either as nitrate, ammonium or amide fertilizer.

Total nitrogen content for certain nitrogen fertilizers is given in Table 23.1.

Table 23.1 Nitrogen content of certain fertilizers

Nitrogenous fertilizers	Nitrogen content (%)
Urea	46.0
Calcium ammonium nitrate (CAN)	26.0
Ammonium sulphates	20.6
Ammonium chloride	25.0

Urea It is an ideal fertilizer and suitable for all crops and soils. It is manufactured by reacting anhydrous ammonia with carbon dioxide at high pressure.

$$2NH_3 + CO_2 \rightarrow NH_2COONH_4 \rightarrow CO(NH_2)_2 + H_2O$$

It is white, crystalline, hygroscopic and freely soluble in water. It has to be stored in moisture-proof bags. Under wet conditions, it is hydrolysed by the enzyme urease as follows:

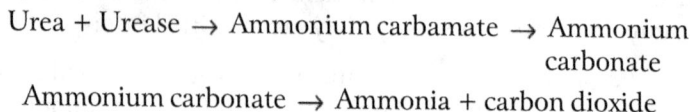

Urea + Urease → Ammonium carbamate → Ammonium carbonate

Ammonium carbonate → Ammonia + carbon dioxide

Ammonium carbonate being less stable in the soil decomposes to liberate ammonia and carbon dioxide. The free liberated ammonia then reacts with water to give ammonium hydroxide, which on ionization yields ammonium ion. The latter then in soil undergoes lot of transformations by the soil microbes initially to ammonium nitrite then to ammonium nitrate. The reaction of ammonium nitrite to nitrate occurs at a fast rate, thus the concentration of toxic nitrite is not allowed to accumulate.

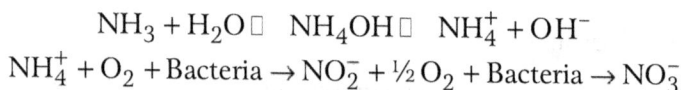

$$NH_3 + H_2O \rightleftharpoons NH_4OH \rightleftharpoons NH_4^+ + OH^-$$
$$NH_4^+ + O_2 + Bacteria \rightarrow NO_2^- + \tfrac{1}{2}O_2 + Bacteria \rightarrow NO_3^-$$

Urea being highly water-soluble, is readily lost by leaching when applied to wet soil or if irrigated immediately. However, under dry conditions urea is lost by volatilization, as ammonia. Thus, urea is to be applied on moderately wet soil.

Calcium ammonium nitrate (CAN)   It is a mixture of ammonium nitrate and powdered chalk which is calcium carbonate. The obtained mixture is called nitro-chalk or nitro-lime. The reaction of ammonia with nitric acid results in ammonium nitrate, the total nitrogen content of which is about 33% and can be used as a good fertilizer, but it is associated with certain disadvantages in being highly hygroscopic. It also increases the acidity of the soil and is explosive. These disadvantages particularly the explosiveness and acidity are reduced amply by mixing with lime.

CAN produced in India contains 25% nitrogen as nitrate and ammonium in equal proportions. The nitrate nitrogen is immediately available and ammonium nitrogen is available on nitrification. CAN is used as a top-dressing fertilizer.

Ammonium sulphate   This is one of the earliest fertilizers. It is also the by-product of coal distillation. The basic constituent of this fertilizer is ammonia. Ammonia is manufactured by Haber's process. Ammonia on reaction with carbon dioxide forms ammonium carbonate, this on further reaction with gypsum produces ammonium sulphate. The physical properties of this fertilizer make it convenient to store, handle and transport. When it is added to soil, the ammonium ions are adsorbed on soil as exchangeable cations resisting loss by leaching. Another essential nutrient, sulphur is also supplied to the crops by this fertilizer.

Demerits of this fertilizer include:

- Ammonium sulphate being an acidic fertilizer enhances the acidity of the soil and needs higher quantity of calcium carbonate to adjust this acidity.
- Lower nitrogen content (20.6%),
- High cost of production.

Ammonium chloride   It is produced as a by-product during the manufacture of soda ash. This is also similar to ammonium sulphate and increases the acidity of the soil.

Phosphatic fertilizers   Commercial phosphatic fertilizers are the compounds of calcium like:

- Monocalcium phosphate $Ca(H_2PO_4)_2$
- Dicalcium phosphate $CaHPO_4$
- Tricalcium phosphate $Ca_3(PO_4)_2$

Some of the commonly used phosphate fertilizers are:

Superphosphate of lime, $Ca(H_2PO_4)_2 \cdot H_2O + CaSO_4 \cdot 2H_2O$ This is suitable for all crops on neutral and alkaline soil. It is a mixture of approximately two parts of monocalcium phosphate and three parts of gypsum, and applied along with nitrogen fertilizer. Super phosphate of lime is ammoniated to neutralize sulphuric acid formed from this fertilizer.

Triple superphosphate, $CaH_4(PO_4)_2$ This has higher phosphate content and is free from gypsum. It is manufactured by treating rock phosphate with phosphoric acid. Its action is similar to that of superphosphate of lime.

Potash fertilizers　Potassium chloride is the most commonly used fertilizer. Others include sulphates and mixed salts.

Muriate of potash (Potassium chloride)　It is manufactured from minerals containing potassium. Being soluble in water, it is directly utilized by the plants in the ionic form. It is suitable for all plants except potato and tobacco because of chloride.

Pupate of potash　This is manufactured from Logebeinite $K_2SO_4 \cdot 2MgSO_4$. It is prepared by treating potassium chloride with sulphuric acid. The action is similar to potassium sulphates but also supplies sulphur additionally.

Schoenite　$K_2SO_4 \cdot 2MgSO_4 \cdot 6H_2O$ is recovered as double salt from marine salt bittern and is a by-product of coastal salt industry. The field trials conducted in India had shown this as a good fertilizer. The presence of MgO does not interfere in the crop production and is found to be good for the soils deficient in magnesium.

Complex Fertilizers

These contain two or more nutrients in one package. This supplies higher concentration of nutrients in one package, reduces the cost of packaging and helps in easy storage, e.g. ammonium phosphate.

EFFECT OF EXCESS FERTILIZATION AND MANURING

Repeated application and larger quantity of fertilizers used are known to alter the chemical and physical nature of soil. Soils with high levels of fertility generally lose more nutrients by leaching than the soils with low fertility. Beyond certain limits the excess fertilizers added to the crop does not necessarily enhance the soil fertility. In fact, some of these added nutrients are lost by many methods. A large loss of nitrogen occurs only when the supply of nitrate nitrogen is enhanced by either fertilization or nitrification process in the soil. The excess of nitrogen fertilizer should be avoided not only for its higher cost but also for the contamination it causes in water bodies either by percolation, natural drainage or by leaching leading to undesirable side effects and toxicity. For instance, contamination of nitrates in drinking water can produce methaemoglobin in babies (blue bodies). In the infant's body, the nitrate ion is converted into nitrite that in turn reacts with the haemoglobin in blood resulting in methaemoglobin. Another major result of contamination of nitrates is eutrophication.

Eutrophication

A lake starts its life cycle as oligotrophic, i.e., a clear water body. When sewage and excess of agricultural products, particularly the phosphates as agricultural chemicals or any other nutrients enter any natural water body, it results in over-nutrition, leading to eutrophication. It is a process of providing a water body with excessive nutrients for the aquatic life it supports. Thus, through the introduction of nutrients, and the growth and decay of aquatic life, the lake collects a good amount of organic substances. Eventually, there is an algal bloom when the lake turns into marsh or debris. This stage is called as *eutrophic*. The aquatic animal life perishes, when the lake is filled with sediments. The lake then becomes a dry land.

The rate of eutrophication strikes a balance between the production of aquatic life and its destruction by bacterial decomposition. With the large input of nutrients, bacterial decomposition does not maintain the balance leading to sedimentation, thus favouring eutrophication. This situation can be prevented by providing proper sewage treatment or preventing the entry of sewage into water bodies.

AGROCHEMICALS

In recent years there has been an increase in use of agricultural chemicals such as pesticides and fertilizers. Pesticide is a general term applied to a substance that kills the disease pests. This includes insecticides, fungicides, nematicides, rodenticides and herbicides (weedicide). The agricultural chemicals increase the crop yields but they contribute to air and water pollution to a significant level. They have many adverse effects on aquatic life and on human beings.

Insecticides

Farmers and gardeners use pesticides to avoid the damage of crops by pests. Insecticides have come into wide use since 1960. DDT was the first chlorinated hydrocarbon to be used as a pesticide. Several such compounds have now been discovered and are classified into stomach poisons, contact poisons and fumigants.

Stomach poisons   They are arsenical and fluoride compounds. Lead arsenate, inorganic fluorides such as cryollite, sodium fluoride, mercury chlorides, boron compounds such as borax and boric acid, etc. are some of the few toxic insecticides affecting both mammals and insects.

Contact poisons   They enter the body when the insect walks or crawls over the treated area. The insecticide is absorbed through the body wall. If the treated surface is a food source (leaf or blossom), then these poisons may enter the gut.

Fumigants   They are insecticides that are gases at normal temperatures. The site of contact is the tracheal (breathing) system. They spread to body tissues through this system where they ultimately kill the insect. These insecticides are applied to enclosures and soils.

Herbicides or Chemical Weed Killers

Herbicides are classified as:

Selective   They are used to kill weeds selectively without harming the crop, e.g. 2,4-D and 2,4,5-T and their esters.

Non-selective   When the purpose is to kill all vegetation, non-selective herbicides like sodium arsenite, sodium chlorate and sulphuric acid, are used.

Both selective and non-selective materials can be applied to weed foliage or to soil containing weed seeds and seedlings, depending on the mode of action. The term *true selectivity* refers to the capacity of a herbicide, when applied at the proper dosage and time, to be active only against certain species of plants but not against others. Extensive use of herbicides like picloram and cacodylic acid has changed the entire ecology of one-third land area; causing defoliation of forest trees and destroyed crops to a greater extent.

Fungicides

They are used to prevent the fungal pathogens of plants; the common chemicals used are sulphur, organic mercurials, formaldehyde and copper sulphate. The pesticides and insecticides used in agriculture which remain in soil are washed away in the environment. Toxicants that are not easily biodegraded enter into food chain, are accumulated in the system and cause undesirable effects, e.g. DDT, BHC, PCBs, etc. They may remain in the soil for 3–15 years or longer. It is well known that many species of birds contain high levels of DDT. The shells of the eggs of these birds are too thin and fragile. This is presumed to be due to the interference of DDT with hormones controlling calcium deposition.

Rodenticides

They are defined as substances that are used to kill rats, mice and other rodent pests. Warfarin, bromodiolone and difenacoum are some examples. These substances kill rodents by preventing normal blood clotting and causing internal haemorrhage. Fumigants such as sulphur dioxide, carbon monoxide, hydrogen cyanide and methyl bromide are also effective rodenticides. In the past, phosphorus paste, barium carbonate salt, and powders such as zinc phosphide, white arsenic, thallium sulphate, strychnine, strychnine sulphate and calcium cyanide were mixed with bait and placed where rodents will find and eat them. All these poisons are toxic to other animals, and mostly cause death by disturbance of nervous system functions. Red squill, a rodenticide derived from the bulbs of a lily-like subtropical plant, is slower-acting and less toxic to animals other than rodents because it is removed from the stomach by vomiting—a reflex that is absent in rodents.

Nematicides

Chemical products used to control nematodes have generally fallen into two major classes, *fumigants* and *non-fumigants* or "contact" nematicides, based on their chemical and physical characteristics.

The first economically effective nematicides were *soil fumigants*. These are the chemicals which when applied to soil, generate fumes that spread through the soil pores to treat a volume of soil reasonably uniformly. Their key characteristic is that they are locally redistributed in the soil by diffusion as gases. Some are economically effective against several kinds of pests (among nematodes, fungi, bacteria, insects and weeds), and are thus often called *multipurpose* or *broad-spectrum* fumigants. Others that have been used primarily for nematode control, lacking significant effects against any other important pest group, are the *fumigant nematicides*. Since the 1960s, *non-fumigant* nematicides have been developed. All are organophosphate or carbamate pesticides; most have significant insecticidal and/or acaricidal properties. Most of them are extremely toxic to humans (acute oral and dermal toxicities). In contrast to the fumigants, these chemicals depend heavily on initial mixing with the soil and local redistribution in solution in the soil water. Some are systemically absorbed and redistributed within plants. They are commonly formulated as granules to which the active ingredient has been adsorbed onto the surface of sand, clay or organic grit of a specific particle size at a rate of 5–20% of the total weight of the formulated product. Many are also available as more concentrated emulsifiable or water-soluble spray liquids. *Land preparation* is important for success with any nematicide. The soil should be turned, tilled or thoroughly disked at least 4–6 weeks before treatment, to encourage decay of roots and other plant trash that could protect nematodes from the chemical and interfere with application equipment. Intact organic matter can also directly adsorb ("tie up").

Multipurpose soil fumigants   The multipurpose soil fumigants include formulations of methyl bromide with varying concentrations of chloropicrin. All methyl bromide products used for soil fumigation should contain a small percentage (0.5–2%) of chloropicrin to warn of free fumigant in the atmosphere, such as from torn plastic tarps, equipment leaks, etc. Methyl bromide is deadly but has no odour; chloropicrin is a powerful tear gas that is easily detected in low concentrations. Higher proportions of chloropicrin are usually chosen to increase fungicidal activity, since it is a far better fungicide than methyl bromide. All methyl bromide products are **extremely toxic** and must be handled with utmost care to avoid injury to those who use them or come near areas where they are used.

Because of their great volatility at normal temperatures, most methyl bromide products and concentrated chloropicrin formulations are sold and handled in gas cylinders. Small quantities (1 or 1.5 lb) of 98% methyl bromide or 2% chloropicrin are sold in sealed cans which are used for treating small areas of seedbed or individual planting sites.

These products are intended to reduce populations of other soil-borne pathogens, insects and weed seeds in addition to nematodes. Offsetting their wider spectrum of activity are greater expenses per acre, more complicated application equipment and methods, and greater potential hazard.

Multipurpose fumigants other than methyl bromide/ chloropicrin mixtures include dazomet, metam and mixtures of 1,3-D with chloropicrin. These products, less volatile than methyl bromide, are usually applied as liquids from drums and bulk containers rather than gas cylinders. Dazomet is a granular product that must be mixed mechanically with the soil.

Fumigant nematicides This group of materials was once the mainstay of soil treatment for nematode control. Three alkyl halides or halogenated hydrocarbons, EDB (ethylene dibromide or dibromoethane), DBCP (1, 2-dibromo-3-chloro-propane) and 1, 3-D (1, 3-dichloropropene), are the principal active ingredients of most of the liquid fumigant nematicides. These compounds are relatively inexpensive and simple to apply, and are also particularly effective in the open-pored sandy soils.

They provide good control of most nematodes and the first good evidence of the economic importance of nematodes as crop pests. However, registrations of DBCP and EDB were cancelled when they were found to present unacceptable health and environmental risks. Only 1,3-D remains as a commercially available fumigant. It is the active ingredient of *Telone II* soil fumigant. The multipurpose soil fumigant, Telone C-17, is the same basic fumigant with 17% chloropicrin added to broaden its spectrum of activity. *Telone II* is recommended for pre-plant nematode control for a wide range of crops. It is especially desirable where root-knot nematodes must be controlled for production of annual crops.

Non-fumigant nematicides   Non-fumigant nematicides are essentially organophosphate or carbamate pesticides, usually developed primarily as insecticides; some are systemic in plants; all are potent cholinesterase inhibitors and hazardous to handle and are water-soluble. They are generally sold in granular formulations, but more concentrated liquid formulations of most are also offered for limited uses.

REVIEW QUESTIONS

1. Define soil.
2. Explain the terms:
   i. Edaphology
   ii. Pedology
3. Discuss the formation of soil.
4. What is a true soil? How is it formed?
5. Give the basis and classification of soil proposed by V.V Dokuchaev.
6. Classify soil based on the origin.
7. What is the USDA classification of soil?
8. What is soil profile?
9. List and discuss the key factors responsible for the soil formation.
10. What is a ternary diagram?
11. Refer to the ternary diagram in Figure 20.2 and identify the texture of a soil for the following compositions

% Clay	% Sand	% Silt
0–12	0–20	85–100
20–40	60–90	65–80
35–55	65–55	55–65

12. What is soil taxonomy?
13. Discuss order and suborder categories of soil taxonomy.
14. What are the features emphasized by great group and sub group of soil taxonomy?
15. Draw and discuss the various horizons of soil profile.
16. Name the three phases of soil.

17. Discuss the following:
    i.   Soil water
    ii. Soil air
    iii. Soil temperature
18. Which type of soil water is the most available one?
19. What is hygroscopic coefficient?
20. Draw a diagram indicating
    i.   Wilt point
    ii. Available water capacity
    iii. Field capacity
21. Why is the subsoil oxygen deficient?
22. How do the seasonal changes alter the soil air composition?
23. Under what condition/s does the soil become a two-phase system?
24. Discuss the solid phase of soil.
25. State and explain the colloidal properties of soil.
26. How is the soil colloid classified?
27. Explain the micellar behaviour of silicate clays.
28. Explain the salient features of the silicate micelle.
29. What constitutes humus?
30. What is cation exchange?
31. Explain the exchange of cations between Ca^{2+} and H^+.
32. What is cation exchange capacity? Describe a suitable method for its determination.
33. Refer table 20.1 and identify the colloids with high CEC values.
34. Why is the CEC determination done at pH of 7?
35. Explain anion exchange capacity.
36. Why do soil colloids exhibit electrophoresis?
37. Discuss the property of dispersion of soil colloids.
38. What is flocculation?
39. Suggest certain methods to flocculate soil.

40. Which phenomenon prevents the soil particles from settling down?

41. What is soil reaction?

42. In which regions are acidic soils found? What are the causes for soil acidity?

43. Discuss the reactions involved in soil acidity.

44. How does the soil acidity influence the availability of plant nutrients?

45. Describe the role of aluminium in plant growth.

46. At what pH ranges are the following minerals abundant?

    i.   Boron

    ii. Copper

    iii. Zinc

47. How does rainfall affect soil acidity?

48. Does the soil pH influence the abundance of microorganism?

49. What is soil alkalinity?

50. What are the factors responsible for soil alkalinity?

51. Discuss the types of alkaline soils.

52. What do the terms exchange and active acidity refer to?

53. What is buffering of soil?

54. How is an acidic soil buffered?

55. Under what conditions is the buffering capacity high?

56. Define amending and reclamation of soil.

57. How is liming effected?

58. Write on reclamation of sodic soil.

59. What is soil fertility?

60. What are the criteria for an element to be termed essential for plants?

61. Discuss permanent and temporary fertility properties of soil.

62. When is soil termed productive?

63. List and discuss the micro- and macronutrients in soil.

64. How do organic manures improve soil fertility?

65. How are the following prepared?
    i.   Farmyard manure
    ii. Compost
66. Enumerate the advantages and disadvantages of green manuring.
67. What are agrochemicals? State their uses.
68. Write a short account on:
    i.   Role of chemical fertilizers
    ii. Advantages and requisites of chemical fertilizers
69. How are fertilizers classified?
70. Explain the application of urea and ammonium carbonate.
71. Discuss CAN.
72. State the demerits of ammonium sulphate.
73. Write a note on phosphatic fertilizers.
74. Discuss the causes and consequences of excessive fertilization.
75. Give a brief account of the following:
    i.   Insecticides   ii. Weedicides
    iii. Fungicides   iv. Nematicides
76. Identify the applications of:
    i.   DDT   ii. PCB
    iii. Warfarin   iv. Methyl bromide
    v.   1,3–D + chloropicrin
77. What are multipurpose soil fumigants?

REFERENCES

Biswas, T.D. and Mukherjee, S.K. *Text Book of Soil Science*, 2nd edn.

David, L. Nelson and Michael, M. Cox. *Lehninger's Principles of Biochemistry*, 2nd edn. W.H. Freeman and Company.

Eric, E. Conn and Stumpf, P.K. *Outlines of Biochemistry*, 3rd edn. Wiley Eastern Pvt. Ltd.

Louis, M. Thompson and Frederick, R. Troeh. *Soils and Soil Fertility.* Tata McGraw-Hill.

Murray, R. K., Granner, D.K., Mayes, P.A. and Rodwell, V.W. *Harper's Biochemistry*, 21st edn. McGraw-Hill.

Nyle, C. Brady. *The Nature and Properties of Soils,* 9th edn. S. Chand and Company Ltd.

Sharma, P.D. *Ecology and Environment.* Rastogi Publications.

Sukumar De. *Outlines of Dairy Technology.* Oxford University Press.

WEB SOURCES

snr.osu.edu/current/courses/ss 300.01

www.rainbowplantfood.com/agronomics/efu/soil

www.uiowa.edu/~mnpcphar/olivo/morphine/morphine.

www.holivo.pharmacy.uiowa.edu/morphine opioids_ringanal.

www.answers.com/topic/nalorphine

www.biology-online.org/dictionary/Nalorphine.

www.physicalgeography.net/fundamentals

www.microsoil.com

INDEX